Springer Series in
MATERIALS SCIENCE 101

Springer Series in
MATERIALS SCIENCE

Editors: R. Hull   R. M. Osgood, Jr.   J. Parisi   H. Warlimont

The Springer Series in Materials Science covers the complete spectrum of materials physics, including fundamental principles, physical properties, materials theory and design. Recognizing the increasing importance of materials science in future device technologies, the book titles in this series reflect the state-of-the-art in understanding and controlling the structure and properties of all important classes of materials.

88 **Introduction to Wave Scattering, Localization and Mesoscopic Phenomena**
By P. Sheng

89 **Magneto-Science**
Magnetic Field Effects on Materials: Fundamentals and Applications
Editors: M. Yamaguchi and Y. Tanimoto

90 **Internal Friction in Metallic Materials**
A Reference Book
By M.S. Blanter, I.S. Golovin, H. Neuhäuser, and H.-R. Sinning

91 **Time-dependent Mechanical Properties of Solid Bodies**
By W. Gräfe

92 **Solder Joint Technology**
Materials, Properties, and Reliability
By K.-N. Tu

93 **Materials for Tomorrow**
Theory, Experiments and Modelling
Editors: S. Gemming, M. Schreiber and J.-B. Suck

94 **Magnetic Nanostructures**
Editors: B. Aktas, L. Tagirov, and F. Mikailov

95 **Nanocrystals and Their Mesoscopic Organization**
By C.N.R. Rao, P.J. Thomas and G.U. Kulkarni

96 **GaN Electronics**
By R. Quay

97 **Multifunctional Barriers for Flexible Structure**
Textile, Leather and Paper
Editors: S. Duquesne, C. Magniez, and G. Camino

98 **Physics of Negative Refraction and Negative Index Materials**
Optical and Electronic Aspects and Diversified Approaches
Editors: C.M. Krowne and Y. Zhang

99 **Self-Organized Morphology in Nanostructured Materials**
Editors: K. Al-Shamery and J. Parisi

100 **Self Healing Materials**
An Alternative Approach to 20 Centuries of Materials Science
Editor: S. van der Zwaag

101 **Organic Nanostructures for Next Generation Devices**
Editors: K. Al-Shamery, H.-G. Rubahn, and H. Sitter

102 **Photonic Crystal Fibers**
Properties and Applications
By F. Poli, A. Cucinotta, and S. Selleri

103 **Polarons in Advanced Materials**
Editor: A.S. Alexandrov

104 **Transparent Conductive Zinc Oxide**
Basics and Applications in Thin Film Solar Cells
Editors: K. Ellmer, A. Klein, and B. Rech

105 **Dilute III–V Nitride Semiconductors and Novel Dilute Nitride Material Systems**
Physics and Technology
Editor: A. Erol

106 **Into The Nano Era**
Moore's Law Beyond Planar Silicon CMOS
Editor: H.R. Huff

Volumes 40–87 are listed at the end of the book.

Katharina Al-Shamery
Horst-Günter Rubahn
Helmut Sitter
Editors

Organic Nanostructures for Next Generation Devices

With 221 Figures and 6 Tables

Professor Dr. Katharina Al-Shamery
Universität Oldenburg, Fakultät 5
Carl-von-Ossietzky-Str. 9–11, 26129 Oldenburg, Germany
E-mail: Katharina.Al.Shamery@uni-oldenburg.de

Professor Dr. Horst-Günter Rubahn
NanoSYD, Mads Clausen Instituttet
Syddansk Universitet, Alsion 2
6400 Sønderborg, Denmark
E-mail: rubahn@mci.sdu.dk

Univ. Prof. Dr. Helmut Sitter
Johannes Kepler Universität, Institut für Halbleiter- und Festkörperphysik
Abteilung Festkörperphysik, Altenberger Str. 69, 4040 Linz, Austria
E-mail: Helmut.Sitter@jku.at

Series Editors:

Professor Robert Hull
University of Virginia
Dept. of Materials Science and Engineering
Thornton Hall
Charlottesville, VA 22903-2442, USA

Professor Jürgen Parisi
Universität Oldenburg, Fachbereich Physik
Abt. Energie- und Halbleiterforschung
Carl-von-Ossietzky-Strasse 9–11
26129 Oldenburg, Germany

Professor R. M. Osgood, Jr.
Microelectronics Science Laboratory
Department of Electrical Engineering
Columbia University
Seeley W. Mudd Building
New York, NY 10027, USA

Professor Hans Warlimont
Institut für Festkörper-
und Werkstoffforschung,
Helmholtzstrasse 20
01069 Dresden, Germany

ISSN 0933-033X

ISBN 978-3-540-71922-9 Springer Berlin Heidelberg New York

Library of Congress Control Number: 2007935922

All rights reserved.
No part of this book may be reproduced in any form, by photostat, microfilm, retrieval system, or any other means, without the written permission of Kodansha Ltd. (except in the case of brief quotation for criticism or review.)
This work is subject to copyright. All rights are reserved, whether the whole or part of the material is concerned, specifically the rights of translation, reprinting, reuse of illustrations, recitation, broadcasting, reproduction on microfilm or in any other way, and storage in data banks. Duplication of this publication or parts thereof is permitted only under the provisions of the German Copyright Law of September 9, 1965, in its current version, and permission for use must always be obtained from Springer. Violations are liable to prosecution under the German Copyright Law.

Springer is a part of Springer Science+Business Media.
springer.com

© Springer-Verlag Berlin Heidelberg 2008

The use of general descriptive names, registered names, trademarks, etc. in this publication does not imply, even in the absence of a specific statement, that such names are exempt from the relevant protective laws and regulations and therefore free for general use.

Typesetting: Data prepared by SPI Kolam using a Springer T$_{\!E}$X macro package
Cover concept: eStudio Calamar Steinen
Cover production: WMX Design GmbH, Heidelberg

Printed on acid-free paper   SPIN: 11735151   57/3180/SPI   5 4 3 2 1 0

Preface

This book is concerned with *organic* nanoaggregates, showing how to master their growth and expanding on their ability to form central building blocks of next generation submicron-scaled devices. It samples reports and views of the central scientific groups working in this field and provides a complete overview of the state-of-the art of basic research and applications.

As compared to their inorganic counterparts, organic materials are by far superior in that they show extremely large design flexibility, very good possibilities for integration into devices, and brilliant performance. Organic molecules exhibit high luminescence efficiencies at low material densities, flexible spectroscopy and are easy and cheap to process.

The growth of organic matter into nanowires, nanotubes, or nanofibers with widths and heights in the lower nanometer regime and lengths up to several microns is a rather recent development. However, the number of studies on these topics has rapidly increased within the last few years. This development is parallel to that happening within the field of *inorganic* nanowires or carbon nanotubes. In both cases, the time span between the first discovery of how to grow those nanowired materials and their prototypical commercial application was no longer than a few years. Numerous examples can be found for carbon nanotubes. In the case of inorganic semiconducting nanowires, the number of examples such as simple logical gates, UV-light emitting nanolasers or sensors to detect explosives is also rapidly increasing. Such tendencies promise a huge consumer market.

In this volume the contributions focus on organic electroluminescent nanostructures, their fabrication as well as properties, use, and implications. Among the basic problems are those concerned with finding appropriate conditions for self-assembled growth, understanding the interaction with the support as well as neighbouring molecules, and detaching the resulting nanostructures from their growth templates. Anisotropic luminescence, waveguiding, nonlinear optical response, random lasing, electrical mobility, and mechanical stability are subsequently discussed, thus paving the route from basic nanomaterials properties to optimized working devices.

The book starts with two tutorials aiming to make it easier for a broad readership to enter the literature of thin organic film growth as well as characterization methods of organic thin films. Part II (Chaps. 3–8) gives an overview on self-assembled growth of nanostructures from rod like molecular building blocks as well as the crystallography of exemplary organic nanowires. This includes a chapter on how to chemically modify molecular building blocks to influence the linear and nonlinear optical properties as well as the fibre shapes. Lasing and nanooptical aspects are discussed in Chaps. 9 and 10 (Part III: Optics). The book ends with a part on first examples of device-oriented studies, enlightening the high potential of this new type of nanoscaled organic material. The editors would like to express their special thanks to Dr. Frank Balzer who did a tremendous work in assembling the manuscripts into a homogeneous book.

Oldenburg, Sønderborg, Linz, *Katharina Al-Shamery*
June 2007 *Horst-Günter Rubahn*
Helmut Sitter

Contents

List of Acronyms and Symbols XVII

Part I Introduction

1 Fundamentals of Organic Film Growth and Characterisation
H. Sitter, R. Resel, G. Koller, M.G. Ramsey,
A. Andreev, and C. Teichert 3
1.1 General 3
1.2 Nucleation Process and Growth Modes 4
1.3 The Surface Science Approach 7
    1.3.1 In Situ UHV MBE 8
    1.3.2 Valence Band Photoemission (ARUPS) 9
    1.3.3 Near Edge X-Ray Absorption
          Fine Structure Spectroscopy (NEXAFS) 12
    1.3.4 Scanning Tunnelling Microscopy 14
1.4 Crystallographic Characterisation 15
1.5 Fundamentals of Atomic Force Microscopy 16
References 18

2 Optical Characterization Methods for Ultrathin Nanoaggregates
H.-G. Rubahn 21
2.1 Dark Field and Fluorescence Microscopy 21
2.2 Two-Photon Microscopy 22
2.3 Scanning Near-Field Optical Microscopy 24
2.4 Two-Photon Near-Field Microscopy 25
References 28

Part II Growth

3 Growth of Oriented Organic Nanoaggregates via Molecular Beam Deposition
F. Balzer 31
3.1 Introduction 31
3.2 Adsorbates 32
3.3 Silicate Substrates 34
3.4 Phenylenes on Muscovite 36
    3.4.1 *Para*-Hexaphenylene 36
    3.4.2 Steps and Defects 43
    3.4.3 Kinetics 45
    3.4.4 Growth Model 51
3.5 Thiophenes on Muscovite 53
3.6 Microrings 56
3.7 Au–Mica Heterostructures 58
3.8 Conclusions 60
References 62

4 Tailored Organic Nanoaggregates Generated by Self-Assembly of Designed Functionalised *p*-Quaterphenylenes on Muscovite Mica Substrates
K. Al-Shamery, M. Schiek, R. Koch, and A. Lützen 67
4.1 Introduction 67
4.2 Design 68
4.3 Synthesis of Oligomers 70
4.4 Vapour Deposition Studies 72
4.5 Nanoaggregates from Symmetrically Functionalised Oligomers 73
4.6 Nanoaggregates from Non-Symmetrically Functionalised Oligomers 78
4.7 Non-Linear Optical Properties 80
4.8 Quantum Chemical Calculations and Optical Properties ... 82
4.9 Conclusion 83
References 84

5 Hot-Wall Epitaxial Growth of Films of Conjugated Molecules
H. Sitter 89
5.1 Introduction: Why Highly Ordered Organic Thin Films? .. 89
5.2 Experimental Setup 90
    5.2.1 Hot-Wall Epitaxy 90
    5.2.2 Source Materials and Substrates 93
    5.2.3 Characterization Methods 94
5.3 Pristine and Ba-Doped C_{60} Layers 95
    5.3.1 C_{60} Films on Mica Substrates 95

| | 5.3.2 | Doping of C_{60} with Ba .... 98 |
5.4 | Highly Ordered Films of *Para*-Sexiphenyl 100
| | 5.4.1 | Needles and Islands of p-6P on KCl Substrates 102
| | 5.4.2 | Islands and Nanofibers of p-6P on Mica Substrates 106
5.5 | Conclusion 114
References 116

6 Crystallography of Ultrathin Organic Films and Nanoaggregates
T. Haber and R. Resel 119
6.1 Overview 120
6.2 Crystal Structure of Rodlike Conjugated Molecules 122
    6.2.1 Oligoacenes 124
    6.2.2 Oligophenylenes 124
    6.2.3 Oligothiophenes 125
6.3 Experimental Methods 126
    6.3.1 Fundamentals 126
    6.3.2 Specular Scans 127
    6.3.3 Rocking Curves 128
    6.3.4 Pole Figure Technique 130
    6.3.5 Surface Diffraction 131
    6.3.6 Line Profile Analysis 132
    6.3.7 Transmission Electron Microscopy 134
6.4 Crystallographic Order within Nanoaggregates 137
    6.4.1 Out-of-Plane Order 138
    6.4.2 In-Plane Order (Organic Epitaxy) 140
    6.4.3 Relation Between Crystal Structure and Film Morphology 145
    6.4.4 Crystallite Size 148
    6.4.5 Polymorphism 152
6.5 Early Stage Growth 153
6.6 Conclusion 158
References 159

7 Growth and Electronic Structure of Homo- and Hetero-epitaxial Organic Nanostructures
G. Koller and M.G. Ramsey 165
7.1 Introduction 165
7.2 Organic Films on Inorganic Substrates 166
    7.2.1 The Substrate as a Template 167
    7.2.2 Structure and Morphology Determinants: Sticking vs. Diffusion Anisotropy 171
    7.2.3 The Electronic Structure 178
    7.2.4 The Electronic Band Alignment on Nanostructured Interfaces 182

7.3　Organic–Organic Heteroepitaxy 185
　　7.3.1　Growth on Closed –CH-terminated Organic Surfaces 186
　　7.3.2　Growth on Open π-terminated Organic Surfaces 187
7.4　Outlook 191
References 191

8 Mechanisms Governing the Growth of Organic Oligophenylene "Needles" on Au Substrates
K. Hänel and C. Wöll 195
8.1　Introduction 195
8.2　Experimental 197
8.3　The Importance of Molecular Conformations in P4P 198
8.4　Molecular Orientation and Conformation
　　within Ultrathin P4P Films Grown on Gold Substrates:
　　Studies using Soft X-ray Absorption Spectroscopy 198
　　8.4.1　Ultrathin Layer Containing only the α-Species 199
　　8.4.2　Full Monolayer Containing α- and β-Species 202
　　8.4.3　Multilayers 202
8.5　The Orientation of Organic Oligophenylene "Needles"
　　on Gold Substrates 203
8.6　Manipulation of Organic Needles
　　Using an STM Operated under SEM Control 206
　　8.6.1　STM Studies of P4P Needles 207
　　8.6.2　Manipulation of STM Needles 211
　　8.6.3　STM Investigations of the Former Contact Area 214
References 215

Part III Optics

9 Nanooptics Using Organic Nanofibers
K. Thilsing-Hansen, S.I. Bozhevolnyi, and H.-G. Rubahn 219
9.1　Morphology and Optical Response 219
　　9.1.1　Static Response 219
　　9.1.2　Dynamic Response 220
9.2　Guiding of Electromagnetic Waves 225
9.3　Spatial Distribution of Molecular Emitters 228
9.4　The Optical Near Field of Nanofibers 231
　　9.4.1　Single Photon Tunneling Microscopy 231
　　9.4.2　Two-photon Near Field Microscopy 234
9.5　Conclusions 236
References 237

10 Optical Gain and Random Lasing in Self-Assembled Organic Nanofibers
F. Quochi, F. Cordella, A. Mura, and G. Bongiovanni 239
10.1 Introduction 239
10.2 Overview on Random Lasing 240
10.3 Experimental Techniques 241
10.4 Random Lasing and Amplified Spontaneous Emission in Close-Packed Organic Nanofibers 242
10.5 Optical Amplification and Random Laser Action in Single Organic Nanofibers 247
    10.5.1 Coherent Random Lasing in Single Nanofibers 247
    10.5.2 Optical Amplification in Single Nanofibers 252
10.6 Potential Applications of Self-assembled Organic Nanofibers 255
10.7 Summary and Conclusions 257
References 258

Part IV Applications

11 Fabrication and Characterization of Self-Organized Nanostructured Organic Thin Films and Devices
A. Andreev, C. Teichert, B. Singh, and N.S. Sariciftci 263
11.1 Introduction 263
11.2 Experimental Methods 265
    11.2.1 Organic Materials and Growth Techniques 265
    11.2.2 OFET: Device Fabrication 266
    11.2.3 Electrical Characterization Using an OFET (Operating Principle) 267
    11.2.4 Morphological Characterization of Organic Thin Films and Devices 270
    11.2.5 Optical and Structural Characterization of Organic Thin Films and Devices 270
11.3 Anisotropy of Self-Organized Organic Thin Films 271
    11.3.1 Anisotropic Epitaxial Growth of p-6P on Mica(001) 271
    11.3.2 Anisotropic Epitaxial Growth of p-6P on KCl(001) 277
    11.3.3 Anisotropic Epitaxial Growth of p-6P and p-4P on TiO_2 and Metal Surfaces 282
11.4 Luminescent and Lasing Properties of Anisotropic Organic Thin Films 285
11.5 Devices Based on Organic Thin Films 288
    11.5.1 OFETs Based on C_{60} Thin Films Grown by HWE 288
    11.5.2 Anisotropic Current–Voltage Characteristics of p-6P Chains on Mica 292
11.6 Conclusions 295
References 296

12 Device-Oriented Studies on Electrical, Optical and Mechanical Properties of Individual Organic Nanofibers

J. Kjelstrup-Hansen, P. Bøggild, H.H. Henrichsen,
J. Brewer, and H.-G. Rubahn 301
12.1 Introduction 301
12.2 Toward Photonic Devices:
      The Optical Properties of Isolated Nanofibers 303
      12.2.1 Preparation and Optical Detection 303
      12.2.2 Nanofiber Tomography and Angular Light Emission 304
12.3 Studies on Electrical Properties 309
      12.3.1 Charge Injection and Transport 309
      12.3.2 Experiments on Single-Nanofiber Devices 311
12.4 Nanofiber Mechanics 315
      12.4.1 2-D Manipulation 316
      12.4.2 3-D Manipulation 320
12.5 Conclusions 322
References 323

13 Device Treatment of Organic Nanofibers: Embedding, Detaching, and Cutting

H. Sturm and H.-G. Rubahn 325
13.1 Introduction 325
13.2 Coating of Organic Nanofibers on Mica 327
      13.2.1 Parameters
             Related to the Embedding of Organic Nanofibers:
             Thermal Conductivity and Thermal Expansion 327
      13.2.2 Evaporation of Silicon Oxide 329
      13.2.3 Antibleaching Effect with SiO_x Coatings 330
      13.2.4 Microscopical Analysis of Nanofibers on Mica,
             Covered by SiO_x 330
13.3 Parameters Related to the Embedding of Organic Nanofibers:
      Preparation of Polymer Films 338
      13.3.1 Motivation for Encapsulation
             of Nanofibers in Polymers 338
      13.3.2 Essential and Desirable Polymer Properties,
             Preparation Strategies 340
13.4 Cutting of Nanofibers 342
13.5 Conclusions 345
References 346

Index 347

List of Contributors

Katharina Al-Shamery
University of Oldenburg
Institute of Pure
and Applied Chemistry
P.O. Box 2503
D-26111 Oldenburg, Germany
katharina.al.shamery@
uni-oldenburg.de

Andrei Andreev
University of Leoben
Institute of Physics
Franz-Josef-Str. 18
A-8700 Leoben, Austria
andrey.andreev@mu-leoben.at

Frank Balzer
University of Southern Denmark
NanoSYD, Mads Clausen Institute
Alsion 2
DK-6400 Sønderborg, Denmark
fbalzer@mci.sdu.dk

Peter Bøggild
Technical University of Denmark
MIC, Department of Micro-
and Nanotechnology
DTU – Building 345east
DK-2800 Kongens Lyngby, Denmark
boggild@mic.dtu.dk

Giovanni Bongiovanni
Università di Cagliari
Dipartimento di Fisica
I-09042 Monserrato (CA), Italy
giovanni.bongiovanni@
dsf.unica.it

Sergey I. Bozhevolnyi
Aalborg University
Department of Physics
and Nanotechnology
Skjernvej 4A
DK-9220 Aalborg East, Denmark
sergey@physics.aau.dk

Jonathan Brewer
University of Southern Denmark
NanoSYD, Mads Clausen Institute
Alsion 2
DK-6400 Sønderborg, Denmark
brewer@ifk.sdu.dk

Fabrizio Cordella
Università di Cagliari
Dipartimento di Fisica
I-09042 Monserrato (CA), Italy
fabrizio.cordella@dsf.unica.it

Thomas Haber
Graz University of Technology
Institute of Solid Sate Physics
Petersgasse 16
A-8010 Graz, Austria
haber@tugraz.at

Kathrin Hänel
Ruhr-University Bochum
Physical Chemistry I
Universitätstraße 150
D-44801 Bochum, Germany

Henrik H. Henrichsen
Technical University of Denmark
MIC, Department of Micro-
and Nanotechnology,
DTU – Building 345east
DK-2800 Kongens Lyngby, Denmark
henrik.henrichsen@mic.dtu.dk

Jakob Kjelstrup-Hansen
University of Southern Denmark
NanoSYD, Mads Clausen Institute
Alsion 2
DK-6400 Sønderborg, Denmark
jkh@mci.sdu.dk

Rainer Koch
University of Oldenburg
Institute of Pure
and Applied Chemistry
P.O. Box 2503
D-26111 Oldenburg, Germany
rainer.koch@uni-oldenburg.de

Georg Koller
Karl-Franzens University Graz
Institute of Physics
Surface and Interface Physics
Universitätsplatz 5
A-8010 Graz, Austria
georg.koller@uni-graz.at

Arne Lützen
University of Bonn
Kekulé-Institute
of Organic Chemistry
and Biochemistry
Gerhard-Domagk-Str. 1
D-53121 Bonn, Germany
arne.luetzen@uni-bonn.de

Andrea Mura
Università di Cagliari
Dipartimento di Fisica
I-09042 Monserrato (CA), Italy
andrea.mura@dsf.unica.it

Francesco Quochi
Università di Cagliari
Dipartimento di Fisica
I-09042 Monserrato (CA), Italy
francesco.quochi@dsf.unica.it

Michael G. Ramsey
Karl-Franzens University Graz
Institute of Physics
Surface and Interface Physics
Universitätsplatz 5
A-8010 Graz, Austria
michael.ramsey@uni-graz.at

Roland Resel
Graz University of Technology
Institute of Solid-Sate Physics
Petersgasse 16
A-8010 Graz, Austria
roland.resel@tugraz.at

Horst-Günter Rubahn
University of Southern Denmark
NanoSYD, Mads Clausen Institute
Alsion 2
DK-6400 Sønderborg, Denmark
rubahn@mci.sdu.dk

Niyazi Serdar Sariciftci
Institute for Physical Chemistry and
Linz Institute for Organic Solar Cells
Altenbergerstr. 69
A-4040 Linz, Austria
serdar.sariciftci@jku.at

Manuela Schiek
University of Southern Denmark
NanoSYD, Mads Clausen Institute
Alsion 2
DK-6400 Sønderborg, Denmark
schiek@mci.sdu.dk

Birendra Singh
Institute for Physical Chemistry and
Linz Institute for Organic Solar Cells
Altenbergerstr. 69
A-4040 Linz, Austria
birendra.singh@jku.at

Helmut Sitter
Johannes Kepler University Linz
Institute of Semiconductor-
and Solid-State Physics
Altenbergerstr. 69
A-4040 Linz, Austria
helmut.sitter@jku.at

Heinz Sturm
Federal Institute
of Materials Research (BAM)
FG VI.2
Unter den Eichen 87
D-12205 Berlin, Germany
heinz.sturm@bam.de

Christian Teichert
University of Leoben
Institute of Physics
Franz-Josef-Straße 18
A-8700 Leoben, Austria
teichert@unileoben.ac.at

Kasper Thilsing-Hansen
University of Southern Denmark
NanoSYD, Mads Clausen Institute
Alsion 2
DK-6400 Sønderborg, Denmark
kth@mci.sdu.dk

Christof Wöll
Ruhr-University Bochum
Physical Chemistry I
Universitätstraße 150
44801 Bochum, Germany
woell@pc.rub.de

List of Acronyms and Symbols

4P	*para*-quaterphenylene
5P	*para*-quinquephenylene
6P	*para*-hexaphenylene, *para*-sexiphenylene
4T	α-quaterthiophene
6T	α-sexithiophene
α-4T	α-quaterthiophene
α-6T	α-sexithiophene
AES	Auger electron spectroscopy
AFM	Atomic force microscopy
ARUPS	Angular resolved UPS
ASE	Amplified spontaneous emission
BCB	Divinyltetramethyldisiloxane-bis(benzocyclobutane)
CG-mode	Columnar growth mode
CLP4	1,4‴-dichloro-4,1′:4′,1″:4″,1‴-quaterphenylene
CNP4	1,4‴-dicyano-4,1′:4′,1″:4″,1‴-quaterphenylene
cw	Continuous wave
DFT	Density functional theory
DPO	2,5-diphenyl-1,3,4-oxadiazine
EBSD	Electron backscatter diffraction
EPD	Enhanced pole density
ESEM	Environmental scanning electron microscope
FET	Field-effect transistor
FH	Fundamental harmonic
FM-mode	Frank-van der Merwe mode
FWHM	Full width at half maximum
GID	Grazing incidence diffraction
HAS	Helium atom scattering
HLG	HOMO–LUMO gap
HOMO	Highest occupied molecular orbital
HOPG	Highly ordered pyrolytic graphite

XVIII   List of Acronyms and Symbols

HRXD	High resolution XRD
HWBE	Hot-wall beam epitaxy
HWE	Hot-wall epitaxy
ILC	Injection limited current
IR	Infrared
ITO	Indium tin oxide
KAP	Potassium acid phthalate
LEED	Low energy electron diffraction
LPA	Line profile analysis
LSM	Laser scanning microscope
LUMO	Lowest unoccupied molecular orbital
MBE	Molecular beam epitaxy
ML	Monolayer
MOCLP4	1-chloro,4‴-methoxy-4,1′:4′,1″:4″,1‴-quaterphenylene
MOCNP4	1-cyano,4‴-methoxy-4,1′:4′,1″:4″,1‴-quaterphenylene
MONHP4	1-amino,4‴-methoxy-4,1′:4′,1″:4″,1‴-quaterphenylene
MOP4	1,4‴-dimethoxy-4,1′:4′,1″:4″,1‴-quaterphenylene
NA	Numerical aperture
NC-AFM	Non-contact AFM
NEXAFS	Near edge X-ray absorption fine structure spectroscopy
NLO	Nonlinear optics
NMeP4	1,4‴-bis(N,N-dimethylamino)-4,1′:4′,1″:4″,1‴-quaterphenylene
OFET	Organic field-effect transistor
OLED	Organic light-emitting diode
OMBD	Organic molecular beam deposition
OMBE	Organic molecular beam epitaxy
P2P	Biphenyl
p4P, P4P, p-4P	*para*-quaterphenylene
p5P, P5P, p-5P	*para*-quinquephenylene
p6P, P6P, p-6P	*para*-hexaphenylene, *para*-sexiphenylene
PAX	Photoemission of adsorbed Xenon
PES	Photoelectron spectroscopy
PL	Photoluminescence
PMMA	Polymethylmethacrylate
PMT	Photomultiplier tube
PS	Polystyrene
PSP	*para*-sexiphenyl
PSTM	Photon-scanning tunneling microscope
PT	Piezoelectric tube
PTCDA	Perylene-3,4,9,10-tetracarboxylic-3,4,9,10-dianhydride
PTFE	Hexafluoroethylene
PTV	Poly(2,5-thienylene vinylene)
RDS	Reflection difference spectroscopy
RFID	Radio frequency identification device

RHEED	Reflection high energy electron diffraction
RT	Room temperature
SAED	Selected area electron diffraction
SCL	Space-charge limited
SCLC	Space-charge limited current
SEM	Scanning electron microscope
SF-mode	Step flow mode
SFG	Sum frequency generation
SHG	Second harmonic generation
SK-mode	Stranski-Krastanov mode
SNOM	Scanning near field optical microscope
SPA-LEED	Spot profile analysis LEED
STM	Scanning tunneling microscope
TED	Transmission electron diffraction
TEM	Transmission electron microscope
TDS	Thermal desorption spectroscopy
TPI	Two-photon intensity
TPI-SNOM	Two-photon intensity near field optical microscope
UHV	Ultra high vacuum
UPS	Ultraviolet photoemission spectroscopy
UV	Ultraviolet
VM-mode	Volmer-Weber mode
XPS	X-ray photoemission spectroscopy
XRD	X-ray diffraction

Part I

Introduction

1

Fundamentals of Organic Film Growth and Characterisation

H. Sitter, R. Resel, G. Koller, M.G. Ramsey, A. Andreev, and C. Teichert

1.1 General

Currently, technical applications of organic nanoaggregates are discussed for various purposes like displays, electronic circuits, sensors and optical waveguides. In all these cases, the optical and electronic properties (and also the combination of both) are decisive parameters. Why are the structural properties of the nanoaggregates so important? The reason is the strong relationship between the structure (arrangement of the molecules within the bulk state) of the nanoaggregates and the application-relevant properties: optical emission and electronic charge transport.

The optical absorption and emission of organic molecules are highly polarised. Therefore, a strong optical anisotropy is observed for ordered bulk materials [1,2]. For example, in case of nanoneedles of the molecule sexiphenyl (or hexaphenyl) light with polarisation along the long molecular axes is emitted [3,4]. Also the electronic charge transport shows anisotropic behaviour within an ordered bulk state [5]. It is generally accepted that the intermolecular charge transport happens along overlapping π-conjugated segments of neighbouring molecules [6,7]. Therefore, specific directions within an ordered bulk state show enhanced ability for charge transport.

Applications based on nanoaggregates which are prepared on surfaces are always associated with specific directions of light absorption/emission and charge transport. The orientation of the molecules as well as the directions of dense π-packing of neighbouring molecules are defined by the crystalline properties of the nanoaggregates. The orientation of the crystallites on the substrate surface is determined by the orientation of the molecules relative to the substrate surface. The alignment of the crystallites along specific surface directions is determined by the in-plane alignment of the molecules relative to the substrate.

One further important parameter for application is the size and shape of the crystalline domains. Generally, large single crystalline domains with regular shape are preferred. Grain boundaries disturb the regular lattice and interrupt the periodic overlapping of neighbouring molecules, this considerably influences the charge transport [8, 9].

1.2 Nucleation Process and Growth Modes

Each crystallisation process can be considered as a phase transition from a mobile phase into the solid phase of the crystal. In the case of layer growth the mobile phase, in our case the vapour is deposited on a solid surface. Consequently, an adsorption process is the first physical step in any deposition process. There are two types of adsorption. The first is physical adsorption, often called *physisorption*, which refers to the case where there is no electron transfer between the adsorbate and the substrate, and the attractive forces are van der Waals type. This type of adsorption is very typical for most of the organic molecules impinging on a substrate surface. The second type of adsorption is *chemisorption*, which refers to the case when electron transfer takes place between the adsorbate and the substrate, which means a chemical bond is formed. The forces are then of the type occurring in the appropriate chemical bond. This case is typical for the deposition of inorganic materials, which are usually at first physisorbed, undergo a dissociation process and are finally chemisorbed and incorporated into the growing crystal. In general, adsorption energies for physical adsorption are smaller than for chemical adsorption, which means that organic molecules are much weaker bound on the substrate in comparison to inorganic materials forming compounds.

It is experimentally well documented that thin films of organic and inorganic materials are formed by a "nucleation and growth" mechanism. Due to their growth, small clusters of atoms or molecules, so-called *nuclei*, are formed which further agglomerate to form islands. As growth proceeds, agglomeration increases, chains of islands are formed and eventually join up to produce a continuous deposit, which still can contain channels and holes. These holes eventually fill up to give a continuous and complete film and further growth leads to smoothening of the surface irregularities or in contrary these defects can be enhanced by further overgrowth. Consequently, understanding and controlling the process of layer formation by epitaxy implies in each case a detailed knowledge of the nucleation and growth modes.

Nucleation is the spontaneous formation of small embryonic clusters with a critical size determined by the equilibrium between their vapour pressure and the environmental pressure. The nuclei are formed in a metastable supersaturated or undercooled medium. Their appearance is a prerequisite for a macroscopic phase transition to take place. This means that nucleation is a precursor of the crystallisation process. Due to their increased surface/volume ratio, the critical clusters have more energy than the bulk phase of the same mass, hence they have a chance to survive and form a macroscopic entity of the new stable phase. Figure 1.1 shows the relationship between the vapour pressure of liquid droplets and their size [10]. If by any chance the critical

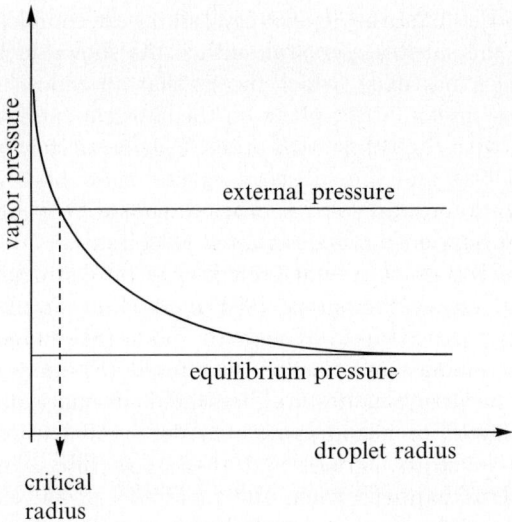

Fig. 1.1. Vapour pressure of small droplets in relation to their size (reprinted with permission from [10])

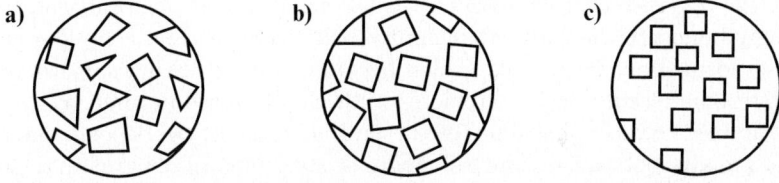

Fig. 1.2. Three intergrowth relations between deposit and substrate: (a) deposit is fully non-oriented, (b) texture orientation where the deposit planes are parallel to the substrate surface and (c) deposit exhibits texture and azimuthal orientation, a so-called epitaxial orientation to the substrate (reprinted with permission from [11])

radius of the droplet, defining its critical size, increases slightly, the droplet or cluster will continue to grow until a macroscopic two-phase equilibrium is attained. In the reverse case the droplet will disappear. This indicates that the energetics of layer formation is governed by the energetics of the critical nucleus.

Epitaxial nucleation is a special case of heterogeneous nucleation in which different kinds of intergrowth relations between the deposit and the substrate may occur. The orientation behaviour of the growing layer on a single crystalline substrate may not only be influenced by the nucleation process of the deposited material, but also by the subsequent growth on the substrate. In Fig. 1.2 three possible, basically different kinds of intergrowth relations between deposit and substrate are schematically shown [11].

According to Fig. 1.2a, the deposit crystallites are completely non-oriented with respect to the substrate crystal surface. As shown in Fig. 1.2b, an orientation of texture may exist, which means that all deposit crystallites grow with the same low index lattice plane on the substrate surface, however they are not oriented with respect to each other. Finally, as shown in Fig. 1.2c, the deposited crystallites and the substrate crystal show both the textural and azimuthal orientation towards one another. It should be emphasised that only these orientation relationship is designated as epitaxial.

Five different modes of crystal growth may be distinguished in epitaxy. These are the Volmer–Weber mode (VW-mode), the Frank–van der Merwe mode (FM-mode), the Stranski–Krastanov mode (SK–mode), the columnar growth mode (CG-mode) and the step flow mode (SF-mode) [12]. These five most frequently occurring modes are illustrated schematically in Fig. 1.3.

In the VW-mode, or island growth mode, small clusters are nucleated directly on the substrate surface and then grow into islands of the condensed phase. This happens when the molecules of the deposit are more strongly bound to each other than to the substrate. This mode is displayed by many systems growing on insulators like mica or organic substrates, because there are no unsaturated dangling bonds on the surface to form a strong bond between impinging molecules and the substrate. The FM-mode displays the opposite characteristics. Because the atoms or molecules are more strongly bound to the substrate than to each other, the first complete monolayer is formed on the surface, which becomes covered with a somewhat less tightly bound second layer. Providing that the decrease in binding strength is monotonic towards the value for the bulk material of the deposit, the layer growth mode is obtained. The SK-mode, or layer plus island growth mode, is an "intermediate" case. After forming the first monolayer, or a few monolayers, subsequent layer growth is unfavourable and islands are formed on top of this

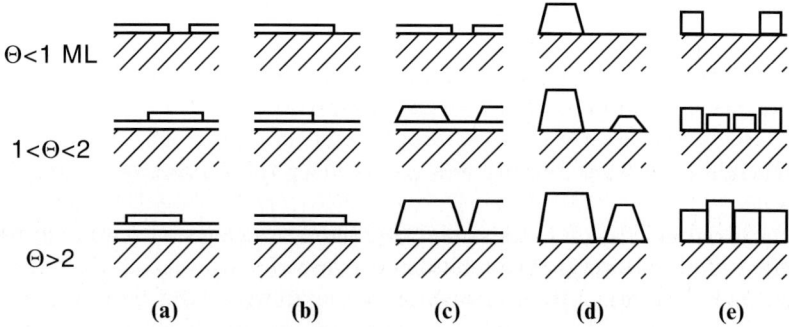

Fig. 1.3. Five crystal growth modes most frequently occurring on flat surfaces of substrate crystals: (**a**) layer-by-layer or Frank–van der Merwe, (**b**) step flow, (**c**) layer plus island or Stranski–Krastanov, (**d**) island or Volmer–Weber and (**e**) columnar growth mode. Θ represents the coverage in monolayers (reprinted with permission from [12])

intermediate layer. There are many possible reasons for this mode to occur and almost any factor which disturbs the monotonic decrease in binding energy characteristic for the layer-by-layer growth may be the cause. It occurs especially in cases when the interface energy is high (allowing for initial layer-by-layer growth) and the strain energy of the film is also high (making reduction of strain energy by islanding favourable).

The fifth mode, the CG-mode, shows some similarities with the SK- and VW-modes, however, it is fundamentally different. In the SK-mode as well as in VW-mode, with increasing thickness, the condensed islands tend to merge and cover the whole substrate surface. Although the grown film may exhibit variations in thickness and presence of structural defects at the interface where the islands merge, it forms a connected structure, in which the density of the film is homogeneous. In contrast, the films grown by CG-mode usually thicken without the merger of columns. As a result, columns usually remain separated throughout the growth process of the film, and the films grown in this mode are easily fractured. The CG-mode occurs where low surface mobility of the impinging molecules leads to the formation of columns of the deposited material.

Beside the described growth modes, the so-called *step flow growth mode* is observed on non-atomically flat substrate surfaces. That can occur, if the substrate surface contains cleaving steps or if the substrate is cut slightly misoriented from a low index plane, however the surface on the planes between the steps, the so-called *terraces*, is atomically flat. Two-dimensional nucleation may occur on the terraces, when the substrate temperature is sufficiently low, or the flux of the constituent molecules of the growing film is to high and therefore prohibits fast surface migration of the molecules. In this case the film may grow on the terraces in the FM- or SK-modes. However, if the substrate temperature is high enough or the flux is sufficiently low, then the impinging molecules can be mobile enough on the terraces that they become incorporated directly into the step edges. In this case, growth of the epitaxial film proceeds by the advancement of steps along the terraces.

1.3 The Surface Science Approach

The surface science approach involves in situ ultrahigh vacuum (UHV) growth on well defined and characterised generally single crystal substrates. The large variety of UHV surface science techniques available have been developed in the last 40 years, because of the needs of the inorganic semiconductor industry on the one hand and the aim to understand catalysis on the atomic/molecular level on the other hand. The blending of surface physics and chemistry into surface science provides the tools of both sciences and the interdisciplinary approach, appropriate for investigating organic semiconductor interfaces, structures and films.

Organic materials have a high propensity to crystallise, as the literature of sexiphenyl films grown on mica or the alkali halides illustrates, where numerous crystallite orientations, often co-existent, are reported (see Chap. 5; [13]). Despite the substrates being single crystals, this (unwanted) richness of organic structures is most likely due to the substrates being cleaved in air and is thus dependent on uncontrolled parameters such as the humidity of the day or the time it took to insert the sample into the growth chamber. The multi-technique approach and the controlled UHV experimental environment for both substrate and organic film preparation appear often rather slow and cumbersome, but the reader should bear in mind that: *You do not need UHV to fabricate organic structures, but you do need it to understand what is happening.*

1.3.1 In Situ UHV MBE

Moderately sized molecules, such as sexiphenyl, sexithiophene, pentacene and other oligomers as base materials for surface science studies, are available in the high purity required for controlled experiments. These can be evaporated from glass or metal crucibles, in a reproducible manner in UHV. Generally several days of thorough out-gassing in UHV are sufficient to purify the materials, here water is the principle problem, after which a background pressure in the 10^{-10} mbar range during evaporation is achievable. The latter is vital for reproducible experiments, particularly for more reactive substrates.

Despite the fact that many oligomers of interest have relatively low vapour pressures and are thus UHV compatible, care should be taken with evaporator design. There is a belief that the molecules "pump" and thus care need not be taken – this is a myth. They can coat chamber walls, etc., and lower the residence time of residual gases (in particular, water) and thus improve pumping rate and the system's base pressure. In standard high vacuum device manufacturing chambers, this can lead to the partial pressures of residual gases being dependent on the chambers history. In UHV chambers, hot filaments in ion gauges, evaporators, sample heaters, etc., crack/desorb the molecules, leading to a partial pressure of molecular fragments, which makes controlled studies difficult. This is a particular problem one often faces in synchrotron end stations and other multiple user facilities. As a consequence, indiscriminate evaporation should be avoided and collimators to direct the evaporant should be used.

The propensity of the molecules to crystallise and grow three-dimensional nanostructures makes the use of surface-sensitive electron spectroscopies, such as X-ray photoemission (XPS) or Auger spectroscopy (AES), to quantify the amount of material applied very unreliable as they are area averaging techniques. Quartz microbalance measurements of the rate of evaporation are necessary. Here, quantities are quoted in equivalent film thicknesses assuming the bulk density of the evaporant. In surface science, a monolayer is defined as the number of adsorbate units equal to the number of substrate surface atoms. It is used more loosely for the large organics as the amount of material required to coat the substrate with a single layer. As the molecules are extremely anisotropic, this varies from $\approx$ 3–4 Å for a layer of lying molecules to $\approx$ 25 Å for a monolayer of standing molecules, for the molecules and substrates considered here.

1.3.2 Valence Band Photoemission (ARUPS)

In photoelectron spectroscopy (PES), electrons are photoemitted by photons of energy $h\nu$ according to the photoelectric effect, and the kinetic energy of the photoelectrons is determined by the conservation law of energy following the modified Einstein relation $E_{\text{kin}} = h\nu - E_{\text{B}} - \phi$, where E_{B} is the binding energy of the electrons in the solid referenced to the Fermi level E_{F} and ϕ, the work function of the sample. As a result of the $h\nu$ dependence of the photoionisation cross-section ultraviolet (UV) photoelectron spectroscopy (UPS) is more suited to probe the valance electronic structure than X-ray photoelectron spectroscopy (XPS), which in turn is used to study core electron states. In addition to the energy, in angular-resolved UPS (ARUPS), the momentum of the photoelectron is measured, which can be related to the momentum of the electron in the solid. This allows the experimental determination of the electronic band structure of solids. Of particular relevance for ARUPS of molecular adsorbates are symmetry-derived polarisation-dependent selection rules for emissions in high symmetry directions of the adsorption complex. These selection rules can be derived from simple group theoretical arguments, and allow a qualitative symmetry analysis of ARUPS data [14–17].

With the example of benzene on Ni(110) in Fig. 1.4, the relationship between the UPS spectra of isolated molecules (gas phase) and those in adsorbed molecular monolayers and multilayer films is illustrated. In the measurement, the substrate and electron spectrometer are in thermodynamic equilibrium, and for the solid state the natural energy reference is the Fermi level (E_{F}). To facilitate comparison to the gas phase spectrum, the solid-state spectra have been referenced to the vacuum level (E_{vac}) by off-setting them by the work function as measured from the secondary electron cut-off in the UPS spectra [18]. In this case, the work function measured for both monolayer and condensed benzene multilayer is 4 eV and thus the Ni substrate Fermi edge is at 4 eV with respect to E_{vac}. In going from gas phase to the molecular solid, there is a rigid shift of all orbital emissions to lower binding energy (higher kinetic energy) due to the extra-molecular relaxation (Δ_{relax}); the electrons of neighbouring molecules are responding to the photohole and screening it. Between second and higher layers and the monolayer, there can be a further rigid shift to lower binding energy due to enhanced screening from the substrate. Additionally, the orbitals that are involved in the bonding to the substrate can have a differential shift to higher binding energy due to bond stabilisation. This is well expressed by the π orbitals involved in the benzene bond to Ni in Fig. 1.4. If, however, there is only an electrostatic bond to the substrate, as

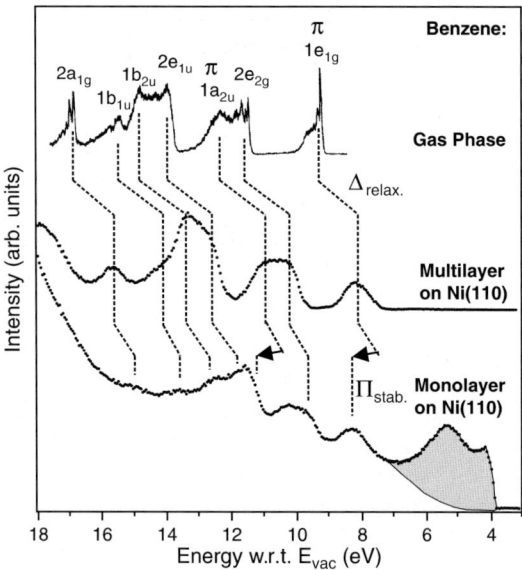

Fig. 1.4. Comparison of UPS spectra of benzene in the gas phase, a condensed film on Ni(110) and a benzene monolayer on Ni(110). The correlation between the orbitals in the three cases is indicated

for instance in the case of benzene on Al(111) [19], not only are no differential shifts observed, there is only a small shift due to substrate screening. Here the weaker bonding results in the molecules being further from the surface and its screening ability. For the larger π-conjugated molecules even when there is strong bonding to the substrate there is almost no screening shift between the monolayer and higher layers [20]. Presumably, the polarisability of the large conjugated systems such as sexiphenyl is as effective as the metal substrate in screening the photohole. Apart from the loss of vibrational fine structure, due to solid-state broadening, the spectral shape of the isolated molecule and the condensed film is very similar. In contrast, the monolayer is somewhat different as it is well ordered and photoemission selection rules are in play making the relative intensities of the orbital emissions very strongly with experimental geometry. To observe all orbitals and their energy positions correctly, spectra need to be taken in a variety of experimental geometries. For the smaller molecules, such as benzene these selection rules have often been exploited to determine the absorbed molecules symmetry and thus the molecular geometry [17, 21]. As will be seen in Chap. 7, strong angular effects in the UPS spectra of monolayers and crystalline structures of the longer oligomers can be used to infer their orientation. A strict group theory selection rule analysis, however, is difficult due to the multiplicity of near degenerate orbitals in the larger molecules.

Figure 1.5 illustrates the development of the valence band structure with increasing chain length by comparing the UPS spectra from films of benzene,

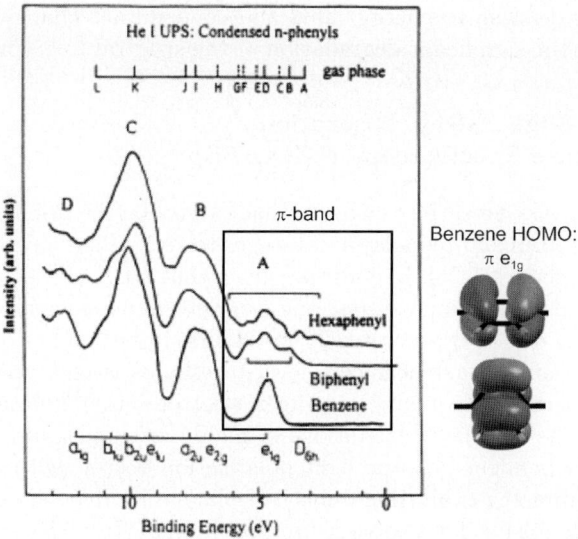

Fig. 1.5. Development of the valence band structure of n-phenyls with increasing chain length by comparing the spectra from films of benzene, biphenyl and sexiphenyl (reprinted with permission from [20])

biphenyl and sexiphenyl. It very clearly shows the development of the all important π band with oligomer length, while the appearance of the higher binding energy bands is essentially the same, as they are derived from intraring orbitals. The lowest binding energy emission of benzene results from the two degenerate π HOMO orbitals shown. In going to biphenyl the degeneracy is lifted, with two inter-ring bonding/anti-bonding pairs resulting. The pair of the first orbital has a high spatial overlap and consequently a large bonding/anti-bonding energy spread of almost 2 eV. The lower spatial overlap of the second results in a small energy separation of the two orbitals. In this seminal gas phase work on the dimer, Turner [22] called these the non-bonding orbitals. In angle-resolved UPS spectra of ordered films of biphenyl and bithiophene, these non-bonding orbitals can be distinguished and are seen to have an energy separation of 0.2–0.3 eV [23–25]. Increasing the oligomer length to sexiphenyl one sees the unresolved peak of the six near degenerate non-bonding π orbitals flanked by three anti-bonding and three bonding π orbitals, with the later merging into band B.

Photoemission requires conductive materials to avoid energy shifts due to charging. With the organics, high nanostructures, which can cause charging shifts, can form even at relatively moderate exposures. Both UV and X-ray radiation can also result in degradation of the molecules either directly or indirectly, via the secondary electrons from the substrate or, in the case of laboratory XPS, from the Al window of the source. The effect on the spectra can be subtle with cationic or anionic defects created effectively doping the materials, leading to energy shifts and concomitant changes in the work function without significant degradation of the spectral fingerprint [26].

1.3.3 Near Edge X-Ray Absorption Fine Structure Spectroscopy (NEXAFS)

NEXAFS is a very versatile technique that can probe the orientation of molecular systems from sub-monolayer coverages to tens of nanometer thick films, if applied correctly [27, 28]. In addition, given that NEXAFS probes the unoccupied molecular electronic states, one can determine changes of these, that may occur, e.g. due to charge transfer in the first layer bond or on doping [29].

The X-ray absorption near core level thresholds is measured via, for example, the yield of secondary or Auger electrons as a function of photon energy [30]. The core level excitations of highly oriented molecules exhibit a pronounced dependence on the light polarisation vector, with excitations of π^* and σ^* symmetry exhibiting opposite polarisation dependance.

Basic prerequisites for a correct interpretation of NEXAFS spectra include normalisation, background subtraction and a clear assignment of the molecular orbital symmetries (π^* or σ^*). A normalisation to the pre-absorption edge intensity is often appropriate. Background subtraction is particularly vital for films of sub-monolayer or monolayer coverage and can be done by either subtraction – or division – by a clean substrate spectrum or division by a simultaneously recorded gold reference spectrum [30]. To facilitate assigning the observed resonances of a NEXAFS spectrum, a simple model the so-called *building block principle* was proposed, which states that, due to the localisation of the excitation at the core hole position, one can describe larger molecules as a superposition of their smaller subunits, e.g. the larger oligomers should appear just as their monomers [30]. While often appropriate for non-conjugated molecules, it can no longer be applied for the conjugated molecules of organic electronics as illustrated in Fig. 1.6. Here the experimental data of thiophene oligomers, together with their corresponding density functional theory (DFT) calculations, illustrate how the carbon K-edge NEXAFS spectra become increasingly complex with increasing oligomer length [31]. The excellent agreement between experiment and theory should be noted, which allows a clear assignment of the orbital symmetries of complex molecules, such as the here displayed sexithiophene. In the top of Fig. 1.6, the azimuthal dependence of uniaxially oriented sexithiophene molecules grown on TiO_2 is shown: if the light polarisation is along the molecules, the π^* orbitals are all but invisible and the σ^* excitation is prominent, while for a polarisation across the molecules, the π^* orbitals are intense. Analysis of the π^* intensity as a function of X-ray incidence angle yields the average angle of the aromatic plane with respect to the substrate.

It is important to note that the measured signal in NEXAFS is an average over all molecules probed and thus one obtains the average orientation over the organic film surface to a depth from $\approx 20\,\text{Å}$ to several hundred Ångstroms

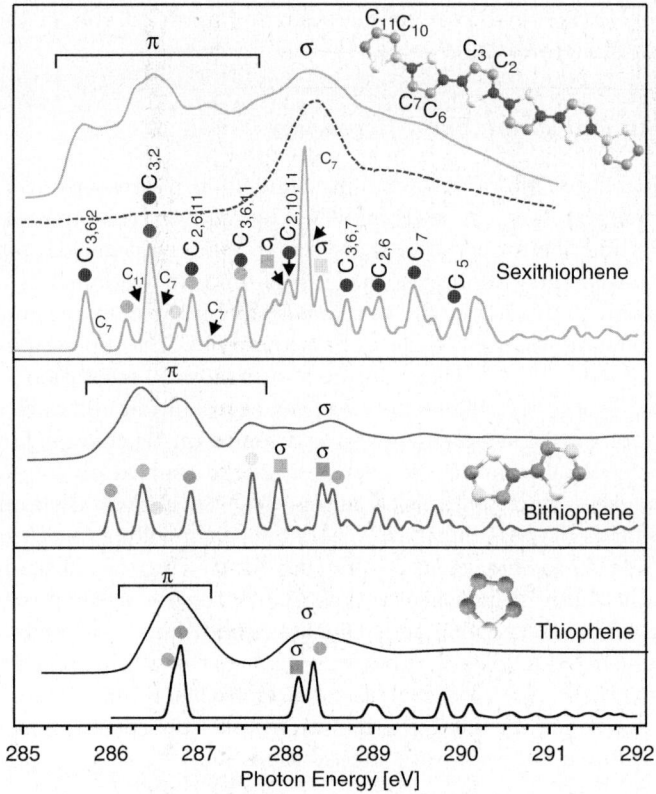

Fig. 1.6. DFT [31] calculated NEXAFS spectra (*bottom curves*) of thiophene, bithiophene and sexithiophene together with the experimental data. A rigid shift of 1.8 eV was applied to match the experiment. In the experimental geometry of the dashed spectrum of sexithiophene, only σ^* symmetric transitions are allowed

(depending on the detection method). This is in contrast to X-ray diffraction, where only molecules in a crystalline arrangement contribute to the signal. The unselective nature of NEXAFS is problematic for non-uniform films with several orientations, since NEXAFS will arrive at an average orientation. Or consider the case of inhomogeneous film growth, where starting from one particular molecular orientation, molecules with a second, different orientation continue to grow, here NEXAFS will wrongly suggest a gradual reorientation in the film. To unambiguously interpret NEXAFS results requires that either a NEXAFS resonance is forbidden for an experimental geometry, as for the sexithiophene example of Fig. 1.6, or the help of a second technique, here atomic force microscopy (AFM), is often quite useful for a first evaluation, as areas with different molecular orientation generally also differ in morphology. A number of synchrotron radiation end stations are being developed around the world that allow spatially resolved photoemission and absorption spectroscopies, these will play an increasingly important role in the study of organic nanostructures [32].

1.3.4 Scanning Tunnelling Microscopy

The physical principle of the scanning tunnelling microscopy (STM) is the tunnelling current that can be established by applying an external bias voltage across a junction formed by a sharp metal tip brought in close proximity to a sample, so that the wave functions of sample and tip overlap. The magnitude of the tunnelling current depends exponentially on the distance, and by scanning the tip along the surface it can be used to obtain a topographical image of clean and adsorbate covered surfaces with atomic resolution. An example is shown in Fig. 1.7, where the exceptional resolution allows to determine the adsorption site of sexiphenyl on an oxygen-reconstructed Ni(110) surface. For the molecule shown, all six phenyl rings are centred on fourfold hollow positions of the underlying nickel atoms [33]. Such clear determination of the molecular registry is only rarely possible as tunnelling conditions appropriate for viewing the molecule are often bad for viewing the substrate and vice versa. Another interesting aspect is the origin of the observed molecular image contrast in STM. By varying the tunnelling conditions, one might expect to tunnel into selected molecular orbitals and image them. This is sometimes the case. However, the observed contrast is often rather robust over a wide bias voltage range, since the images originate from a superposition of individual molecular orbitals and substrate contributions [33, 34].

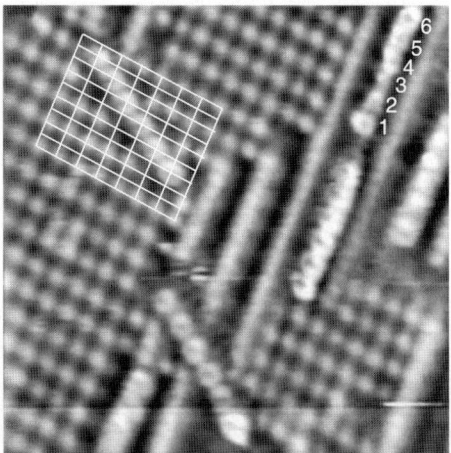

Fig. 1.7. Atomic scale image (80×80 Å^2; 1.5 V, 0.1 nA) of individual sexiphenyl molecules. A (1×1) grid, where the corners match the underlying Ni(110) fourfold hollow positions, has been superimposed on the molecule in the top left corner (reprinted with permission from [33])

The major strengths of STM are the ability to determine the adsorption sites, the two-dimensional periodicity (even for very small domain sizes) and to evaluate changes to the substrate due to molecular adsorption.

A second technique that can evaluate the two-dimensional periodicity of the organic layer is low energy electron diffraction (LEED). While principally very accurate in determining the adsorbate periodicity, with most organic systems it faces the problem of severe damage under the beam of low energy electrons [35]. This can sometimes be improved by using very low electron beam currents or by cooling the sample.

1.4 Crystallographic Characterisation

While crystal structure investigations of inorganic nanoaggregates are frequently performed, organic nanoaggregates are much less studied. The experimental difficulties are small diffraction signals in case of X-ray diffraction and neutron diffraction. Surprisingly, neutron diffraction works for thin organic films [36]. The advantage of electron diffraction studies is the good sensitivity, however the degradation of the organic material due to the electron bombardment is a real problem.

X-ray diffraction studies of thin organic films also show some difficulties. The low scattering probability of the involved atoms (mainly carbon and hydrogen) together with the small scattering volume result in diffraction patterns with small intensities. However, thin films of molecular crystals show acceptable intensity which can be observed by standard laboratory equipment. In case of films with low crystallographic order and also for ultrathin films (with a thickness of few monolayers), synchrotron radiation has to be used. The low crystal symmetry (triclinic, monoclinic or orthorhombic) together with large lattice constants imply a huge number of diffraction peaks which are frequently overlapping.

Good results for structural characterisation of organic nanoaggregates are obtained by combining different diffraction techniques. X-ray diffraction has low sensitivity but high resolution regarding the reciprocal space. It can be combined quite nicely with transmission electron microscopy techniques which have high sensitivity with high spacial resolution. Supplementary information to the crystalline order is the morphology of the nanoaggregates; it can be obtained by diverse microscopy methods. In case of transmission electron techniques both, diffraction and microscopy, can be combined so that the morphology and the crystal structure properties can be related.

The main goals of crystallographic investigations are the determination of basic crystalline properties like crystalline phases, preferred orientations (or epitaxial orders) of crystallites and crystal sizes. The correlation of crystallographic order with the film morphology reveals basic information on the growth mechanisms of organic nanoaggregates on surfaces.

1.5 Fundamentals of Atomic Force Microscopy

AFM is nowadays the favoured imaging technique to investigate organic nanostructures because it is applicable under different environmental conditions and does not require conducting or semiconducting surfaces like STM does. Moreover, due to the imaging principle of AFM, where a sharp probe is scanned across the surface, the resulting image represents a three-dimensional topography $z_{ij} = z(x,y)$. Assuming a proper calibration of the piezoelectric scanner and an infinitely sharp AFM tip, the measured topography corresponds to the true three-dimensional topography of the investigated surface. In the real case of finite dimensions of the probe, the probe geometry has to be taken into account [37]. However, for lateral structure sizes that exceed the tip radius and structure slopes less than half of the opening angle of the AFM tip, this effect can be neglected. Thus, nanostructure size, three-dimensional shape and arrangement can be detected and quantitatively analysed [38]. Moreover, by analysing one-dimensional cross sections through an AFM image, z_{ij}, information on layer distribution and on step heights can be derived. For example, as will be shown in several chapters, terraces and island heights analysis in conjunction with corresponding crystallographic data (X-ray measurements) yields a worthwhile information about the orientation of organic molecules relative to the substrate surface ("lying-down" or "standing-up" growth in organic epitaxy). For three-dimensional islands, straight segments in the cross sections along certain directions reveal the existence of side facets. Analysing the histograms of orientations of local surfaces may yield not only orientation of the facets, but also quantitative information on their area fraction. With respect to self-organised, quasi-periodic structures, two-dimensional power spectrum formalism can be applied to the $z_{ij} = z(x,y)$ matrix, to deduce average real space orientation, symmetry and lateral periodicity of the surface pattern. All these AFM possibilities are of special importance for organic thin films growth, where additional degrees of freedom (orientational and vibrational) of extended organic molecules can result in a large variety of differently oriented three-dimensional morphological features. For further details concerning fundamentals of quantitative analysis of AFM measurements, the reader is referred to Chap. 2.2 of review paper [38].

As shown in Fig. 1.8, three different measurement modes can be generally realised during AFM depending on sample tip separation: contact (C), non-contact (NC) and intermittent contact (IC, often called also as *tapping* mode) modes. In contact mode, i.e. in the repulsive region of the van der Waals forces (at tip sample distances below 1 nm), the interaction between the tip and the surface results in the deflection of the cantilever (on which the interacting tip is mounted), which thus acts as a force sensor. A feedback loop keeps the static deflection constant by adjusting the vertical position of the cantilever base, providing a topographic image. In this mode scanning can be destructive for both the tip and the surface, which is especially true for soft organic samples. This limitation is circumvented in tapping mode AFM

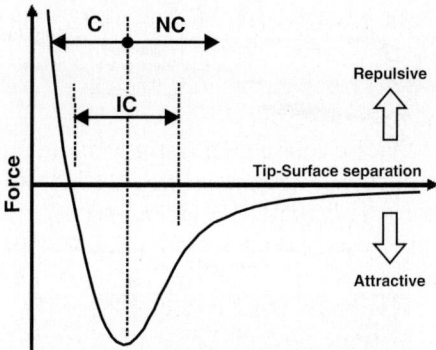

Fig. 1.8. Dependence of van der Waals force on tip sample separation. Contact (*C*), non-contact (*NC*) and intermittent contact (*IC*) modes are shown (reprinted with permission from [39])

(IC in Fig. 1.8). Here, the tip is mechanically oscillated close to its resonance frequency. On approaching the surface, the oscillation amplitude decreases and its phase changes due to the interaction with the surface. Most often the amplitude is then used as the detection signal and scanning is performed so that the oscillation amplitude is kept at predefined set point value smaller than the oscillation amplitude in free space. The phase shift can also be monitored. In this mode, the tip interacts with the surface most strongly in the lower part of the trajectory. Finally, in non-contact AFM (NC-AFM), the tip oscillates in the attractive region of the van der Waals forces and the resonant frequency shift of the cantilever measured by a phase lock loop is used as a feedback signal, providing true non-invasive imaging [41]. Note that high and even atomic resolution is possible in both contact mode AFM and NC-AFM. For further details concerning different variants of the AFM, the reader is referred to a recent paper [39].

One of the advantages of AFM tapping mode is that it allows to measure a phase contrast on the surface, which means that it is possible to distinguish between softer and harder surface areas [42]. Operating principle of such phase contrast imaging is shown in Fig. 1.9. The phase shift between the cantilever oscillating signal and the cantilever output signal from the split photodiode is monitored here simultaneously with topography while operating in tapping mode. As it can be seen in Fig. 1.9, a harder material induces consequently a much stronger phase shift than softer one, where the tip can penetrate a bit into the material, see also Fig. 11.15 in Chap. 11. As a consequence, the phase shift image reflects the changes in the viscoelastic properties of the sample surface [42]. However, in the practical use of this method, it can happen that morphological features with high and sharp edges can influence the phase measurements. Therefore, it is always needed to carefully compare corresponding phase and height images, to be sure that the detected phase contrast is not just a morphological artefact.

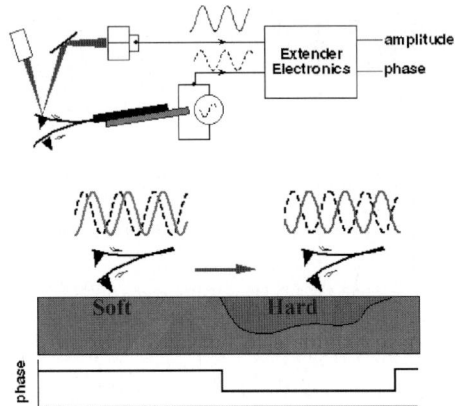

Fig. 1.9. Working principle of the phase shift imaging in tapping mode AFM (reprinted with permission from [40])

For further details concerning fundamentals of AFM measurements, the reader is referred to review papers [38, 39].

References

1. F. Gutmann, L. Lyons, *Organic Semiconductors* (Wiley, New York, 1967)
2. M. Pope, C. Swenberg, *Electronic Processes in Organic Crystals and Polymers*, 2nd edn. (Oxford University Press, New York, 1999)
3. T. Mikami, H. Yanagi, Appl. Phys. Lett. **73**, 563 (1998)
4. F. Balzer, V. Bordo, A. Simonsen, H.G. Rubahn, Appl. Phys. Lett. **82**, 10 (2003)
5. N. Karl, *Charge-Carrier Mobility in Organic Crystals* (Springer, Berlin Heidelberg New York, 2003)
6. E. Silinsh, V. Capek, *Organic Molecular Crystals* (American Institute of Physics, New York, 1994)
7. J. Bredas, J. Calbert, D. de Silva, J. Cornil, Proc. Natl Acad. Sci. USA **99**, 5804 (2002)
8. J. Laquindanum, H. Katz, A. Lovinger, A. Dodabalapur, Chem. Mater. **8**, 2542 (1996)
9. G. Horowitz, M. Hjlaoui, Adv. Mater. **12**, 1046 (2000)
10. M. Gebhart, *Crystal Growth: An Introduction* (North Holland, Amsterdam, 1973), Chap. Epitaxy, p. 105
11. G. Stringfellow, Rep. Prog. Phys. **45**, 469 (1982)
12. M.A. Herman, W. Richter, H. Sitter, *Epitaxy – Physical Principles and Technical Implementation, Springer Series in Material Sciences 62* (Springer, Berlin Heidelberg New York, 2004)
13. E.J. Kintzel, D.M. Smilgies, J.G. Skofronick, S.A. Safron, D.H.V. Winkle, J. Cryst. Growth **289**, 345 (2006)
14. S. Huefner, *Photoelectron Spectroscopy* (Springer, Berlin Heidelberg New York, 1995)

15. S.D. Kevan (ed.), *Angle-Resolved Photoemission* (Elsevier, Amsterdam, 1982)
16. E.W. Plummer, W. Eberhardt, Adv. Chem. Phys. **49**, 533 (1982)
17. H.P. Steinrück, J. Phys.: Condens. Matter **8**, 6465 (1996)
18. K. Jacobi, *Landolt-Börnstein (New Series)*, vol. 24 (Springer, Berlin Heidelberg New York, 1994), Chap. Electronic structure of surfaces, p. 56
19. R. Duschek, F. Mittendorfer, R. Blyth, F. Netzer, J. Hafner, M. Ramsey, Chem. Phys. Lett. **318**, 43 (2000)
20. M.G. Ramsey, D. Steinmüller, M. Schatzmayr, M. Kiskinova, F.P. Netzer, Chem. Phys. **177**, 349 (1993)
21. F.P. Netzer, M.G. Ramsey, Crit. Rev. Solid State Mater. Sci. **17**, 397 (1992)
22. J.P. Maier, D.W. Turner, Faraday Discuss. Chem. Soc. **54**, 149 (1972)
23. M.G. Ramsey, D. Steinmüller, F.P. Netzer, J. Chem. Phys. **92**, 6210 (1990)
24. G. Koller, F.P. Netzer, M.G. Ramsey, Surf. Sci. **421**, 353 (1999)
25. G. Koller, R.I.R. Blyth, S.A. Sardar, F.P. Netzer, M. Ramsey, Surf. Sci. **536**, 155 (2003)
26. D. Steinmüller, M.G. Ramsey, F.P. Netzer, Phys. Rev. B **47**, 13323 (1993)
27. G. Koller, S. Berkebile, J. Krenn, G. Tzvetkov, C. Teichert, R. Resel, F.P. Netzer, M.G. Ramsey, Adv. Mater. **16**, 2159 (2004)
28. B. Winter, S. Berkebile, J. Ivanco, G. Koller, F.P. Netzer, M.G. Ramsey, Appl. Phys. Lett. **88**, 253111 (2006)
29. M.G. Ramsey, F.P. Netzer, D. Steinmüller, D. Steinmüller-Nethl, D.R. Lloyd, J. Chem. Phys. **97**, 4489 (1992)
30. J. Stöhr, *NEXAFS Spectroscopy, Springer Series in Surface Sciences 25* (Springer, Berlin Heidelberg New York, 1992)
31. The Calculations Were Performed Using the StoBe (Stockholm Berlin 2005) Programm Package by K. Hermann, L.G.M. Pettersson, a Modified Version of the DFT-LCGTO Programm Package deMon by A. St.-Amant and D. Salahub (University of Montreal)
32. E.g. the SMART (Spectro-Microscope with Aberration Correction for Resolution and Transmission Enhancement) Project Stationed at the Synchrotron Radiation Facility BESSY II, Berlin
33. G. Koller, F.P. Netzer, M.G. Ramsey, Surf. Sci. Lett. **559**, L187 (2004)
34. P. Sautet, Chem. Rev. **97**, 1097 (1997)
35. B. Winter, J. Ivanco, F. Netzer, M.G. Ramsey, Thin Solid Films **433**, 269 (2003)
36. K. Herwig, J. Newton, H. Taub, Phys. Rev. B **50**, 15287 (1994)
37. J. Villarrubia, J. Res. Natl Inst. Stand. Technol. **102**, 425 (1997)
38. C. Teichert, Phys. Rep. **365**, 335 (2002)
39. S. Kalinin, R. Shao, D. Bonnell, J. Am. Ceram. Soc. **88**(5), 1077 (2005)
40. K.L. Babcock, C.B. Prater, Application Note AN11, Veeco (2004)
41. R. Albrecht, P. Grütter, D. Rugar, J. Appl. Phys. **69**, 668 (1991)
42. S.N. Magonov, V. Elings, M.H. Whangbo, Surf. Sci. **375**, L385 (1997)

2

Optical Characterization Methods for Ultrathin Nanoaggregates

H.-G. Rubahn

2.1 Dark Field and Fluorescence Microscopy

Light-emitting organic nanofibers are an interesting model system for demonstrating the resolution limit of optical microscopy in the border region between micro- and macrocosmos. In Fig. 2.1 dark field and fluorescence images of the same hexaphenylene nanofiber are shown. A standard optical microscope with dia- (for dark field) and epi-illumination (for fluorescence) has been used. Structures with characteristic dimensions of a few ten nanometers such as breaks in the nanofibers (exemplified by an atomic force microscopy (AFM) image as an insert in Fig. 2.1b) are barely visible even in dark field images since the difference in indices of refraction of semiconducting nanofibers and underlying dielectric substrate (mica) is small. In dark field microscopy (Fig. 2.1a) one illuminates the sample under nearly grazing incidence, thus enhancing the visibility for structures on the surface that scatter light. Consequently, such structures appear bright on a dark background. The basic idea for this technique stems from 1903 (H. Siedentopf and R. Zsigmondy). Note that the structures shown in Fig. 2.1 have heights of less than 100 nm, i.e., much smaller than the wavelength of the light used for scattering. The evanescent wave condenser maximizes this principle by ensuring that the illumination occurs *solely* via surface waves, i.e., within the evanescent part of the electromagnetic field. Dark field microscopy is also called "ultramicroscopy" since it allows one to investigate structures with characteristic dimensions smaller than the wavelength of the imaging light (perpendicular plane visibility limit roughly $\lambda/100$).

Even better contrast and visibility of subwavelength structures is obtained for light-emitting objects via epifluorescence microscopy (Fig. 2.1b). In such a setup UV light irradiates the nanofibers under normal incidence and the resulting luminescence is observed under normal incidence, too. Excitation and luminescence light are separated with the help of a wavelength-selective beam splitter and color filters. At the breaks in the needles, the UV-induced luminescence is scattered into the far field and thus submicron structures

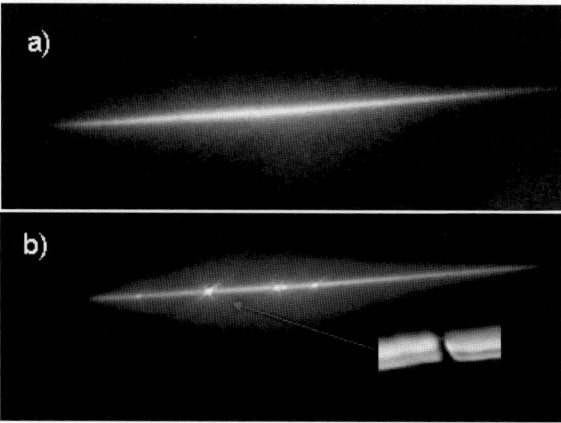

Fig. 2.1. Comparison of dark field (**a**) and epifluorescence (**b**) images (50 μm wide) of the same piece of an organic nanofiber. The inset is an AFM image ($1 \times 1.5\,\mu m^2$) of a break in the nanofiber that gives rise to a bright luminescence spot

become easily visible. The true dimensions of the nanometer-scaled breaks, of course, cannot be determined via optical far-field microscopy.

2.2 Two-Photon Microscopy

A further increase in resolution is obtained if one employs nonlinear optical effects for imaging purposes. This is *in-plane* mainly due to a shortening of the imaging wavelength (by a factor of 2 for second harmonic generation, for example) and *perpendicular plane* because of a better definition of the focal range since the multiphoton effect is observed only above a certain threshold illumination intensity.

Optical second harmonic generation is a nonintrusive, selective, and straightforward method to obtain surface- or interface-specific information. In the absence of noncentrosymmetric scatter sources and given sensitive detectors, it can provide very high signal-to-noise ratio and thus it is a rather obvious idea to obtain also two-dimensional SH or even sum frequency (SF) images [1,2]. Employing the vectorial nature of light or the tensorial nature of the light–matter coupling matrix (viz., susceptibility tensor), two-photon microscopy [3] can be used to obtain the polarized spatial intensity distribution of photons generated in the nanofibers. From a comparison of the intensity distributions at different combinations of polarization of exciting and detected light and employing the tensorial nature of the respective optical response, one can deduce local orientations of the molecules along the nanofibers.

An experimental setup for two-photon microscopy is shown schematically in Fig. 2.2. Here, it consists of a scanning optical microscope in reflection

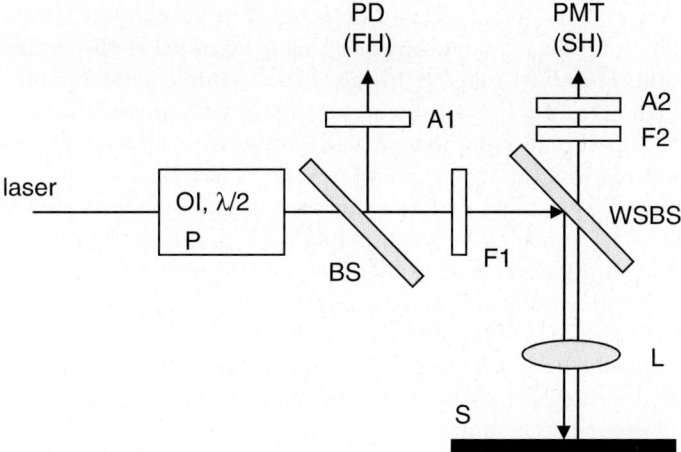

Fig. 2.2. The two-photon microscopy setup in reflection with a Ti-Sapphire laser, *OI* optical isolator, $\lambda/2$ half-wave plate, *P* polarizer, *BS* beam splitter, F_1 and F_2 filters, *WSBS* wavelength-selective beam splitter, *L* objective, *S* sample placed on *XY*-table, A_1 and A_2 analyzers, *PMT* photomultiplier tube, and *PD* photodiode

geometry built on the base of a commercial microscope and a computer-controlled two-dimensional translation stage. The radiation source is a mode-locked pulsed Ti-Sapphire laser with 80 MHz repetition rate, pulse duration of $\tau \approx 200$ fs, and an average output power of 300 mW. The wavelength is tunable from 730 to 920 nm and the radiation line width is $\Delta\lambda = 10$ nm. The linearly polarized light beam from the laser is used as a source of sample illumination at the fundamental harmonic (FH) frequency. An optical insulator is inserted at the laser output to avoid back-reflection into the laser cavity, and the half-wave plate along with the polarizer is used to control the polarization and power of the radiation incident on the sample. The illumination power is kept at a level of about 20 mW to avoid thermal damage and photobleaching of the sample. The red color filter has transmission above 600 nm and absorbs any second harmonic (SH) light generated before the microscope. After passing a wavelength-selective beam splitter, the laser beam is focused at normal incidence on the sample surface with a Mitutoyo infinity-corrected long working distance ×100 objective (NA = 0.7). At the sample surface, the full width at half maximum (FWHM) of the FH spot is about 1 μm giving an intensity at the surface of about $2 \times 10^6 \, \text{W cm}^{-2}$.

The sample is placed on a computer-controlled *XY*-piezoelectric stage, which is capable of moving in steps down to 50 nm with an accuracy of 4 nm over a scanning area of $25 \times 25 \, \text{mm}^2$. The sample reflects the FH and two-photon intensity (TPI) radiation, generated in the direction of reflection, back through the objective. The TPI radiation is transmitted through a beam splitter, FH filter (transmission range 350–550 nm), and an analyzer and detected by a photomultiplier connected to a photon counter. The FH radiation is reflected out of the microscope and detected by a photodiode. Both TPI and FH signals are recorded simultaneously as a function of the scanning coordinate resulting in TPI and FH images of the sample surface. During a normal scan, an area of $50 \times 50\,\mu m^2$ is scanned in $0.5\,\mu m$ steps with a speed of $30\,\mu m\,s^{-1}$, and the TPI photons at each point are counted over a period of 20 ms, resulting in an image acquisition time of 10 min.

On average the number of dark counts is 4, i.e., about 200 counts per second (cps). Special care has to be taken when adjusting to normal incidence and focusing the pump beam. The proper focus adjustment not only decreases the influence of field components other than the one determined by the polarizer, but also increases the field intensity at the surface and improves the spatial resolution in both FH and TPI images. The objective surface distance is adjusted by monitoring the width of the reflected FH beam at a distance of about 2 m from the sample surface. However, even an ideal beam focusing creates a complicated field distribution with all field components being nonzero [4].

2.3 Scanning Near-Field Optical Microscopy

If one is interested in obtaining truly nanoscopic resolution, the limitations given by the dimensions of the propagating optical waves in the far field have to be overcome by investigating objects in the near field at distances much closer than the optical wavelength. Here, a widely distributed and very useful synthesis of atomic force microscopy and conventional optical microscopy is scanning near-field optical microscopy (SNOM), which essentially relies on inspecting the near-field electromagnetic field distribution through a subwavelength aperture, kept at distances of a few nanometers to the field emitter. Scanning principle as well as many advantages and limitations of this new kind of microscopy are very much related to "conventional" scanning microscopies such as scanning tunneling microscopy (STM).

Figure 2.3 shows a setup for investigating the tunneling of photons from UV-excited nanofibers toward a scanning ultrasharp fiber, coupled to a photodetector (i.e., a photon-scanning tunneling microscope or PSTM).

Nanofiber samples were adjusted with respect to both the fiber probe of the PSTM and a fluorescence microscope by monitoring the illuminated area of the sample with the help of a charge-coupled device camera. The illumination source was a high-pressure mercury lamp with a band pass filter centered around 360 nm, which was focused with a 100× objective onto the sample. The spot size of the illumination was of the order of $20\,\mu m$.

The near-field optical probe was produced from a single-mode silica fiber by 2 h etching of the cleaved fiber in 40% hydrofluoric acid with a protective layer of olive oil. The resulting fiber tip has a cone angle of 40° and a radius of curvature of less than 80 nm. The tip was scanned along the sample at a constant distance of a few nanometers maintained by shear-force feedback.

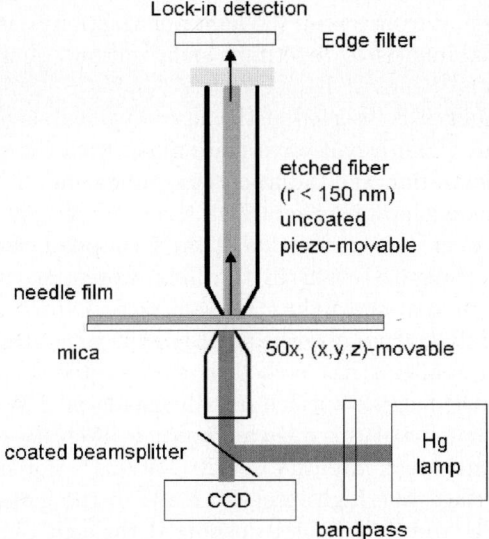

Fig. 2.3. Sketch of an SNOM on top of an inverted fluorescence microscope, allowing one to measure photon tunneling from nanofibers

The radiation scattered by the tip was collected by the fiber itself and propagated in the form of fiber modes toward a photomultiplier. A 420 nm low pass filter blocked the light of the exciting mercury lamp and allowed only fluorescence light to be transmitted. To further reduce background noise a lock-in technique was used.

2.4 Two-Photon Near-Field Microscopy

The above variation of the SNOM, the PSTM, can also be combined with a nonlinear optical excitation technique. This "two-photon SNOM" consists of a scanning XY-table, an optical fiber fabricated for single-mode use at 820 nm and a shear-force system based on a tuning-fork sensor technique. The shear-force system maintains the distance between the sample and the uncoated fiber. The fiber has been electrochemically etched with hydrogen fluoride for 2 h to give a sharp point and then glued to the tuning fork.

The radiation source is a mode-locked pulsed Ti-Sapphire laser with approximately 80 MHz repetition rate, pulse duration $\tau \approx 200$ fs, and an average output power of $P_\omega^{av} = 300$ mW. The wavelength is tunable from 730 to 920 nm and the radiation line width is $\Delta\lambda \approx 10$ nm. The linearly polarized light beam from the laser is used as a source of sample illumination at the FH frequency via the SNOM fiber. The polarization of light incident on the sample is not entirely known since the coupling into the fiber might cause depolarization and furthermore the light exits the sharp fiber point in a light cone. However, during a scan the light polarization is expected to remain constant, thus enabling us to determine polarization ratios as a function of local position.

An optical insulator is inserted at the laser output to avoid back-reflection into the laser cavity, and a half-wave plate along with the polarizer are used to control the polarization and radiation power incident on the cleaved end of the fiber. The incident power is kept at a level of ≈ 40 mW to avoid thermal damage and nonlinear light generation at the focal point of the coupling lens. Such nonlinearly generated light in the fiber would inevitably add to the counts in the SH images and reduce the contrast. A filter with transmission above approximately 600 nm absorbs any SH light generated before the fiber.

Upon exiting the fiber at the etched sharp uncoated point, the laser beam is scattered onto the sample surface usually giving an FWHM resolution of the FH of less than $\approx 0.4\,\mu$m with an average intensity at the surface of ≈ 0.3 MW cm^{-2} and a peak intensity of ≈ 20 GW cm^{-2}. For suitable conditions on the sample surface, the high intensity leads to the generation of SH and TPI light. The FH and TPI light transmitted through the sample are then collected with a lens, passed through the analyzer toward the wavelength-selective beam splitter and onto the respective photomultiplier tubes (PMT). Blue filters (transmission between approximately 325 and 600 nm) are placed in front of the PMT, which measures the SH light.

The sample is placed on a computer-controlled XY-piezoelectric stage, capable of moving in steps down to 1 nm within a scanning area of $50 \times 50\,\mu$m^2. At each step the FH and SH light transmitted through the sample are measured simultaneously with topographic information obtained from the shear-force feedback system. To maintain the distance between sample and fiber, the shear-force system has a piezoelectric tube (PT), which can reduce its length by a maximum of 6 μm if needed. During measurement the fiber is lowered onto the sample until the piezoelectric tube has retracted itself by approximately 2 μm, leaving some distance for the tube to either shorten or extend itself if needed to maintain the appropriate height above the sample. In the case of a rough or slightly tilted sample, the normal scan area is kept below $10 \times 10\,\mu$m^2 or less to stay well within the $\approx 4\,\mu$m of free movement length left in the piezoelectric tube and to prevent the fiber point from crushing onto the sample surface. The computer measures (as a function of the XY-coordinate) the different voltages needed for the piezoelectric tube to maintain the appropriate distance thereby recording a topographic image of the sample surface.

During a normal scan, an area of $10 \times 10\,\mu$m^2 is scanned in 0.1 μm steps and the TPI photons at each point are counted over a period of 100 ms, resulting in an image acquisition time of 33 min. Due to the weak light intensity, no bleaching effect is observed within this time interval. On average, the number of dark counts is less than 10 cps, whereas the signal from the smooth substrate is ≈ 50 cps and the TPI signal from the nanofibers reaches 1,000 cps.

Images obtained using this technique are shown in the chapter about nanooptics. The resolution is limited by (1) the inherent noise given by the fast bleaching of the nanofibers which means that only limited photon intensities for excitation can be used and (2) by the fact that an uncoated, sharpened fiber has been used to increase the detection sensitivity. If one coats the SNOM fiber with a metal and drills a small aperture in the tip via, e.g., ion beam milling, the resolution can be enhanced down to 50 nm or even 20 nm, just limited by damping through the metal coating.

Molecular resolution can be obtained by an "apertureless" near-field technique [5], for example using a PSTM, i.e., the excitation of the sample surface via an evanescent wave and the detection via dipping the fiber tip into the near field of the surface. As an alternative the near-field limited field might also be generated via far-field illumination of a strong scatterer, which is in direct optical contact with the fiber which transmits the light to the detector. In this method fiber and scatterer are scanned in small distance over the surface, which is illuminated in the far field. While the resolution is now no longer restricted by the damping aperture, an obvious disadvantage of this method is that small signal intensities are measured in front of a high background intensity (the illuminated surface). Thus the method works only reasonably well if the scattering rate of the object that has been mounted to the SNOM tip is strongly enhanced, e.g., via plasmon excitation.

Eventually the scattering object at the tip of the fiber could also be a single fluorescing molecule or single excited center [6]. In most cases, one attaches individual molecules at the optical fiber instead of doped host crystals. If it becomes possible to identify spectroscopically individual molecules in this host crystal, one is able to use them as single molecule light sources for optical near-field microscopy [7].

Besides imaging it would be of great help for a thorough understanding of light generation, propagation, and transformation in and along nanofibers if one would be able to perform time-resolved measurements or even spectroscopy in the near field. Of course, besides severe limitations due to the small signal-to-noise ratio given by the inefficient coupling from the near field into propagation modes of an SNOM fiber, the presence of a metallic aperture would affect such spectroscopic information severely. Thus an apertureless approach is more appropriate.

A realization is to scatter light from metallic nanostructures and employ localization and enhancement effects via plasmon coupling ("plasmonics"). Or the metal tip, illuminated by far-field light, can provide a localized excitation source for the sample that is investigated spectroscopically. A variant of that is to laser trap metallic particles of several tens of nanometers width and to use them as near-field probes, which is very much a realization of the basic idea of Synge from 1928 for ultramicroscopy [8]. Such experiments, however, still wait to be performed for organic nanofibers.

References

1. M. Flörsheimer, M. Bösch, C. Brillert, M. Wierschem, H. Fuchs, Adv. Mater. **9**, 1061 (1997)
2. M. Flörsheimer, Phys. Status Solidi A **173**, 15 (1999)
3. K. Pedersen, S. Bozhevolnyi, Phys. Status Solidi A **175**, 201 (1999)
4. M. Mansuripur, J. Opt. Soc. Am. A **3**, 2086 (1986)
5. F. Zenhausern, M. O'Boyle, H. Wickramasinghe, Appl. Phys. Lett. **65**, 1623 (1994)
6. S. Sekatskii, V. Letokhov, JETP Lett. **63**, 319 (1996)
7. J. Michaelis, C. Hettich, J. Mlynek, V. Sandoghdar, Nature **405**, 325 (2000)
8. S. Kawata, Y. Inouye, T. Sugiura, Jpn. J. Appl. Phys. **33**, L1725 (1994)

Part II

Growth

3

Growth of Oriented Organic Nanoaggregates via Molecular Beam Deposition

F. Balzer

3.1 Introduction

The *para*-phenylenes are a class of conjugated rigid rod-like molecules, which have attracted a great amount of interest during the last couple of years. In a series of experiments, it has been shown that needle-like crystals can be grown on various substrates like, e.g., the alkali halides [1,2], muscovite mica [3,4], Au(1 1 1) and Au foil [5–7], potassium acid phthalate KAP [8], and TiO_2 [9]. The needles emit polarized blue light after UV excitation leading to waveguiding [10] and lasing [11], opening a wide range of possible applications in photonics and optoelectronics. On anisotropic substrates like muscovite mica, the needles form large domains of mutually parallel aligned entities. These needles can be transferred onto arbitrary surfaces [12,13] as ensembles and as single aggregates, allowing, e.g., detailed investigations of the mechanical and electrical properties of single needles [14,15] (cf. Chap. 12).

Because the phenylene oligomers, *para*-quinquephenylene and *para*-hexaphenylene, are hardly dissolvable, all the molecules have been deposited by vacuum sublimation from a Knudsen cell. Several few nanometers – measured by a quartz microbalance – are deposited onto a substrate with a deposition rate of typically 0.05–$0.2\,\text{Å s}^{-1}$. The base pressure of the vacuum system is $p \approx 2 \times 10^{-8}$ mbar, which rises to $p \approx 2 \times 10^{-7}$ mbar during deposition; for details see [16]. Such organic molecular beam deposition (OMBD) (see Sect. 1.3.1) is the method of choice if one is interested in high-quality, crystalline films under high- or ultrahigh-vacuum conditions [17,18]. OMBD allows an easy control of the supersaturation and the final thickness of the film, with at the same time a high cleanliness of the deposited material and an independent control of the substrate temperature and the deposition angle. For molecular beam epitaxy (MBE) of an organic material, the growth mechanisms are well known [19]. For OMBD the underlying processes leading to ordered growth are much less understood, due to the lower symmetries of the molecular crystals and, among others, their polymorphism [20]. Here the alignment of the molecules caused by electric surface fields is introduced as an essential driving force for the mutually parallel alignment of the growing entities from rod-like oligophenylenes and oligothiophenes.

3.2 Adsorbates

The blue light-emitting *para*-phenylenes consist of n phenyl groups, forming a linear chain with single bonds in between them. In Fig. 3.1a structural formulas for *para*-phenylene oligomers with $n = 4, 5, 6$ are shown. For *para*-hexaphenylene (p-6P), the length of the molecule is about 2.6 nm. Especially *para*-hexaphenylene has been investigated thoroughly for the last years because of its promising application potential for organic electronical and electrooptical devices like organic light-emitting diodes (OLEDs) [21,22], photovoltaic cells, organic field-effect transistors (OFETs) [23], single- and multimode waveguides [10, 24–26], evanescent wave sensors, and organic nanolasers [27].

The room temperature bulk structure for p-6P is depicted in Fig. 3.1b; it crystallizes in a monoclinic structure with $P2_1/c$ symmetry and two molecules per unit cell (Table 3.1). The long axes of the molecules are all parallel to each other, and the molecular planes are arranged in a herringbone fashion with an angle of approximately 66° in between [32]. The reason for the herringbone packing is atomic charges on the hydrogen atoms [33], which result in a Coulombic quadrupolar interaction between single molecules. That way layers

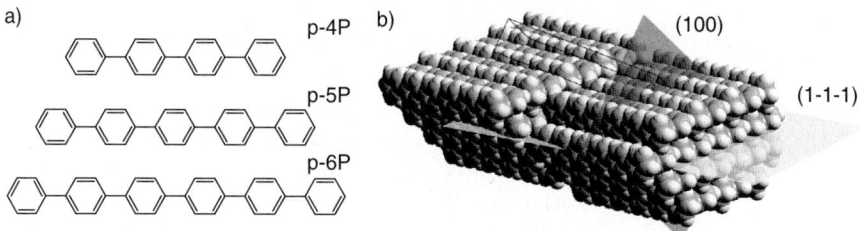

Fig. 3.1. (a) Structural formulas of the *para*-phenylenes: *para*-quaterphenylene (p-4P), *para*-quinquephenylene (p-5P), and *para*-hexaphenylene (p-6P). (b) Bulk herringbone packing of p-6P (β-phase). Part of the bulk unit cell together with the $(1\,0\,0)$ and $(1\,\bar{1}\,\bar{1})$ planes are shown

Table 3.1. Lattice constants for p-6P [28] and α-thiophenes [29–31]

Molecule	Symmetry	a (Å)	b (Å)	c (Å)	β (°)	Step (Å)
p-6P	$P2_1/c$	26.24	5.57	8.09	98.2	26.0
α-4T	$P2_1/c$	6.09	7.86	30.48	91.8	15.2
α-6T	$P2_1/n$	44.71	7.85	6.03	90.8	22.4

The step height is calculated for upright molecules

of parallel molecules form, with only a van der Waals interaction in between, perpendicular to the (1 0 0) planes.

The two planes in Fig. 3.1b resemble two close-packed faces of the bulk crystal. As long as the interaction with the substrate is small, p-6P molecules tend to grow with the (1 0 0) face parallel to the substrate [34, 35], thus forming layers of upright molecules. The molecules include an angle of 73° with the surface normal, the resulting layers having a step height of 2.6 nm. On KCl, NaCl [16] as well as on GaAs [36] at elevated temperatures, also a polymorphic phase has been observed, the so-called γ-phase with oriented molecules along the surface normal. For substrates with an increased interaction between molecules and surface via, e.g., electrostatic interactions or at low substrate temperatures [37], molecules lie on the surface with, e.g., the (1 1 $\bar{1}$) face or the (3 0 $\bar{2}$) face parallel to the substrate [3, 4, 9]. A collection of several low index faces together with the angles β_{Mol} of the molecules with respect to the supposed direction of the needles, and the angles γ_{Mol} of the molecules long axes with the substrate surface is presented in Table 3.2. Some of the realized surface unit cells for lying molecules with close-packed faces parallel to the substrate are depicted in Table 3.3. Assuming that the needle direction is along the layers, i.e., along p-6P [0 1 1], depending on the contact plane different angles between the molecules orientation and the long needle axis are possible. For these low index faces, two different groups of angles arise: 90° and 72°–76°.

Table 3.2. Various hypothetical p-6P needle types together with their contact faces with the substrate

β_{Mol}	Face	γ_{Mol}	β_{Mol}	Face	γ_{Mol}	β_{Mol}	Face	γ_{Mol}
90°	(0 0 1)	25°	74°	(0 1 0)	0°	80°	(3 1 $\bar{2}$)	0°
90°	(1 0 $\bar{1}$)	8°	72°	(1 1 0)	12°			
90°	(2 0 $\bar{1}$)	8°	76°	(1 1 $\bar{1}$)	5°			
90°	(3 0 $\bar{2}$)	0°	75°	(2 $\bar{1}$ $\bar{1}$)	5°			

The angle β_{Mol} denotes the angle of the molecules with the supposed long needle axis and γ_{Mol} denotes the angle of the molecules with the substrate surface

Table 3.3. Surface lattice constants a_s, b_s, and the enclosed angle γ, for some of the low index faces of lying *para*-phenylenes

Molecule	Face	a_s (Å)	b_s (Å)	γ (°)
p-6P	(1 1 $\bar{1}$)	9.82	26.83	76.6
p-6P	(2 $\bar{1}$ $\bar{1}$)	9.82	28.51	70.8
p-6P	(3 0 $\bar{2}$)	5.57	54.59	90.0

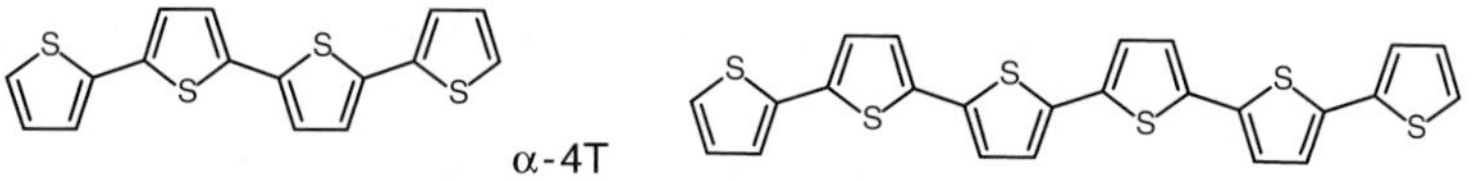

Fig. 3.2. Structural formulas of the α-thiophenes: α-quaterthiophene (α-4T) and α-sexithiophene (α-6T)

Because of the molecules symmetry, the transition dipole for absorption and emission of blue light is parallel to the long molecular axis [38]. This is still valid for crystallized molecules, although packing effects lead to a redshift of the fluorescence. Molecules, which are lying on a substrate can easily be detected by their bright fluorescence after normal incidence UV excitation or by their broad absorption band at 350 nm [39]. Upright molecules on the other hand show only little luminescence after UV excitation with 365 nm light, but feature a characteristic absorption peak close to 300 nm due to absorption along the molecules short axes. That absorption can be detected in transmission for UV-transparent substrates like LiF or NaCl [16]. The electrical conductivity of such organic molecules is largest perpendicular to the molecules long axis, perpendicular to the phenyl rings [14, 21].

Another class of important rod-like molecules for optoelectronic applications is the thiophene oligomers [40, 41]. Quaterthiophene (α-4T) and sexithiophene (α-6T) (see Fig. 3.2) luminesce green and orange after UV excitation. For both thiophenes the most common crystal structures are the high-temperature (HT) and low-temperature (LT) polymorphs. The two polymorphs differ in packing. Molecules lay either on top of each other (LT) or are shifted by one thiophene subunit (HT) [30] – the number of molecules per unit cell being either $Z = 2$ (HT) or $Z = 4$ (LT). For the growth conditions on muscovite, the LT phases will probably be suitable, as has already been reported for the case of α-4T on muscovite [42], KAP [43], and silica [44]. Note that, for α-6T on silicon [45, 46] and silica [47] for upright molecules, various different polymorphial phases have been observed.

3.3 Silicate Substrates

Micas are sheet silicates, consisting of two tetrahedral Al and Si sheets ("T") and an octahedral sheet ("O") in between [48]. Because one out of four Si^{4+} cations in the tetrahedral sheet is replaced by Al^{3+}, the net charge of such a "TOT" layer per unit cell is -1. The charge of the silicon cations is $+4$, therefore a single SiO_4 tetrahedron has a net charge of -2. Every fourth Si cation is replaced by an Al cation with a charge of $+3$, leading to a net charge of one "TOT" layer within the unit cell of -1. Sheets are bound by Al^{3+} and OH^- ions, forming layers. The layers are linked together by interlayer cations such as K^+. Cleavage takes place along the interlayer cations, which then are divided randomly in between the two new surfaces [49]. Cleavage is almost perfect. That way large areas can be prepared which are atomically flat. Because of Löwenstein's rule (avoidance of Al–O–Al linkages) [50] a short-range order of the Si–Al replacement exists, but no long-range order [51]. A freshly cleaved mica surface is positively charged, hydrophilic and polar.

The two micas used for the deposition experiments are muscovite mica and phlogopite mica. Muscovite with the ideal formula $K_2Al_4(Si_6Al_2O_{20})(OH)_4$ (monoclinic lattice, $a = 5.2$ Å, $b = 9.0$ Å, $c = 20.1$ Å, $\gamma = 95.8°$ [52]), is a *dioctahedral* mica. Only two out of three AlO_6 octahedral sites are occupied by a cation. This leads to a tilt of the (Al,Si) oxide tetrahedra and to grooves, which reduce the symmetry of the surface from three- to onefold. For the most common $2M_1$ polytype [48], these grooves alternate by an angle of 120° between consecutive cleavage layers and run along $\langle 1\,1\,0\rangle$, leading to only *two* different groove directions on the $(0\,0\,1)$ surface [53, 54]. To discriminate the grooved $\langle 1\,1\,0\rangle$ from the nongrooved one, the directions are often depicted as $\langle 1\,1\,0\rangle_g$ and $\langle 1\,1\,0\rangle_{ng}$, respectively.

Not only the surface of muscovite is charged due to the K^+ ions, but also the electric fields on the surface possess a component parallel to the substrate and are macroscopically directed, the fields pointing about 15° off one of the sides of the hexagon in a low energy electron diffraction (LEED) pattern. The existence of surface electric dipole fields has been postulated from LEED patterns of freshly cleaved samples [49, 55]. The fields lead to characteristically shaped diffraction spots with a two- or threefold symmetry [56–58]. Angles ±15° off the high-symmetry (*hs*)-directions in the LEED pattern convert to dipole directions ±15° off from the direction perpendicular to a *hs*-direction in real space. LEED cannot distinguish between the three *hs*-directions, but from symmetry arguments (Sect. 3.5), it is reasonable to assign this special direction to the groove direction. The orientation of the $[1\,0\,0]$ direction can be obtained from a Schlagfigur.

The electric fields are supposed to align molecules on the surface, an effect already been observed more than 40 years ago [59–63], even through an up to 20 nm thick layer [64] or through discontinuous Au films [65]. Because of the $2M_1$ polytype stacking of the "TOT" layers on the plane, the direction of the grooves is rotated by 120° between consecutive layers and only two different directions are realized.

As an alternative substrate to elucidate the role of the electric fields and possibly epitaxy, phlogopite mica has been chosen. Phlogopite is a *trioctahedral* mica, with the ideal formula $KMg_3(Si_3AlO_{10})(OH)_2$. It has almost the same surface structure as muscovite mica with a slightly larger lattice constant (monoclinic unit cell, $a = 5.3$ Å, $b = 9.2$ Å, $c = 10.2$ Å, $\gamma = 100.0°$ [54]), but all three octahedral sites are occupied by Mg cations. Therefore, the $(0\,0\,1)$ phlogopite surface does not exhibit grooves [54] and also no uniaxially aligned electric surface fields.

3.4 Phenylenes on Muscovite

3.4.1 *Para*-Hexaphenylene

The deposition of p-6P on muscovite mica at elevated temperatures leads to the formation of long, mutually parallel aggregates, so-called "needles" or "nanofibers." Heights and widths of the needles are several ten nanometers and several hundred nanometers, respectively, whereas their lengths can be adjusted between a few ten nanometers and one millimeter by varying the deposition conditions [8, 66]. In Figs. 3.3a,b the case of deposition at a substrate temperature of 440 K is shown. The mean needle height is 60 nm, whereas the needle widths range between 150 and 600 nm. In a fluorescence microscope (Fig. 3.3a), the needles emit polarized blue light after unpolarized normal incidence UV excitation ($\lambda_{\text{exc}} \approx 365$ nm from a high-pressure Hg lamp), with the plane of polarization almost perpendicular to the needle direction [16, 67] (Fig. 3.3c). Atomic force microscopy (AFM; Fig. 3.3b) finds additional small islands or clusters in between the needles, which are believed to be needle precursors [68].

Usually only two different domains of needles exist on a single sample, with an angle of 120° in between [16] (Fig. 3.4a). Domains can reach sizes up to cm^2 because of the excellent cleavability of muscovite. Needles change orientation when they cross an odd number of mica cleavage steps, but maintain their orientation for an even number of steps (Fig. 3.11). On matching cleavage faces, needles also change directions (Fig. 3.4b,c). On all samples needles grow exactly along the $\langle 1\,1\,0 \rangle$ directions (experimental error: ±3°), never along [1 0 0].

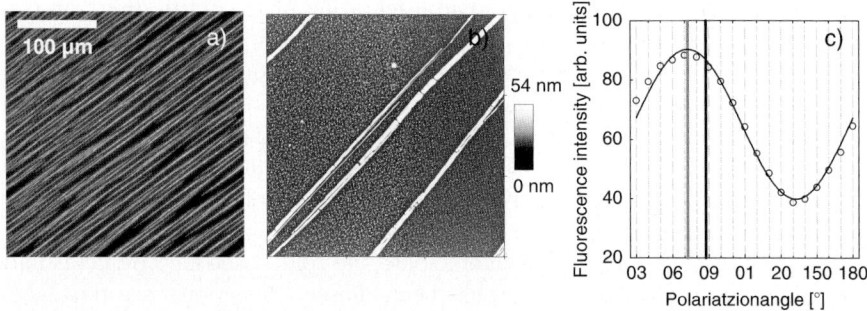

Fig. 3.3. (a) Fluorescence microscope image of p-6P needles grown on muscovite at a substrate temperature of 440 K and a deposition rate of $0.15 \, \text{Å s}^{-1}$. In (b), the sample is imaged by intermittent contact atomic force microscopy ($30 \times 30 \, \mu\text{m}^2$). The mean height of the needles is 60 nm, with needle widths between 150 and 600 nm. A large number of p-6P clusters is located in between the needles, with a denuded zone close to them. (c) For a single fiber a maximum in fluorescence occurs at a polarization angle 14° off the short needle axis (*black vertical line*) (reprinted with permission from [16, 69])

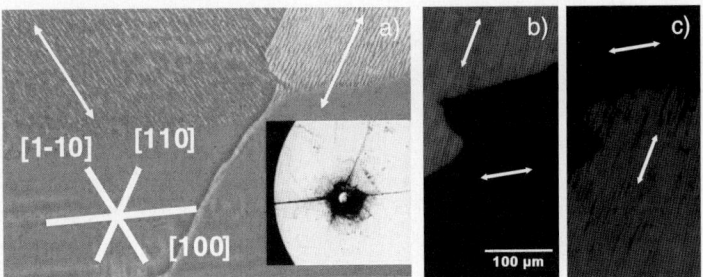

Fig. 3.4. (a) Phase contrast microscope image of p-6P needles on muscovite mica together with uncovered substrate. The needle orientations are clarified by two white arrows. Two domains are visible, with multiple cleavage steps as the domain boundary. As an inset a Schlagfigur from the same surface is presented, giving the three muscovite hs-directions $[1\,0\,0]$, $[1\,\bar{1}\,0]$, and $[1\,1\,0]$. Needles grow along $[1\,\bar{1}\,0]$ and $[1\,1\,0]$, never along $[1\,0\,0]$. Polarized fluorescence microscope images demonstrate that p-6P needles change their orientation by 120° on mirror cleavage faces (b) and (c) (reprinted with permission from [16, 69])

Optical microscopy as well as AFM are real space methods to characterize the grown samples. A diffraction method on the other hand provides information about order and symmetry. Diffraction methods used in the literature on *para*-phenylenes are, e.g., X-ray diffraction (XRD) [70], Helium atom scattering (HAS) [71], transmission electron diffraction (TED) [72], and selected area electron diffraction (SAED) [73]. Here the diffraction of low energy electrons (LEED) has been used to determine the surface unit cells of the organic overlayer with respect to the substrate. The LEED apparatus (Omicron MCP-LEED) is equipped with two channel plates to limit the electron current density on the substrate, minimizing beam damage and reducing charging. Even sensitive self-assembled monolayers of alkanethiols on Au(1 1 1) have been imaged without damage that way [74]. In Fig. 3.5a, a LEED pattern from nominally 7 nm p-6P, deposited on muscovite at a substrate temperature of $T_S = 410$ K, is shown, together with possible reciprocal lattices in (b) and (c). Both the hexagonal pattern from the mica substrate as well as a superstructure from the organic molecules are visible. The diffraction spots along a vertical line along muscovite (00)–(01) are resolved, whereas other spots cannot be distinguished from each other and form vertical streaks. A careful comparison of the hexagonal LEED pattern from the bare mica surface with optical microscope images of grown needles reveals that the needles are oriented exactly perpendicular to the vertical lines (error ±2°).

LEED patterns have been observed both for very thin films down to 0.2 nm nominal thickness as well as for thick films of up to 20 nm thickness. They have been observed for deposition at room temperature as well as for temperatures up to 490 K, where no needles at all grow (Fig. 3.13). The origin for the diffraction patterns can therefore be sought either in the small clusters,

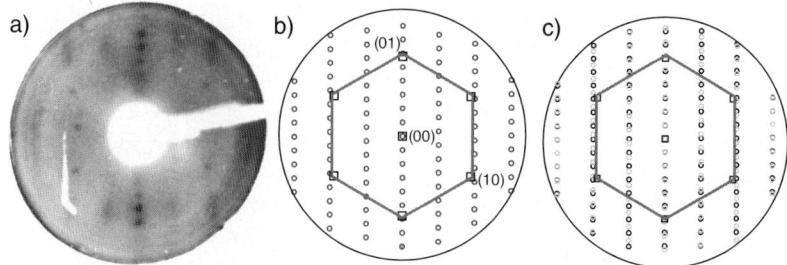

Fig. 3.5. (a) LEED pattern (electron energy $E_{el} = 49\,\text{eV}$) of 7 nm p-6P deposited on muscovite mica at a substrate temperature of $T_S = 410\,\text{K}$, to be compared with the quasihexagonal reciprocal lattice of the mica $(0\,0\,1)$ face (*squares*, forming a centered hexagon) together with the reciprocal lattice of the p-6P $(1\,\bar{1}\,\bar{1})$ face (*circles*) in (b). (c) LEED patterns from both a $(1\,\bar{1}\,\bar{1})$ and a $(2\,\bar{1}\,\bar{1})$ face lead to stripes except for the points along (00)–(01) (from [8])

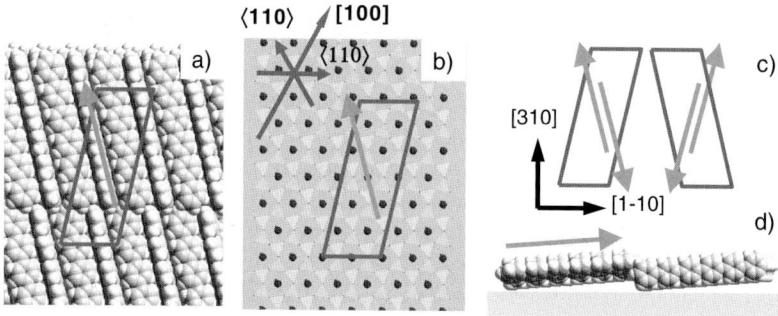

Fig. 3.6. (a) The $(1\,\bar{1}\,\bar{1})$ face of bulk *para*-hexaphenylene together with the surface unit cell. In (b), the unit cell is reproduced together with the mica $(0\,0\,1)$ face. The short axis of the unit cell is parallel to mica $\langle 1\,1\,0\rangle$. The direction of the molecules is symbolized by an *arrow*. (c) Four possible domains within a wetting layer of p-6P molecules. The directions of the molecules are symbolized by *arrows*. (d) The bulk molecules include an angle of $5°$ with the $(1\,\bar{1}\,\bar{1})$ plane, therefore the $180°$ rotational domains are not equivalent

which grow first and have also been observed for high substrate temperatures, or in a wetting layer of organic molecules. Judging from the coverage of the surface and the weakening of the substrate spots by more than a factor of 5, a wetting layer is highly probable. Recently conducted experiments using second harmonic generation (SHG) on grown and transferred fibers [75] and thermal desorption spectroscopy (TDS) [76] also find strong evidence for a wetting layer.

The observed LEED pattern from the wetting layer has been reproduced assuming a single domain with a unit cell as depicted in Fig. 3.6a,b. This unit cell is compatible with the $(1\,\bar{1}\,\bar{1})$ face of p-6P with its bulk crystal structure (Table 3.3), which has been observed via, e.g., XRD for thick films [77] (cf. Chap. 6). The p-6P [0 1 1] direction is parallel to one of the muscovite hs-directions, i.e., parallel to either mica $\langle 1\,1\,0 \rangle$ or mica [1 0 0]. The Schlagfigur technique in determining the optical axis reduces these three directions to two, to the two $\langle 1\,1\,0 \rangle$ directions, [1 1 0] and [1 $\bar{1}$ 0]. In Fig. 3.6a, the arrow marks the molecules orientation, in Fig. 3.6b the muscovite basal plane is sketched. Note that here all K^+ positions are filled, whereas for a real surface only half of the spots are statistically occupied. The only obvious relationship between the substrate and the adsorbate is a coincidence between p-6P [0 1 1] and mica $\langle 1\,1\,0 \rangle$. Assuming a bulk crystal structure of the adsorbate for commensurism or a point-on-line coincidence [78], the lengths of the unit cell axes are either too short or too long by 2–6%. Because of the relatively broad spots within the LEED images, a widening of the adsorbate unit cell cannot be excluded for the wetting layer. The superstructure matrix M, connecting primitive substrate basis vectors a and b with primitive overlayer basis vectors a_s and b_s [78, 79], can be related to either the quasihexagonal surface symmetry of muscovite to emphasize the epitaxial relationship M_{hex}, or to the rectangular surface unit cell to emphasize the absolute directions M_{rect}

$$M_{hex} = \begin{bmatrix} 1.89 & 0.00 \\ 4.09 & 5.79 \end{bmatrix},$$

$$M_{rect}^{a_s||[1\,1\,0]} = \begin{bmatrix} 0.94 & 0.94 \\ -3.75 & 2.04 \end{bmatrix},$$

$$M_{rect}^{a_s||[1\,\bar{1}\,0]} = \begin{bmatrix} 0.94 & -0.94 \\ 4.94 & 0.85 \end{bmatrix}.$$

Only after considering the error bars, a column of the superstructure matrix consists of integer elements, making a point-on-line coincidence probable. However, this possible enlargement is well beyond the resolution of the MCP-LEED apparatus. Future high-resolution SPA-LEED experiments might help to resolve this question. The same holds true for the $(2\,\bar{1}\,\bar{1})$ face, another close-packed face, which is also compatible with the observed LEED pattern and which has been observed by XRD from thick films. The reason for the smeared out lines instead of sharp diffraction spots can be found in different domains or/and in other close-packed contact faces, like the already mentioned $(2\,\bar{1}\,\bar{1})$ face. Both would lead to a shift of the higher-order diffraction spots (Fig. 3.5c).

Assuming that the molecules within the unit cell of the wetting layer are arranged the same way as in the bulk, a twofold rotational axis together with two mirror axes (muscovite [1 1 0] and [3 1 0]) leads to four different domains for the $(1\,\bar{1}\,\bar{1})$ face, for which the p-6P [1 $\bar{1}$ 0] direction is parallel to muscovite $\langle 1\,1\,0 \rangle$. For two orientations (i.e., front- and backside like $(1\,\bar{1}\,\bar{1})$ and $(\bar{1}\,1\,1)$), the molecules directions are also mirrored, whereas for the two rotational domains the molecules directions are the same, but tilted differently with respect of the surface plane (Fig. 3.6c). That way at least four different wetting layer domains can exist within a single needle domain. Such domain

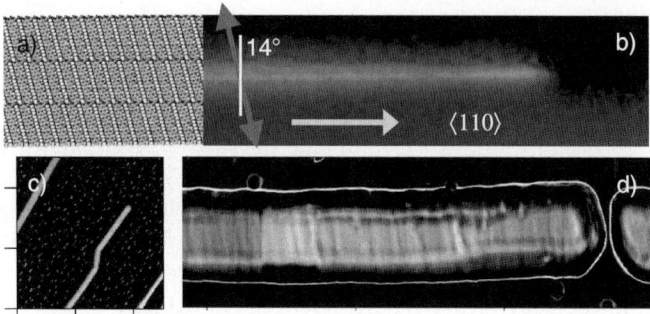

Fig. 3.7. Comparison between the p-6P (1 1̄ 1̄) face (**a**), a fluorescence microscope image of a single fiber (**b**), and an AFM phase contrast image (**d**). The stripes in (**d**) are running almost perpendicular to the long fiber axis (fiber width 500 nm) and correspond to steps in topography. Stripes are parallel to the proposed direction of the p-6P molecules in the fiber (**a**). Within the same domain needles with stripes along +14° and along −14° have been observed. In only a few cases twinned fibers (**c**), $5 \times 5\,\mu m^2$, with an angle of 28° between the two directions grow (**d**) (reprinted with permission from [69])

boundaries could serve as nucleation sites for needles. However, up to now neither such domain boundaries nor different molecular alignments within the wetting layer have been directly observed.

The lowest energy optical transition in p-6P crystals is polarized parallel to the long molecular axis and represents a transition between the highest occupied molecular orbital (HOMO) and the lowest unoccupied molecular orbital (LUMO) [80]. Because the needles are mostly single crystalline with mutually parallel aligned molecules, their optical properties both in absorption as well as in emission are highly anisotropic. The maximum in absorbance as well as in luminescence intensity is shifted by 14° with respect to the short needle axis, and reaches its maximal value for the polarization vector being parallel to the molecules orientation within the fiber (Figs. 3.3c and 3.7). That way the orientation of the molecules within a single fiber has been deduced by far-field optical means.

Many, but not all fibers are single crystalline over their entire length. In Fig. 3.8 the angular dependence of the luminescence intensity has been determined [81] along several p-6P needles. Even for a single needle of this sample, the angles for maximal intensity change continuously between +14° and −14°. An AFM image of a typical needle reveals that needles are not single needles, but that they are bunches of several entities (Fig. 3.8c). Within every single needle the molecules can have an orientation of ±14° with respect to the short fiber axis, thus adding up to an arbitrary angle between +14° and −14° for the fiber bundle. Even within single fibers both orientations have been observed, confirmed by phase contrast AFM image (Fig. 3.8b).

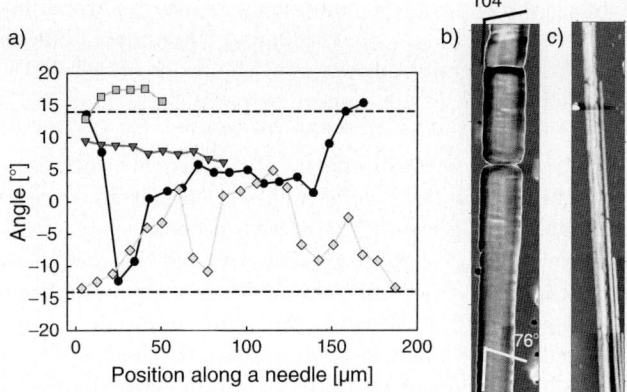

Fig. 3.8. (a) Polarization dependence of emitted fluorescence along four different needles. As a function of position along the fibers, the angle for maximum fluorescence can change from $+14° \pm 3°$ to $-14° \pm 3°$. (b) A $0.75 \times 4\,\mu m^2$ AFM phase contrast of a single p-6P needle. Stripes along two orientations with $\pm 14°$ with respect to the short fiber axis are clearly visible. In (c), a $2 \times 16\,\mu m^2$ AFM image of one of the measured needles from (a) is shown

From the comparison between AFM images, polarized fluorescence microscope images, Schlagfigur images, and LEED patterns, the direction of the needles turns out to be along the $[0\,1\,1]$ direction of the p-6P molecules in the wetting layer, i.e., along $\langle 1\,1\,0 \rangle$ muscovite directions. XRD experiments on thick films as well as on single needles have demonstrated that their contact faces are the close-packed faces $(1\,\bar{1}\,\bar{1})$ and $(2\,\bar{1}\,\bar{1})$ [70], suggesting that the needles grow with the same molecular orientation as the wetting layer. Within a single needle (Fig. 3.7), the long p-6P molecule axis is therefore almost perpendicular to the long fiber axis with an enclosed angle of approximately $14°$. For faceted needles with a flat surface (Fig. 3.7d), AFM is able to image subnanometric steps along the proposed molecule direction (Fig. 3.7a). Steps are resolved well in a phase contrast image (Fig. 3.7b). In a few cases even twinned fibers (Fig. 3.7c) have been detected with an angle of $28°$ in between.

The driving force behind the growth along $[1\,\bar{1}\,0]$ and $[1\,1\,0]$ is a combination of epitaxy and dipole-assisted alignment. From an epitaxial point of view, it is favorable that the molecules are oriented such that p-6P $[0\,1\,1]$ is parallel to one of the mica hs-directions, maybe with an additional enlargement of the unit cell of the wetting layer by, e.g., reducing the tilt of the molecules with respect to the substrate. This alone would give three equivalent growth directions. However, either $[1\,\bar{1}\,0]$ or $[1\,1\,0]$ is energetically favored because of the electric surface fields. Since the average dipole moment of *para*-phenylene oligomers is zero, the confinement of the individual molecules to the surface dipoles is most likely due to a dipole-induced dipole interaction or dipole–quadrupole interaction. The static polarizability of p-6P molecules along their long axis, the largest component of the polarizability tensor, is somewhere between $\alpha_{xx}/4\pi\epsilon_0 \approx 85$ Å^3 [82] and 800 Å^3 [83]. The field E of the individual dipoles on the mica surface is of the order of 10^6–10^8 V cm^{-1} [55]. Hence the interaction energy for molecules parallel to the mica dipoles [84], $w = -1/2\alpha_{xx}E^2$ for an average field of 10^7 V cm^{-1} leads to $|w| \approx 30$ up to 280 meV, well above the thermal energy at a substrate surface temperature of 400 K. Thus this interaction energy defines a preferred orientation on the surface during the initial phase of the growth process.

With an angle of $\phi_{\text{tdm}} = 90° - \beta_{\text{Mol}} = \pm 14°$ of the phenylene molecules with respect to the short fiber axis, the molecules are almost parallel to the dipoles (angle $\phi_D = \pm 15°$). Because the induced dipole moment depends on the component of the electric field parallel to the long molecular axis, and because the interaction of the induced dipole with the surface dipole is proportional to the cosine of the enclosed angle, the interaction energy is

$$w(\phi_{\text{tdm}}) \propto \cos(\phi_D - \phi_{\text{tdm}})^2. \tag{3.1}$$

The two different dipole orientations just favor the two different orientations within a single needle. In Fig. 3.9a, the azimuthal interaction energy from (3.1) is plotted for fixed dipole orientations $\phi_{\text{tdm}} = 30 \pm 15°$, i.e., dipoles corresponding to the $\langle 1\,1\,0 \rangle$ direction at $120°$. The vertical lines denote molecular orientations matching possible needle directions. Realized needle directions within

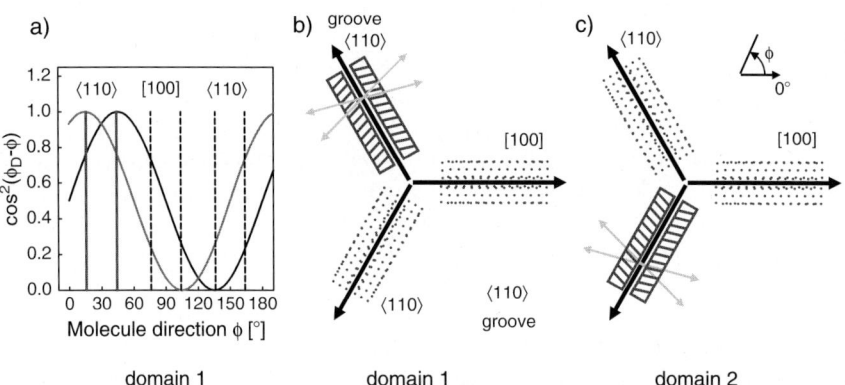

Fig. 3.9. In (a) for the dipole directions along $30° \pm 15°$, the interaction energies $\propto \cos^2(\phi_D - \phi)$ are plotted. The interaction energies reach their maximal values for needles along one of the two $\langle 1\,1\,0 \rangle$ directions, i.e., along the *two solid vertical lines*. In (b) and (c), the orientations of phenylene needles on muscovite mica are sketched for the two mica domains, together with the orientation of the molecules within the needles. The muscovite mica directions are depicted by *black arrows*, the assumed dipole directions by shorter *gray arrows*. Needles not realized on the two domains (b) and (c) are drawn with *dotted lines*, in (a) with *dashed vertical lines*. The calculation in (a) corresponds to the domain shown in (b) (from [8])

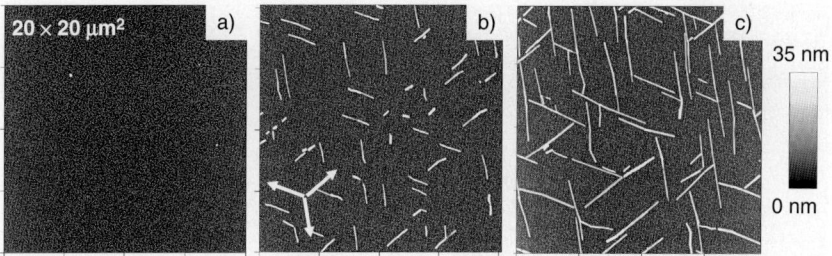

Fig. 3.10. Thickness dependence of p-6P nanofiber growth on phlogopite mica (substrate temperature $T_S = 440$ K). With increasing thickness first clusters grow (**a**) (nominal thickness 0.25 nm), then needles start to form for 0.5 nm (**b**), and increase in length (**c**). The *arrows* in (**b**) depict the phlogopite *hs*-directions

a single domain are emphasized by solid vertical lines, not realized directions by dashed lines. Obviously, the realized directions are the ones with the highest interaction energy. That way on every dipolar domain only a single needle direction exists. The same model is able to explain the roughly three different orientations of thiophene needles on muscovite (Sect. 3.5).

To verify the influence of surface electric fields and of the basal corrugation on the growth direction, phlogopite mica has been chosen as an alternative substrate. Phlogopite possesses almost the same surface structure as muscovite mica. The hexagonal lattice constant is slightly larger, 5.3 Å compared to 5.2 Å, but all three octahedral sites are occupied. Therefore, the (0 0 1) phlogopite surface does not exhibit grooves and also no uniaxially aligned electric surface fields. Microscope images for *para*-hexaphenylene deposited on phlogopite show needles along all three phlogopite *hs*-directions (Fig. 3.10a–c). The morphology (long needles and small clusters in between) resembles the morphology of p-6P on muscovite. Because of the similar lattice constants, epitaxy still favors the *hs*-directions for needle growth. Now $[1\,1\,0]$ and $[1\,\bar{1}\,0]$ are no longer special directions, demonstrating the important role of the vacant octahedral sites for aligned needle growth.

3.4.2 Steps and Defects

A single muscovite step has a height of approximately 1 nm. Most of the steps bunch to several ten nanometer tall entities. As long as the bunches are not taller than a few ten nanometers, needles can pass over them without any disturbance, provided that the total number of steps is even numbered and thus the overall electric field does not change direction (Fig. 3.11b). Along odd-numbered steps with a rotation of the electric field, the needles change direction (Fig. 3.11a,c).

Defects within p-6P fiber films on mica often occur as circular areas with a diameter of 50–100 μm (Fig. 3.12). Within this area no fibers grow, but patches of upright molecules have been found by AFM. Close to the border the needles

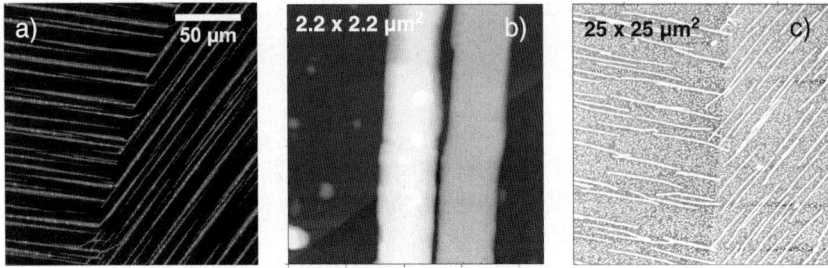

Fig. 3.11. Influence of muscovite steps on the p-6P needle direction. An odd number of 1 nm tall steps changes the needle direction as depicted in the fluorescence microscope image (**a**) and the AFM image (**c**). The step height in (**c**) is 1 nm. An even number of 1 nm steps does not change the needle direction – the step height in (**b**) is 4 nm (from [8])

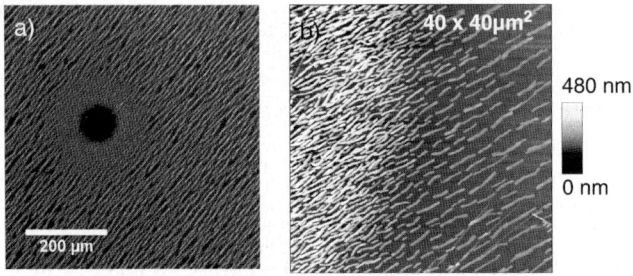

Fig. 3.12. Fluorescence microscope image (**a**) and AFM image (**b**) of a defect on muscovite. The defect disturbs the electric dipole fields on the surface, leading to a spherical disturbance of the p-6P growth. Close to the defect the needles become shorter, taller, and less aligned, in the dark area in (**a**) patches of upright molecules grow

become much shorter but taller and increase in number density, loosing their parallel alignment. That behavior is characteristic for a decreased interaction with the underlaying substrate, like it is expected for an area of muscovite, where the electric fields are disturbed by, e.g., a dust grain or a mica flake sitting in the middle of the circular area, or where the substrate surface energy is reduced. This is additional evidence for electric fields on the surface that are the driving force for molecules lying down on the surface and aligning.

p-6P needles themselves show various types of defects. Most common are a few ten nanometer wide breaks within the needles with mutual distances of a few micrometers (Figs. 3.7d and 3.8b). Twinned fibers as in Fig. 3.7c occur only rarely, whereas defects stemming from only partly agglomerated clusters (Fig. 3.15) are more common. Internal defects like grain boundaries between the forming p-6P clusters have not been quantified, yet, but may be responsible for the granular structure of bleached fibers [85].

3.4.3 Kinetics

The morphology (height, width, length, number density, but not orientation) of the growing p-6P needles and clusters depends strongly on the process parameters during growth. In Fig. 3.13 the temperature dependence is depicted by AFM (upper panel) and fluorescence microscopy (lower panel). At room temperature, (a) and (d), films of densely packed, short fibers form. The optical image (d), where the border between a covered and an uncovered part of the sample is imaged, does not reveal any details on the film. With increasing temperature the needles become longer and their number density decreases, with small clusters appearing in between the needles. For about $T_S > 470$ K no needles at all form, but only the small clusters, which emit considerably less UV light. However, the emitted light is still polarized, meaning the molecules are still aligned.

AFM images for the initial growth stage are presented in Fig. 3.14 for three different substrate temperatures during growth. For the thinnest samples only clusters grow. Only after the number density or the size of the clusters reaches a critical value, needles are formed which then continue growing. A similar behavior has been reported in the literature and has been attributed to accumulating stress in the islands and/or in their environment [76,86]. As soon

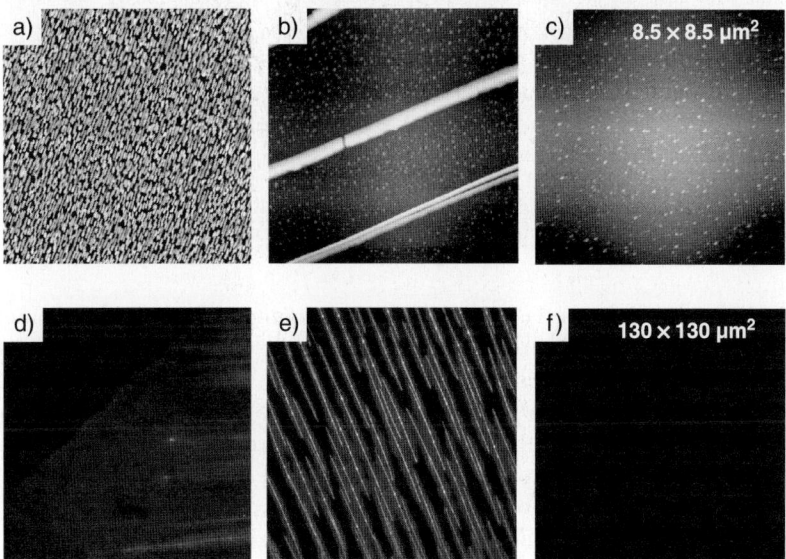

Fig. 3.13. Nominally 5 nm thick p-6P films on muscovite, deposited at different substrate temperatures of $T_S = 300$ K (**a**) and (**d**), 450 K (**b**) and (**e**), and 470 K (**c**) and (**f**). In the upper panel AFM images and in the lower panel fluorescence microscope are shown. At all temperatures p-6P molecules are present on the surface, and the emitted luminescence is strongly polarized (reprinted with permission from [12])

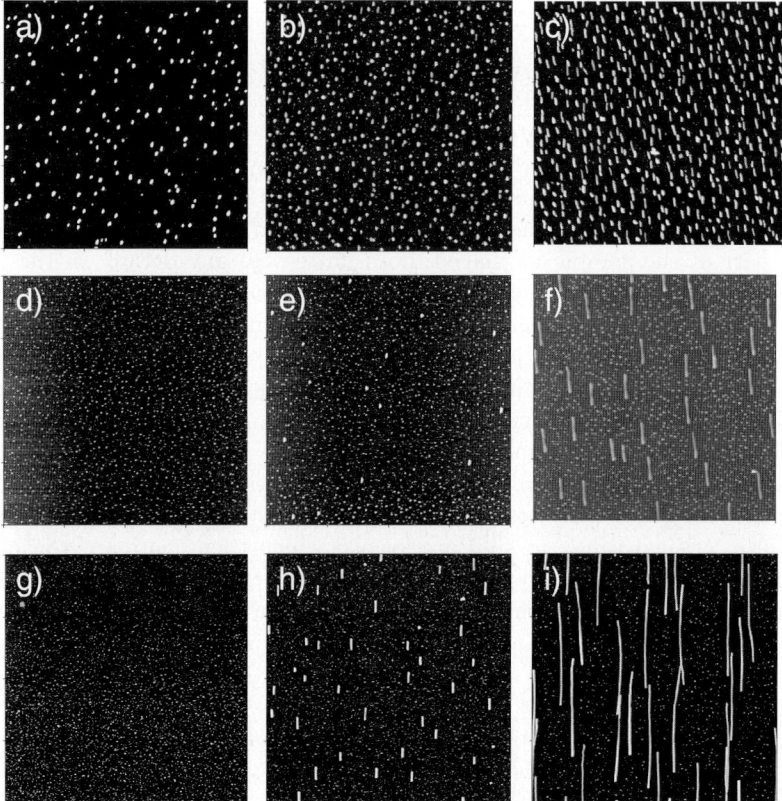

Fig. 3.14. AFM images of p-6P on muscovite for 0.2, 0.7, and 2 nm nominal thickness. (**a**)–(**c**) $6 \times 6\,\mu m^2$ deposited at $T_S = 337\,K$, (**d**)–(**f**) $20 \times 20\,\mu m^2$ deposited at 407 K, and (**g**)–(**i**) $20 \times 20\,\mu m^2$ deposited at 438 K (from [8])

as locally enough stress has built up, single clusters are able to agglomerate into needles. In the very early stages of growth, the clusters within the needles can still be identified [68] as well as on some of the thicker needles (Fig. 3.15).

The dependence of the coverage of the surface with clusters and needles as a function of the overall p-6P thickness for a fixed deposition temperature (here $T_S = 438\,K$), upper panel in Fig. 3.16, clarifies this behavior. First, up to a nominal thickness of about 1 nm, the coverage of the surface with clusters increases. After that, needles start growing, covering more and more of the substrate. The needles are fed by clusters, their number density as a countermove decreasing steadily. However, the number density of the clusters *in between* the needles remains nearly constant (squares). This critical number density depends on T_S, decreasing with increasing substrate temperature. For deposition temperatures close to room temperature, the number of clusters vanishes completely, whereas at higher substrate temperatures many clusters

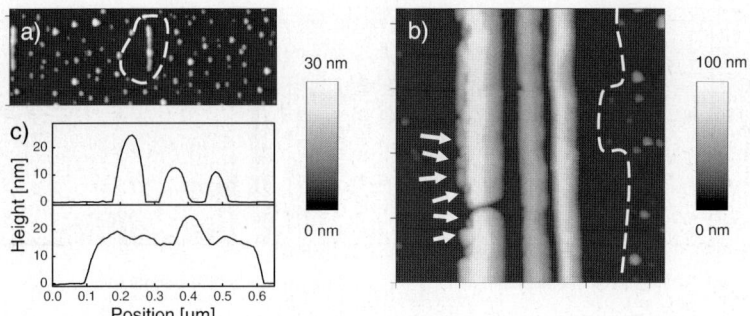

Fig. 3.15. (a) $3\times 1\,\mu\text{m}^2$ and (b) $2\times 2\,\mu\text{m}^2$ AFM images of p-6P needles on muscovite. Both for needles in the very first stage of growth (**a**) as well as for thicker needles (**b**) clusters as their individual building blocks can be identified. The section through the needle in (**a**), lower part of (**c**), shows bumps with dimensions similar to single clusters as can be seen from a section through clusters, upper part from (**c**). The *dashed lines* in (**a**) and (**b**) emphasize 300–400 nm wide cluster depletion zones around the needles. The *arrows* in (**b**) advert to single clusters attached from the left to a needle. Their size is comparable to isolated clusters found outside of the depletion zone to the right

in between the needles survive. In the lower panel of Fig. 3.16, the T_S dependence of the surface coverage with clusters and needles for a constant nominal thickness of 4 nm elucidates this trend. Migrating clusters immediately run into needles at low substrate temperatures, thus keeping their overall number density low and the covered area small. At higher substrate temperatures the distances between needles are getting larger, therefore clusters are able to survive for a longer amount of time. For temperatures higher than $T_S \approx 470\,\text{K}$ no needles form, but clusters are still present.

The height of the clusters ranges between 10 and 50 nm, with a covered area of 3,000–8,000 nm^2. Taller clusters cover a larger area. Needles, however, can reach macroscopic lengths of up to a few hundred micrometers. The temperature dependencies of the length distributions for three different nominal thicknesses and four different temperatures are shown on the right-hand side in Fig. 3.16. From left to right in the upper panel, the nominal thickness increases from 2 and 6 nm to 10 nm for $T_S = 438\,\text{K}$, whereas in the lower panel for a constant thickness of 4 nm the substrate temperature during deposition has been varied from $T_S = 390$ to $458\,\text{K}$. Distributions have been reproduced by Gaussian fits (solid lines) and are getting broader with mean length $\langle l \rangle$. Note that a more careful analysis shows unsymmetric distributions, shining light on the formation mechanism [8].

Not only the mean needle length is affected by the nominal thickness and the substrate temperature, but also the mean needle width and height are influenced. In Fig. 3.17 all these dependencies are concluded. In the upper panel the overall thickness is varied between 0 and 10 nm for three different substrate temperatures during deposition, $T_S = 337\,\text{K}$ (circles), $407\,\text{K}$

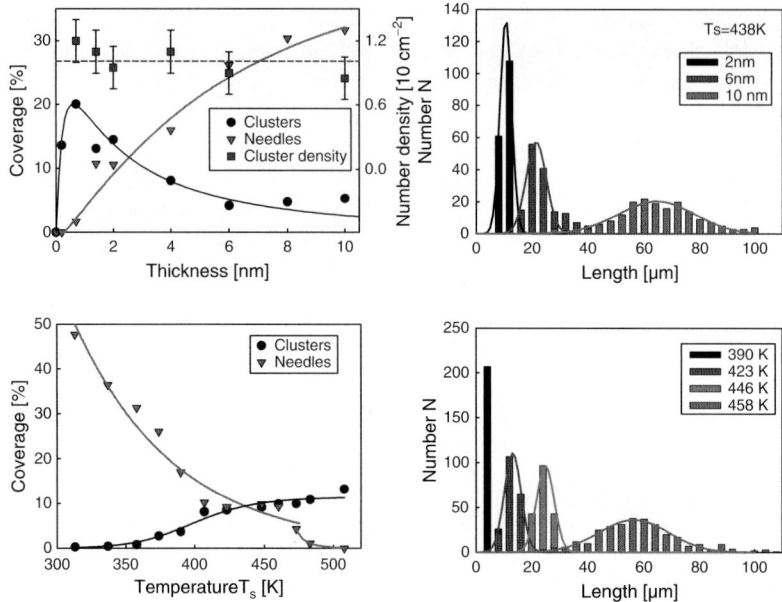

Fig. 3.16. *Upper panel*: the left-hand side depicts, for a substrate temperature of $T_S = 438$ K, the fraction of the muscovite mica surface covered by clusters (*circles*) and needles (*triangles*). Although the coverage of the surface changes with thickness, the number density of the clusters in between the needles remains nearly constant (*squares*). On the right-hand side, the length distribution of p-6P needles as a function of nominal film thickness is shown for the same surface temperature. *Lower panel*: the graph on the left-hand side demonstrates how the coverage of the surface changes with temperature. For higher substrate temperatures less fibers grow, more of the surface being covered by clusters. On the right-hand side, the length distributions for a nominal thickness of 4 nm p-6P are shown for four different values of T_S between 390 and 458 K. Higher substrate temperatures lead to longer needles. On the left-hand side the lines serve to guide the eye, on the right-hand side lines resemble Gaussian fits through the distributions (from [8])

(triangles), and 438 K (squares). In all cases the mean height and mean width first increase with coverage, but after about 2 nm nominal thickness they run into a constant value as emphasized by the solid lines. From that on, needles only grow in length, the length increasing therefore linearly with thickness. The lines in the logarithmic plot are linear fits to the data points, demonstrating the linear behavior for thicknesses larger than 2 nm. For a constant nominal thickness of 4 nm, but increasing substrate temperature the height, width, and length evolution is shown in the lower panel. The trend, already surfacing in the upper panel, is reproduced: both the mean height as well as the mean width and mean length increase with temperature, reaching their maximal values at $T_S \approx 470$ K, i.e., the temperature for optimal needle growth for p-6P. For higher temperatures only clusters grow, which do not aggregate.

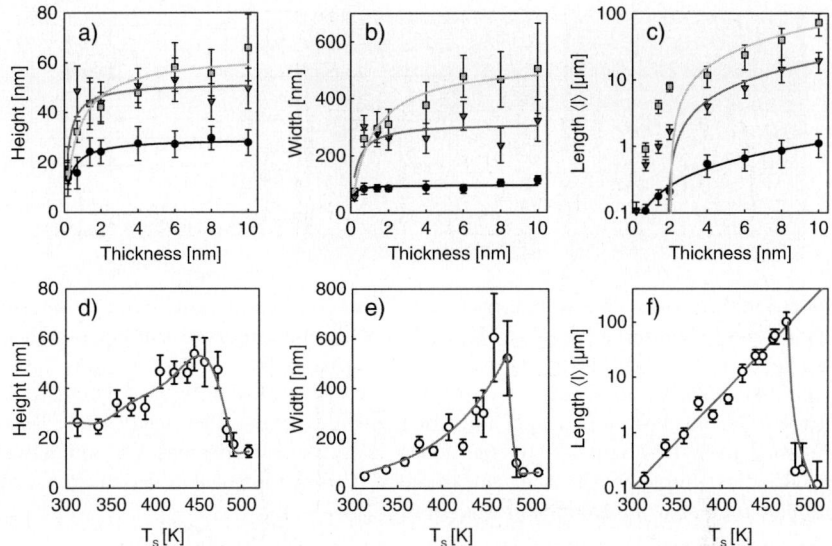

Fig. 3.17. *Upper panel*: mean height, width, and length of p-6P fibers grown at different substrate temperatures ($T_S = 337\,\text{K}$ *circles*, $407\,\text{K}$ *triangles*, and $438\,\text{K}$ *squares*) as a function of nominal thickness. For a nominal film thickness larger than 2 nm width and height saturate. *Lower panel*: mean height, width, and length as a function of the substrate temperature during growth for p-6P fibers (nominal thickness 4 nm) on muscovite mica. The lines serve to guide the eye except for (**c**) and (**f**), where they resemble linear and exponential fits, respectively (from [8])

Compared to the increase in length, the height of the aggregates only changes marginally over the temperature range. Only close to the temperature of optimal needle growth, an enhancement of a factor of about 2–3 is observed. The width of the fibers is also enhanced around this temperature, developing a broad distribution of widths corresponding to the error bars. The main effect of substrate temperature is on the length $\langle l \rangle$ of the fibers. Up to $T_S \approx 470\,\text{K}$ $\langle l \rangle$ increases exponentially with T_S. The increase in needle size together with a decrease in number density is not staggering – such a behavior is predicted by basic nucleation theory. The strong increase in length but not in width and height has its reason in the anisotropic nature of the molecules and in the growth process by attachment of islands. Molecules tend to bind rather via π–π interactions along their long sides than via an interaction of their CH groups, leading to needle-type crystals.

For p-6P on alkali halides either upright or lying molecules can be grown by adjusting the deposition rate. On muscovite mica varying the deposition rate between 0.02 and 0.6 Å s^{-1} for p-6P mainly determines the density of the growing clusters and needles. If long, isolated fibers are desired for waveguiding applications [25] or single organic nanolasers [27, 87], a high substrate temperature together with a low deposition rate is appropriate.

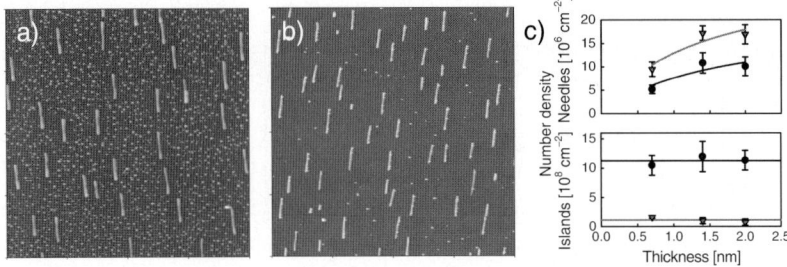

Fig. 3.18. 20×20 μm² AFM images of 2 nm of (**a**) p-6P on muscovite mica, deposited at a substrate temperature of 407 K, and of a similar prepared sample, which has been held at the deposition temperature for 75 min (**b**). Except for an increase in number density the needles do not change considerably, whereas the clusters in between the needles vanish and additional needles grow. The needles increase in number density by about 5×10^6 cm^{-2}, the island density decreases to almost zero (**c**), also for samples with other thicknesses. *Circles* represent films as grown and *triangles* represent annealed samples (from [8])

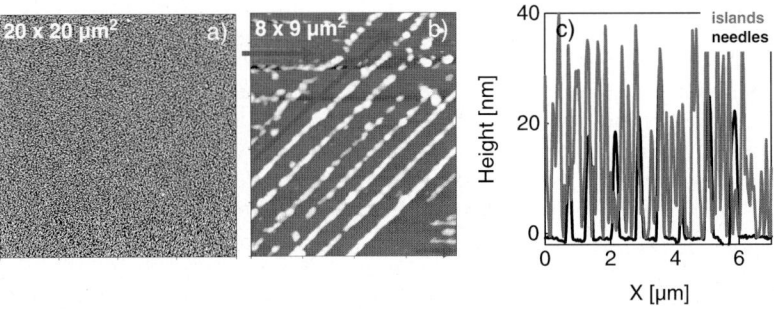

Fig. 3.19. (**a**) p-6P clusters grown on phlogopite mica at room temperature (nominal film thickness 2 nm) can be arranged by an AFM into needles (**b**). Needles form along the scanning direction, i.e., along the *arrows* in (**b**). Heights and widths of the needles (**c**) are comparable to naturally grown ones

Annealing of samples increases the number density of fibers, whereas the clusters vanish. Such a behavior is demonstrated in the AFM images in Fig. 3.18. The as-grown image (a) shows short needles and many clusters in between them. Annealing the sample for 75 min at the growth temperature of 407 K reduces the number of clusters considerably (b). Concurrently new needles take shape, increasing their number density, whereas the islands almost vanish (c).

As long as the number density of the clusters is below the critical number density, no needles form. However, needle formation can be induced by an external force. In Fig. 3.19a, an AFM image of p-6P on phlogopite, deposited at room temperature, is shown. The nominal film thickness is thin enough that no needles have formed, yet. Scanning in intermittent mode with an AFM using a relatively low set point leads to the displacement of the islands, thus islands are forced to agglomerate into needles. That way needles can be created running in any direction on the surface (Fig. 3.19b). Heights and widths of the artificial needles are comparable to ones grown naturally (Fig. 3.19c).

3.4.4 Growth Model

The basic idea for modeling the growth process is to divide it into three parts, which can be treated independently of each other (1) the initial formation of a wetting layer, (2) the nucleation and growth of clusters up to the critical number density or up to a critical cluster size, and (3) the assembly of needles from single clusters. Growth of needles from isolated molecules is supposed to play only a minor role and therefore is neglected.

First, a wetting layer of lying p-6P molecules on the polar muscovite surface grows. The layer forms domains of aligned molecules, where the unit cells are eventually broadened compared to the bulk value to increase the overlap between the overlayer and substrate lattice. After the wetting has been completed, clusters' growth sets in and the system follows the Stranski–Krastanov growth mode. In this second stage of growth only clusters form, their number density increasing with the amount of deposited material. The clusters consist of aligned molecules, with the alignment being inherited from the wetting layer. Stress builds up in the clusters and/or in their environment during their growth [76] until, from a critical size and therefore from a critical number density n_{crit}, needle growth is favored, the third stage of the growth process. The critical number density of the clusters depends slightly on temperature and changes from $n_{\text{crit}} = 3.4 \times 10^9 \, \text{cm}^{-2}$ at $T_S = 337 \, \text{K}$ to $n_{\text{crit}} = 0.9 \times 10^9 \, \text{cm}^{-2}$ at $T_S = 438 \, \text{K}$. n_{crit} might decrease with T_S because from capillarity theory [88] the clusters are expected to become larger with increasing substrate temperature, but with a reduced number density. Therefore, the critical stress is reached for a smaller value of n_{crit} and a lower nominal film thickness.

From basic nucleation theory [88, 89] a power law is predicted for the number density n of the clusters

$$n \propto R^\kappa \exp\left(E_\text{N}/kT\right). \tag{3.2}$$

For a higher deposition rate R, a larger island density n is expected, with the exponent κ and the activation energy E_N being functions of the type of condensation, of characteristic energies like the energies for adsorption and diffusion and of the critical nucleus size of the system [89]. In the literature this activation energy has been determined for several organic molecules. For the formation of three-dimensional PTCDA islands on Ni, a value of $E_\text{N} = 0.37 \, \text{eV}$ has been reported [90], for PTCDA islands on Ag (1 1 1) $0.6 \pm 0.2 \, \text{eV}$ [91], for chlorinated p-4P $0.56 \pm 0.1 \, \text{eV}$ [92], for p-6P islands on GaAs $0.90 \pm 0.04 \, \text{eV}$ [93], and for sexithiophene islands on mica $0.36 \pm 0.04 \, \text{eV}$ [94]. Determining the cluster distributions is difficult because of their intermediate nature, but as long as the size distribution of the needles is fairly homogeneous, such an activation energy can also be determined by an Arrhenius plot of the *needle number density*. For films with a constant nominal p-6P thickness (here 4 nm), a value of $E_\mathrm{N} = 0.8 \pm 0.1\,\mathrm{eV}$ results, which is close to the value obtained for p-6P on GaAs. Although an understanding of the physical meaning of this activation energy is still lacking, this value will be used in future Monte Carlo simulations of the growth process.

Most of the obtained results for the temperature dependence like the decrease in needle number density and the increase of needle size with deposition temperature are in consensus with predictions from nucleation theory [88, 95]. Using simple thermodynamics, islands are expected to be larger at higher substrate temperatures. The cause for the minimal needle length, which exists for every distribution, is probably related to the agglomeration process by single clusters. As soon as the critical number density n_crit of the clusters is reached, clusters agglomerate to needles and reduce the mean number density below n_crit. Therefore only needles forming at the very beginning continue growing, resulting in a minimal value for their length. To include more subtle details like anisotropic growth, depletion zones, and shading of areas by already grown needles like in Fig. 3.15b, and to test possible mechanisms a simple kinetic Monte Carlo model [96, 97] for the growth kinetics of the third growth stage has been developed [98], which is able to explain qualitatively the effects observed by AFM. The basis of the model is clusters of organic molecules, which diffuse freely on the surface of a wetting layer of aligned molecules. At the beginning a certain number of clusters is dropped onto the substrate, which then are followed over an appropriate number of time steps with additional clusters arriving continuously to simulate a deposition experiment. The clusters are placed on the surface in a way that their distribution is random, but keeping a minimal distance from each other to mimic the observed cluster distributions below the critical density (Fig. 3.14).

For every cluster, the rate for a jump along a unit distance into a random direction is calculated for every time step, depending on the substrate temperature T_S and on the activation energy for diffusion. If two clusters come closer than a critical distance and if the local number density of the clusters is larger than the critical density n_crit, two clusters might agglomerate if they overcame an activation energy. Two clusters sticking together form a stable needle nucleus [95, 99], which does not diffuse on the surface any more but can decompose again into two single clusters. This decay is thermally activated by a temperature-dependent nucleation barrier, assuring that at high substrate temperatures no needles form. All forming needle nuclei are artificially aligned, pointing into the same direction. These prerequisites can of course be adapted to substrates with lower symmetry like, e.g., phlogopite mica with a threefold symmetry or to the alkali halides exhibiting a fourfold surface symmetry. Further diffusing clusters are able to attach to these nuclei as soon as they come closer than the critical distance and overcome the thermal activation

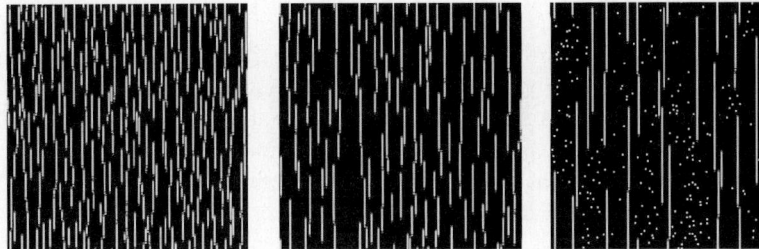

Fig. 3.20. Monte Carlo simulation of the temperature dependence of needle growth. The substrate temperature increases from left to right, all other parameters are constant. The size of the images corresponds to approximately $20 \times 20\,\mu m^2$

barrier. Probabilities are calculated according to either activation energies for attachment along the side of the needles, or at the ends of the needles.

Very first results of the Monte Carlo simulations are presented in Fig. 3.20. The same initial distribution of clusters has been followed over time with increasing temperature. Qualitatively, AFM images like the ones shown in Fig. 3.14 are reproduced. For low substrate temperatures, densely packed needles with a narrow length distribution grow. No clusters can be found in between the needles. Increasing the substrate temperature leads to more isolated and longer fibers with surviving clusters in between. Depletion zones around the needles are visible.

3.5 Thiophenes on Muscovite

On muscovite the thiophenes show a much richer morphology after vacuum deposition than the phenylenes [94, 100, 101]. For α-4T only needles of a few micrometers length grow (Figs. 3.21 and 3.22). Under UV illumination, the needles fluoresce green (Fig. 3.21b). Together with these short needles, which have heights of around 100 nm, dendritic islands of upright molecules in between the needles form. Step heights on the islands are usually multiples of a single step height of molecules with the (0 0 1) plane parallel to the substrate; the overall height of the islands is in the range of 10–20 nm. These islands mostly form at already existing needles and do luminesce only weakly. Their area increases with nominal film thickness (Fig. 3.22) and decreases with substrate temperature. From $T_S \approx 320\,K$ almost no islands form, from $T_S \approx 335\,K$ neither needles nor dendritic islands grow. Clusters with heights of approximately 10 nm can be found in between the dendritic islands and the needles.

For samples prepared at constant substrate temperature, the mean heights and widths of the fibers increase first with increasing nominal thickness, then at about 3–6 nm run into saturation at values of 150 and 600 nm, respectively. Opposite to p-6P the mean length also saturates for thicknesses larger than

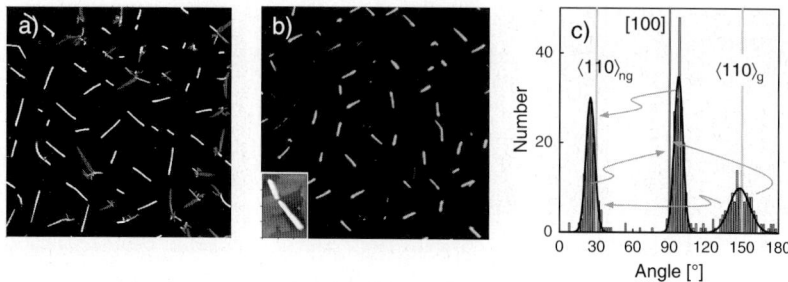

Fig. 3.21. (a) $75 \times 75\,\mu m^2$ AFM image, (b) $80 \times 80\,\mu m^2$ fluorescence microscope image of α-4T deposited at a substrate temperature of $T_S = 320\,K$ on muscovite. In (c), the needle orientations together with the muscovite crystallographic orientations are shown. The *lines* are Gaussian fits to the experimental distributions, the *arrows* point to the directions where the different needle orientations emit their maximum in fluorescence, with *vertical lines* depicting the hs-directions. The $3 \times 4\,\mu m^2$ inset in (b) is an AFM image of a needle pair along $\langle 1\,1\,0 \rangle_g$ with an angle of 15° in between (from [102])

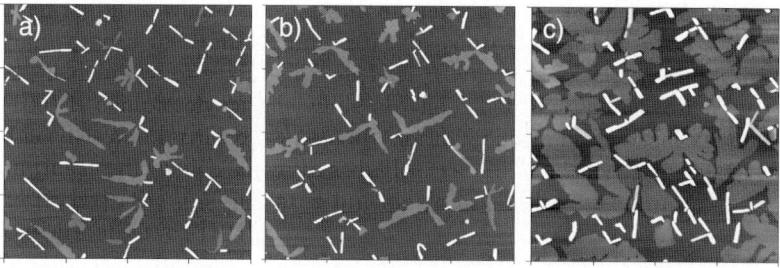

Fig. 3.22. $20 \times 20\,\mu m^2$ AFM images of α-4T on muscovite mica, deposited at $T_S = 310\,K$. The nominal thicknesses are 2 nm (a), 4 nm (b), and 10 nm (c). The height scales are 70 nm for (a) and (b), and 150 nm for (c) (from [102])

about 5 nm, because most of the new material is used for the growth of islands instead of needles. Reducing the deposition rate and increasing the substrate temperature to values larger than approximately 325 K might diminish this growth behavior. An increase of the deposition rate to about $1\,\text{Å}\,s^{-1}$ on the other hand leads to the growth of islands only [103]. A phase contrast exists in AFM images between the islands and the area in between the islands, suggesting that this area is either the bare substrate or covered by a wetting layer of lying molecules.

The next longer molecule with an even number of thiophene rings is sexithiophene, α-6T, its polarizability being considerably larger than that for α-4T [104]. Together with branched needles and small clusters of lying molecules, two types of dendritic islands of upright molecules grow (Fig. 3.23). Fibers show a bright orange, polarized fluorescence, whereas again the islands of upright molecules only luminesce weakly [105]. The main difference between

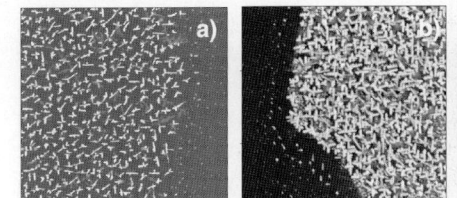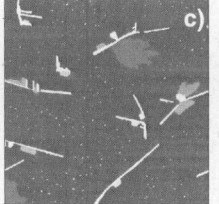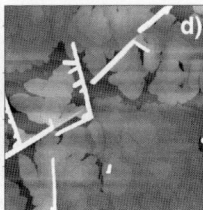

Fig. 3.23. Thickness and substrate temperature dependence of α-6T on muscovite. For a substrate temperature of 310 K and thicknesses of (**a**) 1 nm and (**b**) 6 nm, *partly shadowed areas* are imaged for comparison with the bare mica substrate. The size of the images is $15 \times 15\,\mu m^2$. For a higher substrate temperature of 370 K, (**c**) 1 nm and (**d**) 8 nm, initially needles grow, then islands of upright molecules start to form. The size of the two images is $9 \times 9\,\mu m^2$ (from [102])

the two types of dendritic islands is their height. The shallow type is only around 2.9 nm tall, the platelet type shows multiple steps with step heights of 2.4 nm, which is close to the 2.24 nm for upright α-6T molecules with their (1 0 0) face parallel to the substrate. In [94] it has been speculated that the shallow islands consist of a layer of lying molecules and, on top, a layer of upright molecules.

At room temperature a close-packed film of short α-6T fibers forms, whereas at higher substrate temperatures the number density of the fibers decreases, and their length increases (Fig. 3.23). This effect is well known from the *para*-phenylenes and is expected from nucleation theory. The dendritic islands start their growth at fibers, their area increasing with nominal film thickness. The increase of width and height of the needles saturates at about 4 nm nominal thickness, but the height of the islands continues to increase.

The most striking difference between phenylene and thiophene growth is the many different orientations of the fibers. From large-scale fluorescence microscope images, the distribution of α-4T and α-6T needle orientations reveals three preferred orientations. A rather broad distribution around one of the $\langle 1\,1\,0\rangle$ directions, $\langle 1\,1\,0\rangle_g$ in Fig. 3.21c, and two narrower distributions close to the other $\langle 1\,1\,0\rangle$ direction, $\langle 1\,1\,0\rangle_{ng}$, and to $[1\,0\,0]$. The two narrow distributions are not exactly centered around the high-symmetry directions. Polarization-dependent images reveal that the orientations of the transition dipoles are along $[1\,0\,0]$ and $\langle 1\,1\,0\rangle_{ng}$, not along $\langle 1\,1\,0\rangle_g$. Note that for the thiophenes the transition dipoles are a few degrees off the long molecules axis [106]. Fibers along $\langle 1\,1\,0\rangle_{ng}$ emit light polarized along $[1\,0\,0]$, fibers along $[1\,0\,0]$ emit light polarized along $\langle 1\,1\,0\rangle_{ng}$. Fibers along $\langle 1\,1\,0\rangle_g$ have their emission maximum either along $[1\,0\,0]$ or along $\langle 1\,1\,0\rangle_{ng}$. So, as observed for the case of the *para*-phenylenes, a single high-symmetry direction of the muscovite substrate is a special direction. For p-6P this special direction is $[1\,0\,0]$, along which no needles grow. For the thiophenes the broad distribution is never along $[1\,0\,0]$, along $\langle 1\,1\,0\rangle_g$ none of the needles has its maximum in polarization. Obviously, thiophene molecules like to be oriented along [1 0 0] or $\langle 1\,1\,0\rangle_{ng}$, not along $\langle 1\,1\,0\rangle_g$. As for the phenylenes two domains are found, where on the other domain the asymmetries are reversed.

Assuming that the contact face between the α-4T needles and the substrate is a close-packed face with the bulk crystalline structure and that the needle direction is along the molecular lamella structure, three types of needles can be predicted. The first type is the (1 0 0)-type with the (1 0 0) face or similar faces like $(1\,0\,\bar{1})$, $(1\,0\,1)$, $(1\,0\,\bar{2})$, $(1\,0\,2)$, etc., parallel to the substrate. For this type the angle between the long molecular axis and the needle axis is $\beta_{\text{Mol}} = 90°$. The second type is the (0 1 0)-type (similar contact faces (0 1 1), (0 2 0), (0 2 1)) with $\beta_{\text{Mol}} \approx 67°$, the third type is the (1 1 0)-type ($(\bar{1}\,1\,0)$, $(1\,\bar{1}\,0)$, (1 1 1), etc.) with $\beta_{\text{Mol}} \approx 74°$. The same considerations lead to very similar angles for α-6T needles.

That way the asymmetry of the distribution of orientations can be easily explained. Because the angle of the molecules with respect to the long fiber axis is larger than 60° and because the thiophene molecules tend to aline along the nongrooved hs-directions, the needle directions have to be off by 7°, i.e., the (0 1 0)-type, and 14° from the mica high-symmetry directions, i.e., the (1 1 0)-type of needles. From high-resolution AFM images, angles between roughly parallel entities along $\langle 1\,1\,0\rangle_g$ of (6°, 14°, 20°, and 26°) 2° are observed, which fit the predicted values very well. Note that the distributions in Fig. 3.21 stem from optical images, and therefore average over needle orientations. The fine structure can only be observed by AFM images [102]. The broader distribution along $\langle 1\,1\,0\rangle_g$ is symmetric, because here the molecules can be aligned along two different directions.

3.6 Microrings

In Sect. 3.4.2 the role of the surface energy on aggregate morphology has already been mentioned. A deliberate way of changing the surface energy of the substrate muscovite is rinsing it for a few minutes with deionized water or with methanol. The surface changes from hydrophilic to hydrophobic, the surface energy is getting reduced to a value probably close to $250\,\text{mJ}\,\text{m}^{-2}$ [107].

The morphology of growing phenylene and thiophene films changes drastically by the preceding water treatment (Fig. 3.24). Luminescing aggregates are still visible, but the shape of the aggregates varies between needles, which are partly bent, and closed rings. The minimal diameter of the rings is in all cases around 2–5 µm. Larger rings up to a few ten micrometers in diameter, which not necessarily have to be circular in shape, can also form. Sometimes rings are isolated, sometimes form at the end of fibers or even agglomerate to multirings, form rackets or loops. AFM reveals that rings possess heights of the order of 500 nm, i.e., reach heights comparable to the needles close to defects (Sect. 3.4.2), and to the case of p-6P on Au (Sect. 3.7). In between the

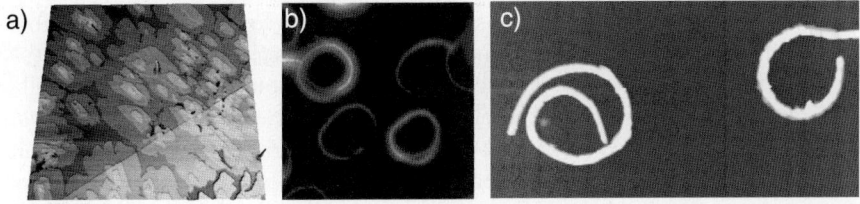

Fig. 3.24. (a) $12 \times 12\,\mu m^2$ AFM image of patches of upright p-6P molecules on a water treated muscovite surface, height scale 20 nm. The step heights of the islands are 2.5 nm. (b) $20 \times 20\,\mu m^2$ fluorescence microscope image of p-6P microrings from p-6P. (c) $20 \times 11\,\mu m^2$ AFM image of α-4T microrings, height scale 250 nm (b) (reprinted with permission from [108])

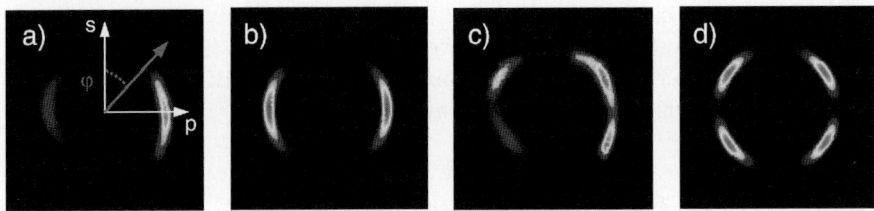

Fig. 3.25. Polarized two-photon images $10 \times 10\,\mu m^2$ of a selected microring for pp-polarization (a) and ps-polarization (c). In (b) and (d), simulated images are shown (reprinted with permission from [109])

rings, layers of upright molecules are visible (Fig. 3.24a). The step height is 2.5 nm, i.e., the step height of (0 0 1) layers (cf. Table 3.1).

Polarized fluorescence microscopy suggests that these rings are made of lying radially oriented molecules [109], i.e., they resemble bent fibers. These findings from linear optics are confirmed by polarized two-photon measurements (Fig. 3.25). The investigations are similar to previous two-photon measurements on straight organic nanofibers [110], where the orientation of molecular building blocks was determined as a function of position along the nanofibers. Two-photon microscopy is performed with linearly polarized 200 fs IR pulses from a Ti-Sapphire laser in reflection geometry (cf. Chap. 2).

For pp-polarization (i.e., p-polarized excitation and p-polarized detection) and the main transition dipole moment parallel to the long molecular axis, the two-photon signal is expected to vary like $I^{pp} \propto \sin^6 \varphi$ along the ring [110], and for ps-polarization like $I^{ps} \propto \cos^2 \varphi \sin^4 \varphi$. Measured two-photon fluorescence images are that way reproduced very well (Fig. 3.25), for pp- and ps-polarization. Differences in relative intensities along the rings are attributed to local height variations as confirmed by AFM.

Thickness-dependent experiments (not shown here) demonstrate that rings already form for very thin films of nominal 2 nm thickness, and that both open rings, agglomerated rings, as well as rings of various sizes and shapes exist.

Multiple-curled fibers (Fig. 3.24c) suggests that rings form from straight fibers by bending, which thereafter might continue growing in width and height.

Ring formation due to van der Waals interactions might be a possible scenario, but electrostatic interactions as for counterion-induced actin ring formation [111] are also possible because of incorporation of, e.g., potassium cations. It is not clear, yet, at which size and why the rings start to bend. First a large energy barrier due to the strain energy has to be overcome, which will be compensated by cohesion only if the two ends are already close together. The persistence length ℓ_p even for fibers in the very beginning of their growth process is large, about $700\,\mu$m at 440 K for fibers with 10 nm width and 20 nm height. Bending due to thermal fluctuations only can probably be excluded that way. In accordance to this, drifting fibers in water or in methanol appear stiff. Stress relief like described in [112] for inorganic nanobelts from ZnO and SiC might be a possible factor. However, formation of organic microrings by or after OMBD is rather scarce, but obviously is most advantageous because even molecules, which are insoluble, can be used.

3.7 Au–Mica Heterostructures

A more systematic way of changing the substrate surface energy implies covering it with various amounts of a different material. For this task, ultrathin films of Au are deposited on mica by an electron beam evaporator at a pressure of $p = 1 \times 10^{-7}$ mbar. Thereafter the Au/mica sample is transferred through ambient into the deposition chamber for OMBD. For very thin films gold does not form a continuous layer on mica, but shows Volmer–Weber growth (Sect. 1.2). In Fig. 3.26 the growing Au clusters are clearly distinguishable from the overgrowing p-6P aggregates.

AFM images (Figs. 3.26 and 3.27) together with fluorescence micrographs [26, 66] demonstrate the influence of Au layers of thicknesses between 1 and 17 nm on the overall morphology of the growing organic aggregates. Fluorescing fibers are still formed, but their mutually parallel alignment

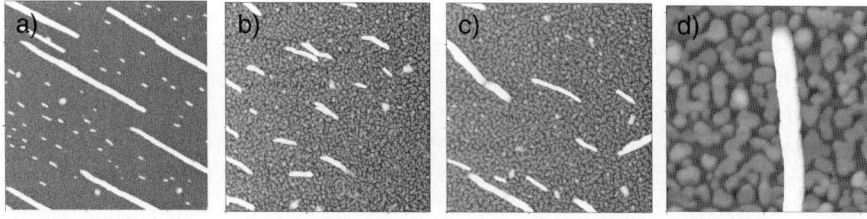

Fig. 3.26. $5 \times 5\,\mu$m^2 AFM images of p-6P on muscovite covered with (**a**) 1 nm, (**b**) 5 nm, and (**c**) 8 nm Au, before 3 nm p-6P were deposited. In (**d**), an $1 \times 1\,\mu$m^2 AFM image of 3 nm p-6P on 8 nm Au, it can clearly be seen that the p-6P fibers lie on top of the Au islands (reprinted with permission from [26])

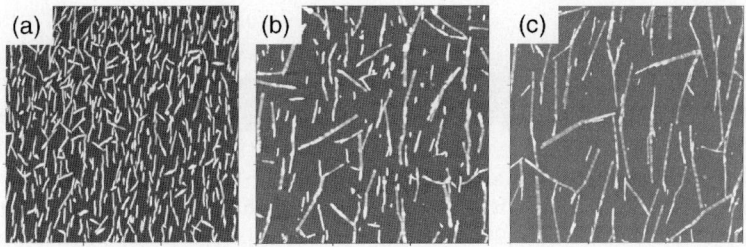

Fig. 3.27. $30 \times 30\,\mu m^2$ AFM images of p-6P needles (nominal p-6P thickness 5 nm, substrate temperature during deposition 400 K) on muscovite covered with (**a**) 8 nm, (**b**) 11 nm, and (**c**) 17 nm Au, before p-6P deposition. The height scale for all images is 330 nm (reprinted with permission from [66])

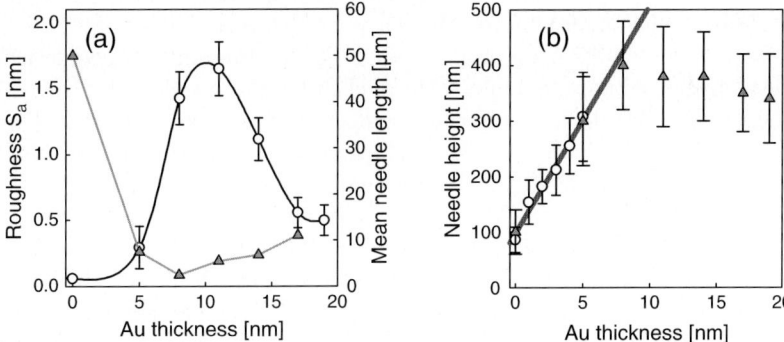

Fig. 3.28. (**a**) Mean roughness S_a over $1 \times 1\,\mu m^2$ of the Au film as a function of the Au film thickness (*open circles*), together with the mean length of the p-6P fibers (*filled triangles*). The lines serve to guide the eye. (**b**) Mean needle height as function of Au film thickness for two comparable samples (*open circles* and *filled triangles*). The *solid line* resembles a linear regression for the data points up to 8 nm Au thickness (reprinted with permission from [66])

decreases with Au thickness. But even when the Au cluster for a 8 nm thick Au film covers a large fraction of the substrate surface, the fibers still follow a preferred orientation. This supports the assumption that it is a long-range interaction which determines the alignment, i.e., electric fields on the surface. These fields are just weakened by the screening due to the Au islands, resulting in less alignment. The growth of ammonium chloride on mica covered with up to 20 nm thick crystalline and amorphous coatings shows a similar alignment due to mica electric fields [64]. However, for thicker Au films, three preferred needle orientations evolve (Fig. 3.27b).

The mutual alignment not only changes with increasing Au thickness but also aggregates mean lengths and heights (Fig. 3.28). The mean needle length decreases strongly to values below 5 μm, until, from a Au thickness of 5–8 nm on starts to increase again. The mean height increases linearly up to values

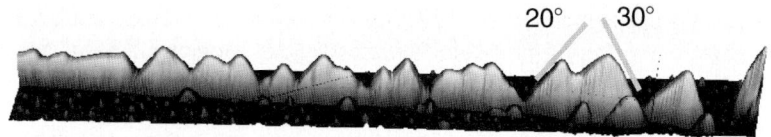

Fig. 3.29. Three-dimensional representation ($1.3 \times 10\,\mu m^2$, 350 nm height range) of a single p-6P needle on a thick Au film. The width of the needle is 250 nm. Two common slopes of the facets (20° and 30°) are emphasized (reprinted with permission from [66])

of 400 nm, until from about 10 nm Au thickness the value remains approximately constant. The mobility determines the mean length and the number density of the needles. The mean roughness S_a of uncovered Au films (circles in Fig. 3.28a) reaches a maximum value close to 10 nm Au thickness. This maximum in Au roughness corresponds well with the minimum in needle length. The Au clusters hinder agglomeration of clusters into needles, thus increasing the needle number density and decreasing the mean needle length. As soon as the roughness decreases due to the formation of a continuous Au film, the mean needle length starts to increase again.

For the surface energy of mica values between 5,000 and 250 mJ m^{-2} have been reported [107,113], depending on the cleanliness of the substrate. Freshly cleaved mica is expected to possess surface energies closer to 5,000 mJ m^{-2} than to 250 mJ m^{-2}. Depositing a thin layer of Au will result in a linear decrease of the surface energy with increasing Au thickness, until eventually the value for a Au (surface free energy 1,400 mJ m^{-2} [114], but probably smaller due to contaminations [115] during transfer) is reached. For equilibrium crystal shapes on a flat surface, the crystal height h decreases with an increasing ratio of adhesion energy β and surface energy of the adsorbate crystal γ [116], $h \propto 1 - \beta/\gamma$. A linear decrease of the average adhesion energy with increasing Au thickness from the mica to the Au value thus leads to the observed linear increase in needle height.

From the point of maximum surface roughness where the organic aggregates reach maximum height, the surface morphology of single needles changes too. For thin films the needle surfaces are either flat and show facets along the short fiber axis, or are rounded. As soon as the Au thickness is larger than approximately 8 nm, facets along the long needle axis show up. In Fig. 3.27b,c, these facets manifest themselves as a knob-like structure. The most common slopes for the facets are 10°, 20°, and 30° (Fig. 3.29).

3.8 Conclusions

Fabrication of organic nanofibers by OMBD on anisotropic silicate substrates is a well-suited method to produce high-quality entities. For the *para*-phenylenes – here demonstrated for *para*-hexaphenylene – the fibers grow along the two high-symmetry directions ⟨1 1 0⟩ on different rotational substrate domains, never along [1 0 0]. Needles are assembled from nanoscale clusters, which form after the completion of a wetting layer. The molecules within the fibers are oriented approximately perpendicular to the two ⟨1 1 0⟩ directions, i.e., almost perpendicular to the actual groove direction. No obvious match with the substrate lattice has been found except a possible point-on-line coincidence for the preceding wetting layer and the alignment of the [1 1 0] direction of the needles with the ⟨1 1 0⟩ mica direction. Electric dipole fields on the muscovite mica are crucial for the alignment of the molecules within the wetting layer and, further on, within clusters and needles. With an angle of ±14° of the phenylene molecules with the short fiber axis, the molecules are almost parallel to the dipoles and thus the interaction energy is minimized.

For the thiophenes the molecules long axes tend to orient along the non-grooved hs-directions. That way roughly three needle directions are realized on a single substrate domain, with only two mean molecule orientations. The alignment and growth for both families of molecules are determined by two processes (1) an epitaxy-related alignment of the molecules on the surface and (2) a selection of otherwise symmetry-allowed domains. The knowledge of the basic principles behind the growth mode opens up the road to custom-made nanofibers. As long as, e.g., the basic phenylene chain and the packing of the bulk organic crystal are not changed significantly, even functionalized phenylenes will adopt needle growth with mutually parallel aligned entities. This has already been demonstrated for methoxylated as well as for chlorinated phenylenes [92, 117] (cf. Chap. 4). On the other hand needle films with specific symmetries are approachable by tailoring the angle between the long molecules axis and the needle axis, as already shown for the case of thiophene/phenylene cooligomers [118].

Reducing the surface energy of muscovite by either rinsing it with water or covering it with an ultrathin Au film leads to much taller organic aggregates as for the case of a clean substrate: either rings with radially oriented molecules or facetted needles evolve. That way a wide range of aggregate morphologies is experimentally accessible. In the future such building blocks might be used for the bottom-up design of new photonic devices.

Acknowledgments

The author would like to thank Horst-Günter Rubahn from Syddansk Universitet for a very fruitful cooperation on the topic of organic nanofibers; Laxman Kankate, Horst Niehus, and Karin Braune from Humboldt-University Berlin; and Manuela Schiek from University Oldenburg for help in preparation and characterization of the samples, and for many valuable discussions. Norbert Koch, Humboldt-University Berlin, is acknowledged for providing α-6T. The Hanse Institute for Advanced Study (HWK), Delmenhorst, is thanked for financial support.

References

1. T. Mikami, H. Yanagi, Appl. Phys. Lett. **73**, 563 (1998)
2. T. Haber, A. Andreev, A. Thierry, H. Sitter, M. Oehzelt, R. Resel, J. Cryst. Growth **284**, 209 (2005)
3. A. Andreev, G. Matt, C. Brabec, H. Sitter, D. Badt, H. Seyringer, N. Sariciftci, Adv. Mater. **12**, 629 (2000)
4. F. Balzer, H.G. Rubahn, Appl. Phys. Lett. **79**, 3860 (2001)
5. R. Resel, M. Oehzelt, T. Haber, G. Hlawacek, C. Teichert, S. Müllegger, A. Winkler, J. Cryst. Growth **283**, 397 (2005)
6. S. Müllegger, S. Mitsche, P. Pölt, K. Hänel, A. Birkner, C. Wöll, A. Winkler, Thin Solid Films **484**, 408 (2005)
7. S. Müllegger, O. Stranik, E. Zojer, A. Winkler, Appl. Surf. Sci. **221**, 184 (2004)
8. L. Kankate, F. Balzer, H. Niehus, H.G. Rubahn (2007) (submitted)
9. G. Koller, S. Berkebile, J. Krenn, G. Tzvetkov, G. Hlawacek, O. Lengyel, F. Netzer, C. Teichert, R. Resel, M. Ramsey, Adv. Mater. **16**, 2159 (2004)
10. F. Balzer, V. Bordo, A. Simonsen, H.G. Rubahn, Appl. Phys. Lett. **82**, 10 (2003)
11. F. Quochi, A. Andreev, F. Cordella, R. Orru, A. Mura, G. Bongiovanni, H. Hoppe, H. Sitter, N. Sariciftci, J. Lumin. **112**, 321 (2005)
12. F. Balzer, H.G. Rubahn, Adv. Funct. Mater. **15**, 17 (2005)
13. J. Brewer, H. Henrichsen, F. Balzer, L. Bagatolli, A. Simonsen, H.G. Rubahn, Proc. SPIE **5931**, 250 (2005)
14. J. Kjelstrup-Hansen, H. Henrichsen, P. Bøgild, H.G. Rubahn, Thin Solid Films **515**, 827 (2006)
15. J. Kjelstrup-Hansen, O. Hansen, H.G. Rubahn, P. Bøggild, Small **2**, 660 (2006)
16. F. Balzer, H.G. Rubahn, Surf. Sci. **548**, 170 (2004)
17. S. Forrest, Chem. Rev. **97**, 1793 (1997)
18. A. Sassella, M. Campione, M. Moret, A. Borghesi, C. Goletti, G. Bussetti, P. Chiaradia, Phys. Rev. B **71**, 201311(R) (2005)
19. M. Herman, W. Richter, H. Sitter, *Epitaxy, Springer Series in Materials Science*, vol. 62 (Springer, Berlin Heidelberg New York, 2004)
20. L. Kilian, E. Umbach, M. Sokolowski, Surf. Sci. **573**, 359 (2004)
21. H. Yanagi, S. Okamoto, Appl. Phys. Lett. **71**, 2563 (1997)
22. G. Leising, S. Tasch, C. Brandstatter, F. Meghdadi, G. Froyer, L. Athouel, Adv. Mater. **9**, 33 (1997)
23. D. Gundlach, Y.Y. Lin, T. Jackson, D. Schlom, Appl. Phys. Lett. **71**, 3853 (1997)
24. H. Yanagi, T. Morikawa, Appl. Phys. Lett. **75**, 187 (1999)
25. F. Balzer, V. Bordo, A. Simonsen, H.G. Rubahn, Phys. Rev. B **67**, 115408 (2003)
26. F. Balzer, L. Kankate, H. Niehus, H.G. Rubahn, Proc. SPIE **6117**, 20 (2006)
27. F. Quochi, F. Cordella, A. Mura, G. Bongiovanni, F. Balzer, H.G. Rubahn, J. Phys. Chem. B **109**, 21690 (2005).
28. K. Baker, A. Fratini, T. Resch, H. Knachel, W. Adams, E. Socci, B. Farmer, Polymer **34**, 1571 (1993)
29. L. Antolini, G. Horowitz, F. Kouki, F. Garnier, Adv. Mater. **10**, 382 (1998)
30. T. Siegrist, C. Kloc, R. Laudise, H. Katz, R. Haddon, Adv. Mater. **10**, 379 (1998)

31. G. Horowitz, B. Bachet, A. Yassar, P. Lang, F. Demanze, J.L. Fave, F. Garnier, Chem. Mater. **7**, 1337 (1995)
32. G. Heimel, P. Puschnig, M. Oehzelt, K. Hummer, B. Koppelhuber-Bitschnau, F. Porsch, C. Ambrosch-Draxl, R. Resel, J. Phys.: Condens. Matter **15**, 3375 (2003)
33. D. Williams, T. Starr, Comput. Chem. **1**, 173 (1977)
34. R. Resel, Thin Solid Films **433**, 1 (2003)
35. G. Koller, S. Berkebile, J. Krenn, F. Netzer, M. Oehzelt, T. Haber, R. Resel, M. Ramsey, Nano Lett. **6**, 1207 (2006)
36. R. Resel, G. Leising, Surf. Sci. **409**, 302 (1998)
37. Y. Yoshida, H. Takiguchi, T. Hanada, N. Tanigaki, E.M. Han, K. Yase, J. Cryst. Growth **198/199**, 923 (1999)
38. C. Ambrosch-Draxl, J. Majewski, P. Vogl, G. Leising, Phys. Rev. B **51**, 9668 (1995)
39. A. Niko, F. Meghdadi, C. Ambrosch-Draxl, P. Vogl, G. Leising, Synth. Met. **76**, 177 (1996)
40. C. Ziegler, in *Handbook of Organic Conductive Molecules and Polymers. Vol. 3. Conductive Polymers: Spectroscopy and Physical Properties*, ed. by H. Nalwa (Wiley, New York, 1997)
41. D. Fichou, J. Mater. Chem. **10**, 571 (2000)
42. W. Schoonveld, Transistors based on ordered organic semiconductors. Ph.D. Thesis, Rijksuniversiteit Groningen (1999)
43. V. Marcon, G. Raos, M. Campione, A. Sassella, Cryst. Growth Des. **6**, 1826 (2006)
44. A. Sassella, D. Besana, A. Borghesi, M. Campione, S. Tavazzi, B. Lotz, A. Thierry, Synth. Met. **138**, 125 (2003)
45. B. Servet, S. Ries, M. Trotel, P. Alnot, G. Horowitz, F. Garnier, Adv. Mater. **5**, 461 (1993)
46. B. Servet, G. Horowitz, S. Ries, O. Lagorsse, P. Alnot, A. Yassar, F. Deloffre, P. Srivastava, R. Hajlaoui, P. Lang, F. Garnier, Chem. Mater. **6**, 1809 (1994)
47. M. Campione, A. Sassella, M. Moret, A. Thierry, B. Lotz, Thin Solid Films **500**, 169 (2006)
48. D. Griffen, *Silicate Crystal Chemistry* (Oxford University Press, New York, 1992)
49. K. Müller, C. Chang, Surf. Sci. **9**, 455 (1968)
50. E. Palin, M. Dove, S. Redfern, A. Bosenick, C. Sainz-Diaz, M. Warren, Phys. Chem. Miner. **28**, 534 (2001)
51. C. Herrero, J. Sanz, J. Serratosa, J. Phys. C: Solid State Phys. **18**, 13 (1985)
52. R. Knurr, S. Bailey, Clays Clay Miner. **34**, 7 (1986)
53. Y. Kuwahara, Phys. Chem. Miner. **26**, 198 (1999)
54. Y. Kuwahara, Phys. Chem. Miner. **28**, 1 (2001)
55. K. Müller, C. Chang, Surf. Sci. **14**, 39 (1969)
56. S. Dorel, F. Pesty, P. Garoche, Surf. Sci. **446**, 294 (2000)
57. S. Dorel, Nanostructuration de la muscovite: Une étude par diffraction d'électrons lents en mode oscillant. Ph.D. Thesis, Université de Paris-Sud, Centre D'Orsay (2000)
58. F. Pesty, P. Garoche, Surf. Sci. **580**, 153 (2005)
59. N. Uyeda, M. Ashida, E. Suito, J. Appl. Phys. **36**, 1453 (1965)
60. M. Ashida, Bull. Chem. Soc. Jpn **39**, 2625 (1966)

61. G. Distler, Kristall Technik **5**, 73 (1970)
62. S. Kobzareva, G. Distler, J. Cryst. Growth **10**, 269 (1971)
63. M. Metsik, L. Golub, J. Appl. Phys. **46**, 1983 (1975)
64. C. Motoc, M. Badea, J. Cryst. Growth **17**, 337 (1972)
65. S. Folgueras, F. Da Costa, M. Hoyos, Thin Solid Films **113**, 257 (1984)
66. F. Balzer, L. Kankate, H. Niehus, R. Frese, C. Maibohm, H.G. Rubahn, Nanotechnology **17**, 984 (2006)
67. J. Beermann, S. Bozhevolnyi, F. Balzer, H.G. Rubahn, Laser Phys. Lett. **2**, 480 (2005)
68. A. Andreev, C. Teichert, G. Hlawacek, H. Hoppe, R. Resel, D.M. Smilgies, H. Sitter, N. Sariciftci, Org. Electron. **5**, 23 (2004)
69. F. Balzer, L. Kankate, H. Niehus, H.G. Rubahn, Proc. SPIE **5925**, 31 (2005)
70. H. Plank, R. Resel, S. Purger, J. Keckes, A. Thierry, B. Lotz, A. Andreev, N. Sariciftci, H. Sitter, Phys. Rev. B **64**, 235423 (2001)
71. E. Kintzel, Jr., D.M. Smilgies, J. Skofronick, S. Safron, D. Van Winkle, T. Trelenberg, E. Akhadov, F. Flaherty, J. Vac. Sci. Technol. A **19**, 1 (2001)
72. R. Resel, K. Erlacher, B. Müller, A. Thierry, B. Lotz, T. Kuhlmann, K. Lischka, G. Leising, Surf. Interface Anal. **30**, 518 (2000)
73. H. Plank, R. Resel, H. Sitter, A. Andreev, N.S. Sariciftci, G. Hlawacek, C. Teichert, A. Thierry, B. Lotz, Thin Solid Films **443**, 108 (2003)
74. F. Balzer, R. Gerlach, G. Polanski, H.G. Rubahn, Chem. Phys. Lett. **274**, 145 (1997)
75. J. Brewer, M. Schiek, A. Lützen, K. Al-Shamery, H.G. Rubahn, Nano Lett. **6**, 2656 (2006).
76. C. Teichert, G. Hlawacek, A. Andreev, H. Sitter, P. Frank, A. Winkler, N. Sariciftci, Appl. Phys. A **82**, 665 (2006)
77. H. Plank, R. Resel, A. Andreev, N. Sariciftci, H. Sitter, J. Cryst. Growth **237–239**, 2076 (2002)
78. D. Hooks, T. Fritz, M. Ward, Adv. Mater. **13**, 227 (2001)
79. S. Mannsfeld, T. Fritz, Phys. Rev. B **71**, 235405 (2005)
80. E. Zojer, N. Koch, P. Puschnig, F. Meghdadi, A. Niko, R. Resel, C. Ambrosch-Draxl, M. Knupfer, J. Fink, J. Brédas, G. Leising, Phys. Rev. B **61**, 16538 (2000)
81. M. Abramoff, P. Magelhaes, S. Ram, Biophot. Int. **11**, 36 (2004)
82. V. Mochalkin, Teoret. Eksp. Khim. **3**, 587 (1967)
83. Y. Verbandt, H. Thienpont, I. Veretennicoff, G. Rikken, Phys. Rev. B **48**, 8651 (1993)
84. J. Israelachvili, *Intermolecular and Surface Forces*, 2nd edn. (Academic, London, 1991)
85. C. Maibohm, J. Brewer, H. Sturm, F. Balzer, H.G. Rubahn, J. Appl. Phys. **100**, 054304 (2006)
86. J. Tersoff, R. Tromp, Phys. Rev. Lett. **70**, 2782 (1993)
87. F. Quochi, F. Cordella, A. Mura, G. Bongiovanni, F. Balzer, H.G. Rubahn, Appl. Phys. Lett. **88**, 041106 (2006)
88. M. Ohring, *The Materials Science of Thin Films* (Academic, San Diego, 1992)
89. J. Venables, G. Spiller, M. Hanbucken, Rep. Prog. Phys. **47**, 399 (1984)
90. M. Tiba, Organo-metallic structures for spintronic applications. Ph.D. Thesis, Technische Universiteit Eindhoven (2005)
91. B. Krause, A. Dürr, K. Ritley, F. Schreiber, H. Dosch, D. Smilgies, Phys. Rev. B **66**, 235404 (2002)

92. M. Schiek, A. Lützen, K. Al-Shamery, F. Balzer, H.G. Rubahn, Cryst. Growth Des. **7**(2), 229 (2007).
93. B. Müller, T. Kuhlmann, K. Lischka, H. Schwer, R. Resel, G. Leising, Surf. Sci. **418**, 256 (1998)
94. F. Biscarini, R. Zamboni, P. Samori, P. Ostoja, C. Taliani, Phys. Rev. B **52**, 14868 (1995)
95. I. Markov, *Crystal Growth for Beginners*, 2nd edn. (World Scientific, New Jersey, 2003)
96. H. Brune, Surf. Sci. Rep. **31**, 121 (1998)
97. D. Landau, K. Binder, *A Guide to Monte Carlo Simulations in Statistical Physics* (Cambridge University Press, Cambridge, 2000)
98. U. Wilensky, *NetLogo* (Center for Connected Learning and Computer-Based Modeling, Northwestern University, Evanston, IL, 1999) (http://ccl.northwestern.edu/netlogo/)
99. A. Pimpinelli, J. Villain, *Physics of Crystal Growth* (Cambridge University Press, Cambridge, 1998)
100. C. Taliani, F. Biscarini, M. Muccini, in *Semiconducting Polymers*, ed. by G. Hadziioannou, P. van Hutten (Wiley-VCH, Weinheim, 2000), pp. 149–188
101. F. Balzer, L. Kankate, H. Niehus, H.G. Rubahn, Proc. SPIE **5724**, 285 (2005)
102. F. Balzer, L. Kankate, H. Niehus, H.G. Rubahn (2007) (in preparation)
103. L. Kankate, Light emitting organic nanofibers from *para*-phenylene and α-thiophene oligomers. Ph.D. Thesis, Humboldt-Universität zu Berlin (2007)
104. B. Champagne, D. Mosley, J.M. Andre, J. Chem. Phys. **100**, 2034 (1994)
105. J. Vrijmoeth, R. Stok, R. Veldman, W. Schoonveld, T. Klapwijk, J. Appl. Phys. **83**, 3816 (1998)
106. H. Sun, Z. Zhao, F. Spano, D. Beljonne, J. Cornil, Z. Shuai, J.L. Brédas, Adv. Mater. **15**, 818 (2003)
107. H. Christenson, J. Phys. Chem. **97**, 12034 (1993)
108. F. Balzer, H.G. Rubahn, Opt. Photon. News **14**, 27 (2003)
109. F. Balzer, J. Beermann, S. Bozhevolnyi, A. Simonsen, H.G. Rubahn, Nano Lett. **3**, 1311 (2003)
110. J. Beermann, S. Bozhevolnyi, V. Bordo, H.G. Rubahn, Opt. Commun. **237**, 423 (2004)
111. J. Tang, J. Käs, J. Shah, P. Janmey, Eur. Biophys. J. **30**, 477 (2001)
112. L. Zhang, E. Ruh, D. Grützmacher, L. Dong, D. Bell, B. Nelson, C. Schöneberger, Nano Lett. **6**, 1311 (2006).
113. M. Odelius, M. Bernasconi, M. Parrinello, Phys. Rev. Lett. **78**, 2855 (1997)
114. D. Porter, K. Easterling, *Phase Transformations in Metals and Alloys* (CRC, Boca Raton, 1992)
115. S. Wu, *Polymer Interface and Adhesion* (Dekker, New York, 1982)
116. B. Mutaftschiev, *The Atomistic Nature of Crystal Growth, Springer Series in Materials Science*, vol. 43 (Springer, Berlin Heidelberg New York, 2001)
117. M. Schiek, A. Lützen, R. Koch, K. Al-Shamery, F. Balzer, R. Frese, H.G. Rubahn, Appl. Phys. Lett. **86**, 153107 (2005)
118. F. Balzer, M. Schiek, A. Lützen, K. Al-Shamery, H.G. Rubahn, Proc. SPIE **6470**, 647006 (2007)

4

Tailored Organic Nanoaggregates Generated by Self-Assembly of Designed Functionalised p-Quaterphenylenes on Muscovite Mica Substrates

K. Al-Shamery, M. Schiek, R. Koch, and A. Lützen

4.1 Introduction

Materials built up of π-conjugated molecular building blocks have seen a tremendous development recently due to their interesting optical, electrical and optoelectrical properties [1]. Studies of single molecule properties on the Ångstrom length scale have demonstrated the feasibility of molecular electronics [2] as bulk polymeric systems have proven the practicability of plastic electronics on the micrometer scale, which resulted in the first commercially available polymer-based light-emitting diodes already [3]. However, it is still a challenge to develop structures that are confined to the (sub)-wavelength nanometer regime in at least one or two dimensions [4]. Promising bottom-up approaches have been developed using inorganic nanowires [5] or carbon nanotubes [6], however, the use of supramolecular self-assembly processes [7] of π-conjugated organic molecules is an attractive alternative to get access to defined shape-persistent objects under thermodynamic control. Although inorganic materials are usually chemically more stable, the use of organic rather than inorganic compounds to fabricate nanostructures usually has the advantage of higher luminescence efficiency at the same material density, more flexible spectroscopic properties, and in general easier and cheaper processing. Especially rod-like molecules [8] like oligothiophenes [1]b, [9], perylenes [10], pentacenes [11] and oligo-p-phenylenes [12, 13] have been studied in this context [14]. Among those the latter ones and especially the p-hexaphenylene has been found to form very interesting nanoaggregates upon vapour deposition on various solid supports using different deposition techniques [15–24].

However, the controlled generation of mutually aligned single or even polycrystalline nanoaggregates from organic molecules is by no means a trivial process. In fact, the kind of nanofibres we are dealing with can only be produced in a rather sophisticated high-vacuum surface growth process so far. Previous growth studies using various types of oligomers and different dielectric substrates led to even more restrictions, namely an almost unique combination of muscovite mica as the growth substrate and p-phenylene oligomers as molecular building blocks, especially p-hexaphenylene [13]b,c, [24]. The nanofibres grown that way via dipole-assisted self-assembly possess typical widths and heights on the nanometer scale, but their lengths can reach several hundred micrometers, depending on the growth conditions [25]. From coverage-dependent atomic force microscopy (AFM) images of p-hexaphenylene on mica, it has been concluded that three-dimensional islands form first on top of a wetting layer of lying molecules, which, after reaching a critical number density and/or a critical aggregate size, result into long fibres [26]. The nanofibre morphology (length, width, height and number density) can be tailored within certain limits by modifying the growth conditions such as substrate temperature, deposition rate, amount of deposited material, substrate roughness and free surface energy [27].

These nanofibres show high quantum yields of anisotropic blue luminescence, which makes them promising candidates to become new building blocks for optoelectronic devices like organic light-emitting diodes (OLEDs). Waveguides [28, 29] as well as ultraviolet light-pumped organic nanolasers [30] are other potential applications, if the fibres possess sufficient widths, heights and optical gain. In addition electrical properties have been investigated [31–33] in view of potential applications of nanofibres in organic field-effect transistors (OFETs) or electroluminescent nanolasers.

As one changes the molecular building blocks, shape and alignment of the nanofibres also change. It has been noted earlier for oligo-p-phenylenes that the nanoaggregates become shorter and less well oriented with decreasing chain length of the molecular building blocks [34]. To some degree this is attributed to the decrease of polarisability of the molecules.

To further explore the limits of this approach, we were wondering if chemically functionalised derivatives would still undergo similar self-assembly processes and allow the creation of quantitative amounts of self-similar nanoshaped aggregates with tailored morphologies and optical, electrical, mechanical and even new properties like non-linear optical (NLO) activity from designed molecules with specific properties in the sense of a *bottom-up nanoengineering* approach.

4.2 Design

As revealed by previous studies with p-hexaphenylene, a coplanar arrangement of the phenylene rings is important both for the self-assembly process to ensure optimum CH–π interactions between the individual molecules within the crystalline nanoaggregates, as it is also found in the bulk crystal structure [35], as well as for their optical properties because of a better conjugation of the π-electron systems. Therefore, the only positions suitable for carrying functional groups seemed to be the two *para*-positions because any other position would lead to a significant out-of-plane orientation of the phenylene rings and probably also prevent the self-assembly for steric reasons. Thus, one could think of two different classes of compounds, symmetrically and non-symmetrically (i.e. differently) substituted derivatives. The last one is certainly more interesting because it offers more possibilities to fine-tune the desired molecular properties and also to address aspects like NLO activity. However, the symmetrically functionalised compounds also often show interesting optical behaviour like strong luminescence and they should, of course, be easier to synthesise than their non-symmetrically substituted analogues. Thus, these compounds seemed to be ideal candidates to proof whether the introduction of functionalities is possible and if they would still allow these compounds to form self-assembled nanofibre-like structures.

At first we thought to synthesise and study functionalised p-hexaphenylenes due to their superior properties in comparison to shorter oligo-p-phenylenes. However, we quickly had to realise that the solubility of these compounds is so low that their synthesis and especially their isolation and purification are severely hampered. This is a big obstacle against high-vacuum nanofibre growth which asks for quantitative amounts of substances with high purity.

Thus, we started some modelling studies first to look for alternative targets that would be easier to access but still promise interesting optical properties. One of these target properties is undoubtedly the generation of non-linearly optically active nanofibres and the most promising approach to it is to grow them from organic molecules with optimised hyperpolarisabilities. Hence, we performed a theoretical study on several functionalised oligo-p-phenylenes as single molecules in the gas phase using ab initio and density functional theory (DFT) methods to get a better impression of the influence of the number of phenylene rings and different substituents at the *para*-positions of p-oligophenylenes on the HOMO–LUMO gap (HLG) and hyperpolarisability β. Generally, relatively large basis sets and post-SCF methods are a necessity to get accurate quantitative β values [36]. Since we were only interested in relative trends at this stage, however, we used a moderate 6-31G(d) basis set throughout as suggested by recent literature [37, 38]. Employing this methodology the geometries of oligophenylenes with a focus on different p-methoxy-substituted oligophenylenes as model systems with 2, 4, 6 and 8 aromatic rings have been optimised at a DFT level (B3LYP) and second-order susceptibilities β were evaluated at the Hartree–Fock level of theory [37]. Figure 4.1 gives an overview of the properties dependent on the chain length of the phenylenes.

As expected the hyperpolarisabilities become larger the longer the oligo-p-phenylenes get. However, the increase is found to be the highest when the chain length is changed from two phenylene units to four and becomes less pronounced when it is further increased. The same holds true for the decrease of the HLGs of the several p-methoxy-substituted oligophenylenes as a measure for the band gap. One can clearly see from the data above that the gap is largely reduced when going from biphenyls to p-quaterphenylenes, the influence of even larger chain lengths is lessened. This trend is independent of the functional groups and is in agreement with experimental data

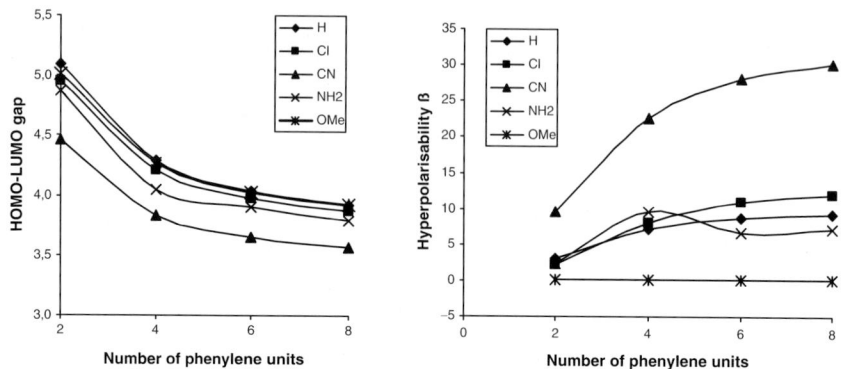

Fig. 4.1. Calculated HOMO–LUMO gap (eV; *left*) and hyperpolarisabilities β (10^{-30} esu cm^{-5}; *right*) for several *p*-methoxy-substituted oligophenylenes (MeO–(Ph)$_n$–X) as a function of the chain length n

obtained for 1,4-substituted mono-, 4,4'-substituted bi- and 1,4''-substituted *p*-terphenylene carrying amino or methoxy on one end and nitro groups on other end [39]. Thus, the calculations indicate that – although the absolute values improve – longer chains do not lead to significantly better (molecular) properties. Thus, it seemed reasonable to focus on rather short functionalised *p*-quaterphenylenes first.

4.3 Synthesis of Oligomers

Although the non-substituted *p*-quaterphenylene is known for more than 125 years now [40] and like its higher oligomers, it is even commercially available, 1,4'''-disubstituted derivatives and especially non-symmetrically functionalised compounds are still rare. This is mainly due to the notoriously low solubility of these compounds that almost prevents a (regio-)selective functionalisation [41]. To get access to substances with a defined substitution pattern, it is therefore mandatory to introduce the desired functional groups into smaller building blocks and use these precursors to establish the synthesis of the *p*-quaterphenylene scaffold. In the past this was achieved by cyclotrimerisations of acetylenes [42], Diels-Alder reactions of cyclopentadienones and subsequent aromatisation [43], Wittig reactions of cinnamaldehydes followed by Diels-Alder reactions with acetylenic dicarboxylates and subsequent aromatisation [44], addition of Grignard reagents to arines [45], Grignard reactions with *p*-quinones and subsequent dehydratisation [46] or Ullmann-type coupling reactions [47].

With the advent of modern transition metal-catalysed homo- and cross-coupling reactions over the last 30 years [48], however, these methods became more and more popular also for the synthesis of oligophenylenes and dominate this area today. Especially Kharash- and Suzuki-type couplings using

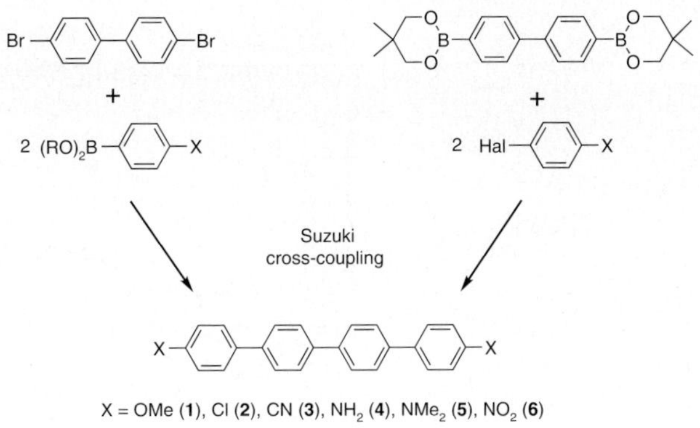

4 Functionalised Nanofibres   71

Scheme 4.1. Synthesis of symmetrically substituted p-quaterphenylenes **1–6** via Suzuki cross-coupling reactions

Grignard reagents or arylboronic acids or esters have been used very successfully in this context [49–51]. Although most of the molecules prepared in this way carry long alkyl or alkoxy groups that ensure solubility in common organic solvents, 1,4'''-disubstituted p-quaterphenylenes could also be prepared [49]b,c, [52, 53].

Thus, we decided to follow a similar approach using Suzuki cross-coupling reactions as the key steps in the synthesis of our target compounds. Symmetrically substituted compounds were synthesised in a twofold Suzuki cross-coupling reaction from commercially available p-substituted phenylboronic acids or esters and 4,4'-dibromobiphenyl or 4,4'-biphenyl-bisboronic acid ester and a respective p-substituted arylhalide as shown in Scheme 4.1. The desired products were usually obtained in very good to excellent yields.

The preparation of non-symmetrically disubstituted derivatives is somewhat more challenging and was achieved in a multi-step synthesis involving three Suzuki cross-coupling reactions and an iodination reaction as the key steps. It starts with building blocks consisting of a single phenyl ring substituted with a protective group or a desired functional group and a reactive group in *para*-positions. Further phenyl rings are added stepwise at reactive groups using Suzuki cross-coupling reactions to finally give the p-quaterphenylene core bearing functional groups at the 1,4'''-positions (Scheme 4.2). Following this procedure we prepared three different compounds **21–23** carrying a methoxy group at the 1-position and a chloro, amino or cyano substituent in the 4'''-position as a first set of non-symmetrically substituted p-quaterphenylenes in excellent yields of more than 90% each [53].

This strategy has the advantage of allowing access to a variety of p-quaterphenylenes with different combinations of functional groups from the same precursors. Furthermore, it is flexible in the sense that the sequence of Suzuki coupling and iodination reactions can be changed or additional functional group manipulations like palladium-catalysed borylations of

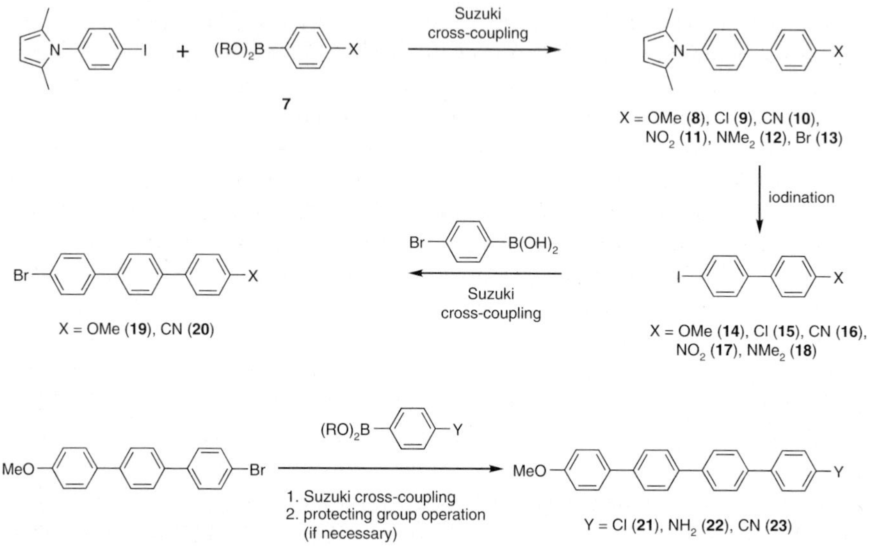

Scheme 4.2. Synthesis of non-symmetrically *p*-functionalised oligo-*p*-phenylene building blocks via Suzuki cross-coupling and iodination reactions (R = H or alkyl)

halogenated compounds can be performed to synthesise other functionalised oligo-*p*-phenylenes.

As in the case of the symmetrical analogues, the final product precipitated from the reaction mixture and was washed repeatedly with water and organic solvents for purification. Residual water and organic solvents were then removed by outgassing in vacuo to give the desired functionalised compounds in high purity.

4.4 Vapour Deposition Studies

As mentioned above the use of muscovite mica as solid support has proven to be mandatory to get the desired mutually parallel nanofibres so far. It is a sheet silicate (monoclinic lattice, $a = 5.2$ Å, $b = 9.0$ Å, $c = 20.1$ Å, $\gamma = 95.8°$) [54] consisting of octahedral Al–O layers sandwiched between two tetrahedral Si–O layers. One out of four Si atoms in the tetrahedral layers is replaced by an Al atom. The resulting charge is compensated by an intercalation of potassium ions in between two tetrahedral sheets. The cleavage of the substrate occurs along these interlayers and is almost perfect, resulting in large atomically flat areas on the surface. Each cleavage face possesses half of the potassium cations, which leads to the formation of surface dipoles [55]. Muscovite mica with the ideal formula ($K_2Al_4[Si_6Al_2O_{20}](OH)_4$) is a dioctahedral mica [56], which means that only two out of three octahedral sites are occupied by a cation. This leads to a tilt of the Al/Si oxide tetrahedra and to the formation of grooves along a ⟨1 1 0⟩ direction of the surface in case of the most common $2M_1$ polytype. These grooves alternate by an angle of 120° between consecutive cleavage layers and determine together with the oriented electric fields the orientation of the resulting nanoaggregates.

For the studies with our functionalised p-quaterphenylenes, muscovite mica was cleaved in air and transferred immediately into a high-vacuum chamber with a base pressure of about 5×10^{-8} mbar. The sample was heated for at least 2 h for outgassing purposes and for reaching the desired growth temperature, which is different for different molecules. The deposition of the p-quaterphenylenes was carried out via sublimation from a Knudsen cell. The molecules were kept in high vacuum for at least 10 h while the oven was heated slowly to a temperature below the sublimation temperature of the respective molecules to remove residual contaminations from the solvents. After that time the oven was heated to about 530–580 K to evaporate the p-quaterphenylenes and to deposit them onto the solid substrate. During evaporation the pressure in the chamber increased to about 5×10^{-7} mbar. For monitoring of the deposition rate and the final nominal mass thickness of the organic molecule films on the mica support, a water-cooled quartz microbalance was used.

4.5 Nanoaggregates from Symmetrically Functionalised Oligomers

The deposition of functionalised p-quaterphenylenes on a muscovite mica substrate resulted in the well-organised formation of light-emitting nanoaggregates whose optical properties are basically determined by the functional group of the molecular building block [57]. The morphology also varies with changing functional groups, but also depends on other factors such as the growth temperature or growth rate. In Fig. 4.2, fluorescence microscopy images are shown of the nanoaggregates from 1,4‴-dimethoxy-4,1′:4′,1″:4″,1‴-quaterphenylene (**1**, **MOP4**), 1,4‴-dichloro-4,1′:4′,1″:4″,1‴-quaterphenylene (**2**, **CLP4**), 1,4‴-dicyano-4,1′:4′,1″:4″,1‴-quaterphenylene (**3**, **CNP4**) and 1,4‴-bis(N,N-dimethylamino)-4,1′:4′,1″:4″,1‴-quaterphenylene (**4**, **NMeP4**). All aggregates strongly emit polarised blue light after normal incidence excitation with non-polarised UV light ($\lambda_{\text{exc}} = 365$ nm). Two different domains of aggregates exist on a single sample with an angle of 120° in between. The mean orientation of the fibres (in the case **3** and **4** the fibre-like structures) is along the muscovite ⟨1 1 0⟩ and never along the [1 0 0] direction.

Deposition of molecule **1** (Fig. 4.2a) on muscovite mica at elevated temperatures leads to mutually aligned almost parallel fibres with mean widths and heights of several hundred and several ten nanometers, respectively, and lengths up to several hundred micrometers [57]. Interestingly, a lateral instability leads to sawtooth-like aggregates with one straight and one rugged side, if the fibres reach a critical width of 700 nm. Within each of the two domains two orientations with an angle of about 14° in between can be observed with

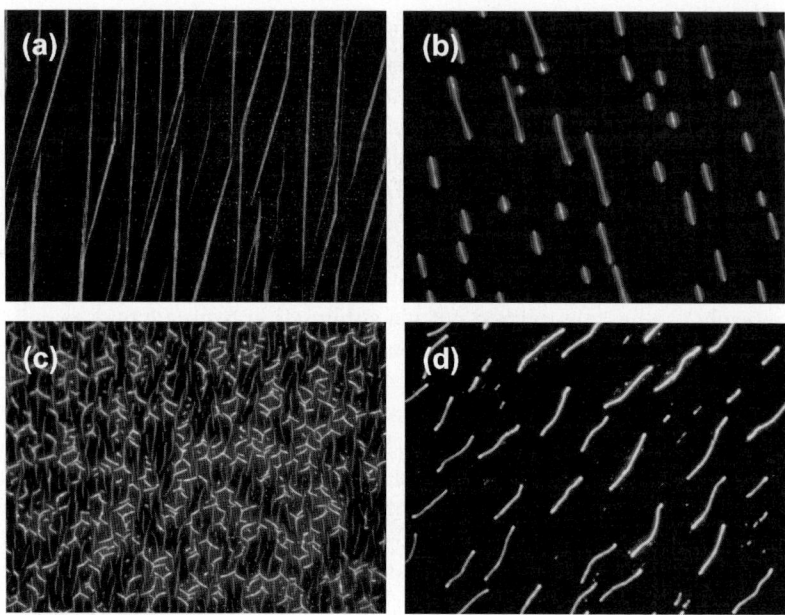

Fig. 4.2. Fluorescence microscopy images of aggregates from **1** (a) ($113 \times 85\,\mu m^2$, $T_S = 340\,K$), **2** (b) ($70 \times 53\,\mu m^2$, $T_S = 380\,K$), **3** (c) ($124 \times 94\,\mu m^2$, $T_S = 300\,K$) and **4** (d) ($56 \times 42\,\mu m^2$, $T_S = 340\,K$) on muscovite mica after non-polarised UV excitation ($\lambda_{exc} = 365\,nm$). Temperatures indicate the substrate temperatures during deposition (from [58])

the $\langle 1\,1\,0\rangle$ muscovite direction being the bisecting line of the angle [57]. The fibres form kinks to switch between these growth directions. It can be concluded from the measured polarisation patterns (not shown) that the molecules are oriented with their long axes almost perpendicular to the long axes of the fibres (for details see [57]a). In general, fibres from **1** are wider and flatter compared to those obtained from p-hexaphenylene under similar growth conditions. Again, small islands are located in between the needles, which are believed to be needle precursors. A depletion zone of several hundred nanometers up to a few micrometers around the needles exists. This depletion zone is larger as compared to p-hexaphenylene, indicating a higher mobility of the islands and/or the molecules on the surface. The effect of surface mobility on the growth mechanism becomes more evident if one deposits the material at lower temperatures. In that case the average number density of needles becomes larger, but the mean height decreases. Figure 4.3a–c shows needle growth for an increasing amount of organic material deposited at room temperature. Initially short needles are formed, followed by a close-packed film of short fibres. Two orientations within a single domain still exist, and the same instability occurs for wider needles as has been observed at elevated surface growth temperatures.

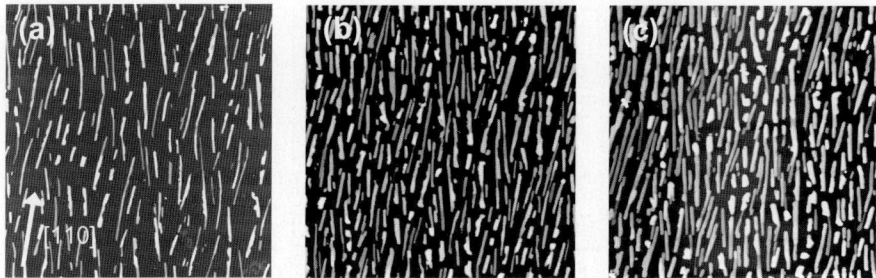

Fig. 4.3. AFM images ($20 \times 20\,\mu m^2$) of aggregates of **1** grown on mica at $T_S = 300\,K$. Parts (**a**)–(**c**) show images with increasing nominal thickness of deposited organic material (2, 4 and ≈ 6 nm, respectively). The height scale are 50, 100 and 100 nm, the mica $\langle 1\,1\,0 \rangle$ direction is shown in (**a**)

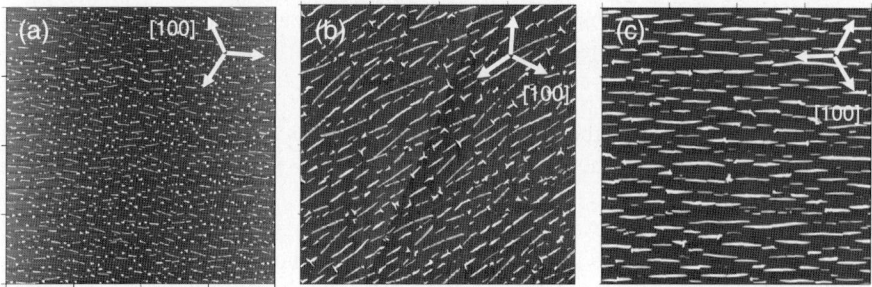

Fig. 4.4. AFM images ($80 \times 80\,\mu m^2$) of aggregates of **2** grown on muscovite mica with a nominal thickness of deposited material of 5 nm. The substrate temperatures were (**a**) $T_S = 310\,K$, (**b**) $T_S = 330\,K$ and (**c**) $T_S = 370\,K$. The height scale for all images is 180 nm

The nanofibres from **2** exhibit widths and heights comparable to the fibres from **1** but they are significantly shorter with a length up to 30 µm at maximum. Figure 4.4b shows parallel, well-separated nanofibres, which emit polarised blue light in a fluorescence microscope with the plane of polarisation almost perpendicular to the needle directions. Similar to the case of the p-hexaphenylene and **1** the needles are presumably not the initially formed aggregates. Three-dimensional islands in between the needles have been observed on AFM images for the samples obtained from **2** grown at elevated substrate temperatures. Needles probably grow by agglomeration of such islands. The shape of the aggregates from **2** is strongly depending on the substrate temperature during deposition [57]. For substrate temperatures close to room temperature aggregates consist of straight needles and up to 500 nm tall islands of µm^2 area. The islands are usually situated at the end of fibres and form star-shaped entities as can be shown in Fig. 4.4b. The mean height of the needles increases with substrate temperature from 80 to 150 nm, but the number of star-shaped entities decreases with temperature, until from $T_S \sim 350\,K$ on

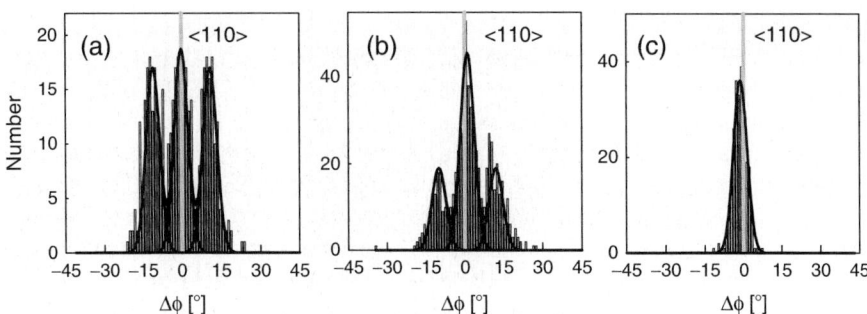

Fig. 4.5. Distributions of needle orientations obtained from AFM and fluorescence microscope images of **2** for three different substrate temperatures: (**a**) $T_S = 310$ K, (**b**) $T_S = 330$ K and (**c**) $T_S = 350$ K. The *vertical line* at $\Delta\phi = 0°$ indicates the $\langle 1\,1\,0 \rangle$ mica orientation. The main needle direction is along this orientation, with two additional directions ("off-needles") at $\Delta\phi = \pm 11°$ at lower temperatures (reprinted with permission from [57]c)

mostly needles with homogeneous heights form as shown in Figs. 4.2b and 4.4c. For somewhat higher substrate temperatures no nanofibres from **2** are observed at all.

Interestingly, not only the size of the fibres but also their mutual alignment strongly depends on the growth temperature in the case of **2**. At relatively low surface temperature of $T_S = 310$ K the needles exhibit three orientations within one domain. Needles grow along $\langle 1\,1\,0 \rangle$ and along two directions about $\Delta\phi = \pm 11°$ off $\langle 1\,1\,0 \rangle$, which we call "off-needles." For increased substrate temperature during deposition the number of *off-needles* decreases, until almost all needles form along $\langle 1\,1\,0 \rangle$ at $T_S \geq 350$ K (Fig. 4.5).

The heights of the needles are between 60 and 120 nm. These two needle types have indeed been observed for needles from *p*-hexaphenylene before [21], [13]a. An AFM image of an off-needle and two $\langle 1\,1\,0 \rangle$ needles (Fig. 4.6) discloses that the two fibre types have quite different morphologies. The $\langle 1\,1\,0 \rangle$ needles are straight and show clear facets, whereas the *off-needles* have a rugged morphology. Heights and widths, however, are similar to the straight fibres. The reason for the higher luminescence efficiency of the *off-fibres* therefore is not due to more material within the needle, but might result from a different crystal structure with the same lateral orientation of the molecules with respect to the substrate. This would, e.g. imply that the molecules are more parallel to the substrate for the rugged than for the straight fibres.

Figure 4.2c displays nanofibre growth from **3**, showing diverse structures within a single domain. Beside mutually aligned fibre-like structures, aggregates with increased heights are visible with a shape that reminds of swallow wings. The fibres have a mean length of about 7 μm as well as a typical width of several 100 nm and a typical height of several 10 nm, respectively. A single swallow wing exhibits a mean length of 4 μm and a width in the same

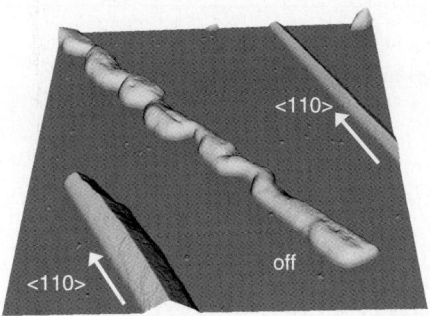

Fig. 4.6. Typical ($5 \times 5\,\mu m^2$) AFM image of fibres of **2** grown at lower substrate temperatures. Two fibres grow along the $\langle 1\,1\,0\rangle$ muscovite mica direction and a single "off-fibre" in between. Whereas the main needles look well defined and rather single crystalline the off-needle is much less defined. The heights of the needles are between 60 and 120 nm (reprinted with permission from [57]c)

dimension as the fibre-like structures but the height is about twice as high compared to those. The fibres grow along the $\langle 1\,1\,0\rangle$ muscovite direction whereas the swallow wings grow along $\langle 1\,1\,0\rangle$ (but rotated by 120° with respect to the fibres) as well as along $[1\,0\,0]$. Both nanofibre types emit polarised blue light in the fluorescence microscope with the polarisation vector pointing in the same direction, i.e. perpendicular to the long axes of the $\langle 1\,1\,0\rangle$ fibres.

Figure 4.2d shows nanoaggregates from **4**, which look slightly bent like worms. They possess a mean length of about 10 μm as well as a typical width of a few hundred nanometers and height of several ten nanometers, respectively. The growth direction is along muscovite $\langle 1\,1\,0\rangle$ as usual. The nanoaggregates emit polarised blue light, with the plane of polarisation almost perpendicular to the growth direction, after excitation with unpolarised UV light. In between the nanoworms small fluorescent aggregates are visible. AFM images (not shown here) reveal that the space between the nanoworms is densely filled with small elongated aggregates. These hardly fluoresce due to their littleness compared to the nanoworms. The shape of the nanoaggregates is strongly depending on the substrate temperature during deposition (not shown here). The *low-temperature case* leads to the needle-like aggregates in this case in opposition to **2**. However, the generation of well-defined nanoaggregates from **4** demonstrates that even the comparatively bulky N,N-dimethyl amino group is accepted within the growth process.

Fluorescence spectra of the nanoaggregates are plotted in Fig. 4.7. As seen, the fluorescence is tuneable within the blue depending on the functional group attached to the oligomer.

The spectra were obtained by recording the fluorescence after continuous UV irradiation (325 nm) of a sample area (roughly $1\,mm^2$) of aggregates on the mica substrate. Well-resolved excitonic transitions are observed, peaked at (0-1), i.e. per definitionem zero quanta in the excited state, one vibrational

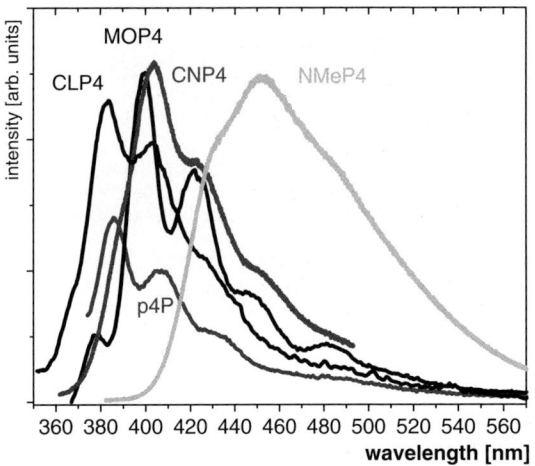

Fig. 4.7. Fluorescence spectra of **1** (MOP4), **2** (CLP4), **3** (CNP4), **4** (NMeP4), and non-substituted p-quaterphenylene (p4P) after continuous wave UV excitation (325 nm). The intensities are arbitrary

quantum in the ground state. Apparently the spectrum of **2** is blue-shifted in comparison to all other spectra because of the electron withdrawing chloride atoms. In contrast the spectrum of **1** is red-shifted in comparison to the parent p-quaterphenylene spectrum due to the electron-pushing methoxy groups. The cyano groups of **3** are strongly electron withdrawing but the triple bond within the cyano group also causes an elongation of the conjugated π-electron system. Therefore the fluorescence is red-shifted in this case. The spectrum of **4** is strongly red-shifted due to electron donating N,N-dimethyl amino group. The lone pair of the nitrogen conjugates into the aromatic system of the phenyl rings, which causes a pronounced increase in electron density. Thus, the fluorescence exhibits the largest red-shift in this case compared to the non-functionalised p-quaterphenylene.

4.6 Nanoaggregates from Non-Symmetrically Functionalised Oligomers

Vapour deposition of 1-chloro,4'''-methoxy-4,1':4',1'':4'',1'''-quaterphenylene (**21**, **MOCLP4**), 1-cyano,4'''-methoxy-4,1':4',1'':4'',1'''-quaterphenylene (**23**, **MOCNP4**) and 1-amino,4'''-methoxy-4,1':4',1'':4'',1'''-quaterphenylene (**22**, **MONHP4**) onto muscovite mica substrates at elevated surface temperatures also leads to nanofibre growth as shown in Fig. 4.8. In fact the fluorescence of the fibres from **22** is so weak (about 100 times weaker than the fluorescence of the fibres from **21** and **23**) that it is not possible to obtain a fluorescence microscopy image. Instead Fig. 4.8b is a dark field microscopy image.

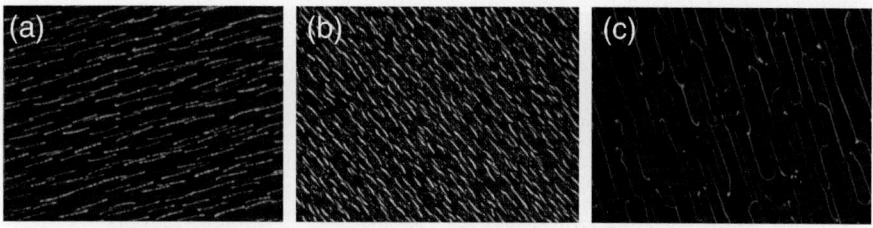

Fig. 4.8. Fluorescence microscopy images of aggregates from **21 (MOCLP4)** (a) ($56 \times 42\,\mu m^2$, 400 K) and **23 (MOCNP4)** (c) ($65 \times 48\,\mu m^2$, 380 K) after non-polarised UV excitation ($\lambda_{\text{exc}} = 365$ nm) and dark field microscopy image of aggregates from **22 (MONHP4)** (b) ($113 \times 85\,\mu m^2$, 380 K) on muscovite mica

Figure 4.8 shows fluorescence microscopy images from all three nanoaggregates obtained after 325 nm excitation. Again the fluorescence is tuneable by the functional groups implemented into the oligomers. Due to the weak fluorescence in case of **22**, no measurements of polarisation of the fluorescence in comparison to the needle orientation have been performed. Like in all other fibres, aligned along muscovite $\langle 1\,1\,0 \rangle$, the plane of polarisation of the emitted blue light of the nanofibres from **21** and **23** is almost perpendicular to the long axis of the needles. The "knob" of the nanoaggregates from **23** whose morphology reminds of a walking-stick consists of a bent fibre piece: the polarisation vector is always almost perpendicular to the needle direction.

In the case of **21** there is again a strong dependence on the growth temperature: clearly a high surface temperature case (Fig. 4.9a) and a low surface temperature case (Fig. 4.9b,c) can be distinguished as it has already been the case with **2** and **4**.

If the substrate temperature is kept at 400 K (*high surface temperature case*), mutually aligned fibres are formed with a mean length of about 6 μm, a typical width of a few hundred nanometers and a height of a few ten nanometers, respectively. Hence, width and height are about the same as in the case of **1** and **2**, but the length is comparable to the needles of **2**. The aggregates from **21** always grow along the $\langle 1\,1\,0 \rangle$ muscovite direction and form two different domains within a single sample, which are rotated by 120° to each other such as usual for the *para*-phenylenes. More details of the morphology were obtained by means of AFM. As deduced from those morphologically well-resolved studies, a single fibre is made up of smaller aggregates. In between the fibres small clusters are visible which are not seen in the light microscope image. This reflects the bottom-up fashion of the growth process. The fibres form kinks which have already been observed in the case of **1**. Thus it seems that the trend to form kinks is an intrinsic property due to the substitution with methoxy groups. If the surface temperature is kept at 360 K during the deposition (*low surface temperature case*), mutually aligned branches are formed. The growth direction is again the $\langle 1\,1\,0 \rangle$ muscovite direction before the fibre splits. The $\langle 1\,1\,0 \rangle$ muscovite direction is then the bisecting line of the angle between the branches.

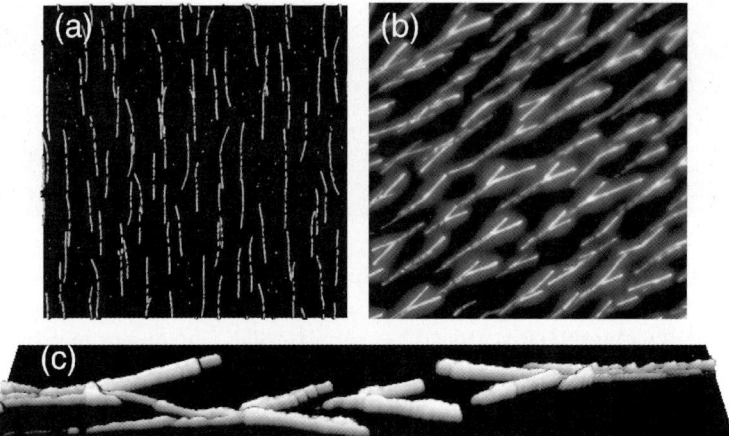

Fig. 4.9. At an elevated substrate temperature of 400 K during deposition (*high surface temperature case*), the aggregates from **21 (MOCLP4)** form domains of mutually parallel aligned needles as shown in the atomic force microscopy (AFM) image (**a**) ($25 \times 25\,\mu m^2$, height scale 80 nm). The needles are built up from smaller clusters and grow always along the $\langle 1\,1\,0 \rangle$ high-symmetry direction of the muscovite mica substrate. At a low substrate temperature of 360 K during deposition (*low surface temperature case*), the aggregates from **21** form branches on a muscovite mica substrate as shown in the fluorescence microscopy image (**b**) ($66 \times 66\,\mu m^2$) and the AFM image (**c**) ($48 \times 8\,\mu m^2$, height scale 136 nm). The fibres grow along the $\langle 1\,1\,0 \rangle$ high-symmetry direction before they split and form kinks

4.7 Non-Linear Optical Properties

Non-symmetrically substituted oligomers should not only allow an accurate tuning of the luminescence behaviour but also give access to systems that show NLO properties. **21**, **22** and **23** are (to a different extent) push–pull substituted systems that can act as NLO prototypes. Due to the non-symmetrical functionalisation the oligomers are not centrosymmetric anymore. Theoretical calculations have shown that this leads to an increased hyperpolarisability. Because of this intrinsic property of the molecular building blocks and the highly crystalline order of the aggregates, the resulting nanofibres are expected to possess large non-linear susceptibilities. Such a property is of obvious importance for future integrated optical circuits, but optical second-harmonic generation (SHG) is also a powerful technique for understanding the correlation between morphology and optoelectronic response of nanoaggregates.

It has been reported before that nanofibres from *para*-hexaphenylene exhibit NLO activity, but this can mainly be assigned to *surface* SHG from a continuous organic film (wetting layer) on the growth substrate and from the nanofibre surfaces [59]. The amount of true SHG from the nanofibres of *p*-hexaphenylene is reported to be in order of a few percent of the two-photon luminescence intensity at infrared femtosecond excitation around 800 nm.

It is important to note that the investigations reported in [59] have been performed with nanofibres on their original growth substrate mica. From our recent investigations on *transferred* nanofibres, we assign the reported SHG mainly to *surface* SHG from a wetting layer on the growth substrate and from the nanofibre surfaces. The non-symmetrically functionalised *p*-quaterphenylenes, however, behave different. For the observation of the fibres' intrinsic properties, we first transferred the nanofibres obtained from **22** from mica to glass using a standard procedure [60] to avoid SHG from the underlying substrate. Upon irradiation with femtosecond pulses, a strong second-harmonic peak was observed which shifted as expected with excitation wavelength, thus clearly proving true SHG from the *bulk* of the fibre [61]. This is the first example of freely transferrable nanofibre frequency doublers from designed organic molecules.

Figure 4.10 shows the second-harmonic signal from **21 (MOCLP4)** obtained after excitation with femtosecond pulses at three different wavelengths within the near infrared. For comparison the fluorescence spectrum after cw UV excitation is also shown. An accompanying fluorescence signal appears for excitation wavelengths shorter than 800 nm. This fluorescence spectrum is the same as the one obtained after cw UV excitation.

On individual fibres, a local variation of the SH intensity in transmission and reflection has been observed simultaneously and compared directly to morphological variations of the very same fibres. In contrast to two-photon luminescence, the second-harmonic intensity does not correlate simply to the amount of light-emitting material but depends sensitively on local surface

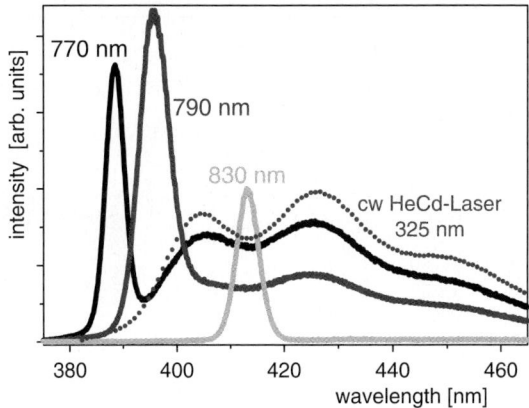

Fig. 4.10. Emission of second-harmonic signal from the bulk of the fibres of **21 (MOCLP4)** after excitation with femtosecond near-infrared laser pulses (100 fs) of different wavelength (770, 790 and 830 nm, respectively). For comparison the fluorescence spectrum after cw UV excitation is shown as well (from [58])

morphological variations. Hence, although the generation of second-harmonic intensity by itself is due to the hyperpolarisability of the molecular building blocks of the nanofibres and would be close to zero for symmetrically functionalised *p*-quaterphenylenes, its local variation along individual nanofibres appears to be dominated by local field enhancement of the responsible effective dielectric function of the nanofibres. The local field enhancement, in turn, is due to local changes in the fibre's morphology.

4.8 Quantum Chemical Calculations and Optical Properties

To get a better understanding of the experimental findings, the influence of functional groups in the *para*-positions of *p*-quaterphenylenes on the second-order susceptibility β and the luminescence of the fibres, the hyperpolarisabilities and the HLGs of the constituting molecules have been calculated with ab initio and DFT methods. All structures were optimised at the DFT level B3LYP/6-31G(d) and β values were calculated with the Hartree–Fock approach using the same basis set. HLGs are derived from the DFT data.

Figure 4.11 displays the calculated values of the HLG (left-hand side) and β (right-hand side), while luminescence spectra and second-harmonic response are shown in Fig. 4.12 for fibres of the non-symmetrically functionalised *p*-quaterphenylenes. As seen, the order of measured second-harmonic intensities and the relative intensities, obtained under similar excitation conditions, are in good agreement with the calculations. A quantitative comparison is

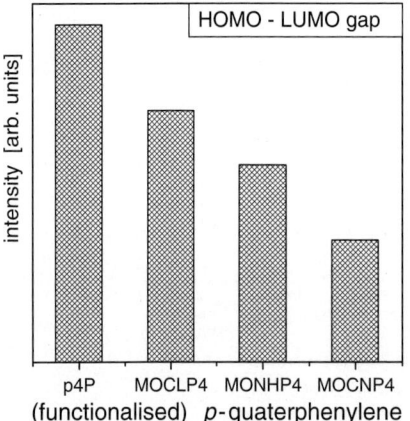
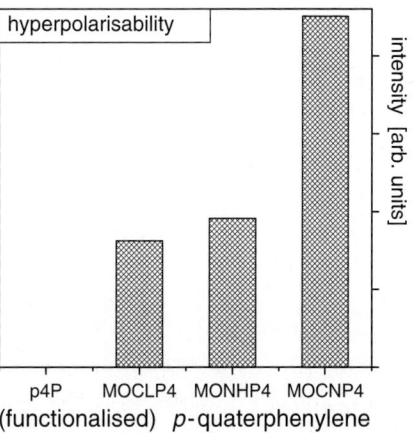

Fig. 4.11. Calculated HOMO–LUMO gaps (B3LYP/6-31G(d)) and hyperpolarisabilities (HF/6-31G(d)//B3LYP/6-31G(d)) for the functionalised compounds **21 (MOCLP4)**, **22 (MONHP4)** and **23 (MOCNP4)** in comparison with the parent *p*-quaterphenylene (p4P)

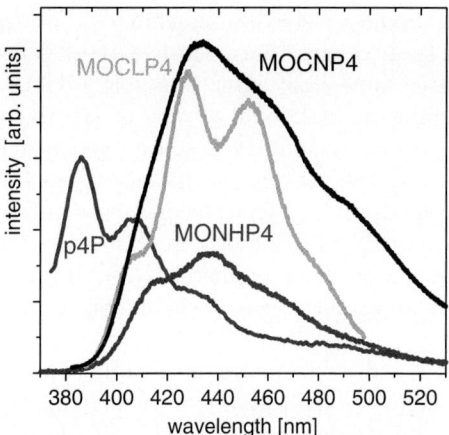 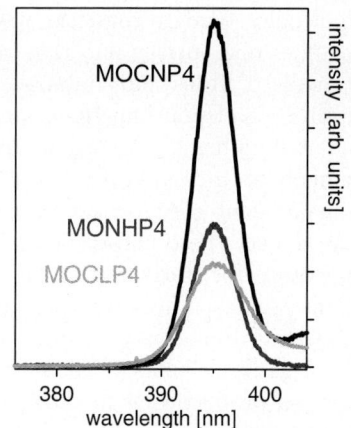

Fig. 4.12. Measured luminescence spectra (*left-hand side*) and second-harmonic intensities (*right-hand side*) after excitation with UV light (325 nm, *left-hand side*) and femtosecond laser light (790 nm, 100 fs, *right-hand side*), respectively

difficult since it depends on, e.g. fibre morphology, excitation wavelength and corresponding field enhancement effects. High-level hyperpolarisability calculations will be topic of future studies.

As for the linear spectra, the calculations demonstrate that the introduction of different functional groups allows the control of the band gap [62]. The decrease of the HLG is reflected also in an increase of the hyperpolarisability. However, while the linear spectra are clearly red-shifted compared to the non-functionalised *p*-quaterphenylene (as expected from theory), the positions of the maxima are similar for **21–23**. This reflects the fact that the exact position of the optical transitions is not simply given by the HLG frequency.

The qualitative agreement between calculations and measurements shows that the employed approach is suitable to reflect the trend found by the experiments. Calculations can now be applied as a tool to predict which functionalisation provides, e.g. optimal NLO properties. This will allow an even better design on the basis of theoretical calculations to focus on the most promising molecular building blocks concerning the desired property and application.

4.9 Conclusion

In conclusion we have established a reliable strategy for the synthesis of both symmetrically and non-symmetrically 1,4'''-disubstituted *p*-quaterphenylenes in a very flexible manner. These molecules were shown to form well-defined mutually aligned fibre-like nanostructures similar to that of the non-functionalised *p*-hexaphenylenes upon vapour deposition on muscovite mica substrates. This is a non-trivial finding since rod-like functionalised molecules tend to interfere with the dipole-assisted self-assembly processes on the solid surface and thus do not lead to nanofibres. Most of these functionalised fibres emit intense, polarised blue light after non-polarised UV excitation. Upon implementing two different functional groups to the molecular building block, a new property of the nanofibres has been generated, namely an increased NLO response. Thus, the nanofibres can act, e.g. as frequency doublers after excitation with near-infrared femtosecond laser pulses. We expect these nanoscaled frequency doublers to be of potential great importance for upconversion of the light from submicrometer-scaled infrared light sources. The second-harmonic signal could be used to image fibres with a spatial resolution of about 500 nm.

Quantum chemical calculations of the individual molecular building blocks' optical properties were shown to be in good qualitative agreement with the experimental data. Thus, it seems possible to design functionalised molecules with tailor-made optical properties for the generation of well-ordered and stable nanofibres with a distinct profile of desired properties. That way the fibres bridge the gap between microscopic and macroscopic dimensions and are an important step forward towards integrated nanoscopic optoelectronics in a bottom-up fashion. Thus, such tailor-made molecules are expected to develop into an exciting new group of functionalised nanomaterials.

Acknowledgements

The authors are grateful to the DFG for financial support. M.S. and A.L. also thank the Fonds der Chemischen Industrie for further financial support. The authors would like to thank Prof. Dr. H.-G. Rubahn, Dr. F. Balzer, R. Frese, and J. Brewer for helping to prepare and characterise the nanoaggregates. M.S. and A.L. are indebted to Prof. Dr. Peter Köll for providing them with excellent working conditions. Generous allocation of computing time by the HLRN, Hannover-Berlin and the CSC Oldenburg is gratefully acknowledged.

References

1. Some recent books: (a) K. Müllen, G. Wegner, *Electronic Materials: The Oligomer Approach* (Wiley-VCH, Weinheim, 1998). (b) D. Fichou, *Handbook of Oligo- and Polythiophenes* (Wiley-VCH, Weinheim, 1999). (c) H.S. Nalva, *Handbook of Advanced Electronic and Photonic Materials and Devices* (Academic, San Diego, 2000). (d) K. Müllen, U. Scherf, *Organic Light Emitting Devices – Synthesis, Properties and Applications* (Wiley-VCH, Weinheim, 2006). (e) H. Klauk, *Organic Electronics – An Industrial Perspective* (Wiley-VCH, Weinheim, 2006)
2. R.L. Carroll, C.B. Gorman, Angew. Chem. **114**, 4556 (2002); Angew. Chem. Int. Ed. **41**, 4378 (2002)
3. (a) A. Kraft, A.C. Grimsdale, A.B. Holmes, Angew. Chem. **110**, 416 (1998); Angew. Chem. Int. Ed. **37**, 402 (1998). (b) S.R. Forrest, Nature **428**, 911 (2004)

4. Special issues on nanowires: (a) Adv. Mater. **15**(5), 341 (2003). (b) L. de Cola (ed.), Top. Curr. Chem. **257** (2005). A recent review: (c) S.J. Hurst, E.K. Payne, L. Qin, C.A. Mirkin, Angew. Chem. **118**, 2738 (2006); Angew. Chem. Int. Ed. **45**, 2672 (2006)
5. Y. Xia, P. Yang, Y. Sun, Y. Wu, B. Mayers, B. Gates, Y. Yin, F. Kim, H. Yan, Adv. Mater. **15**, 353 (2003)
6. A special issue on carbon nanotubes: Acc. Chem. Res. **36**(12), 997 (2002)
7. A.P.H.J. Schenning, E.W. Meijer, Chem. Commun. 3245 (2005)
8. (a) P.F.H. Schwab, M.D. Levin, J. Michl, Chem. Rev. **99**, 1863 (1999). (b) P.F.H. Schwab, J.R. Smith, J. Michl, Chem. Rev. **105**, 1197 (2005)
9. G. Ziegler, in *Handbook of Organic Conductive Molecules and Polymers*, vol. 3, ed. H.S. Nalwa (Wiley, New York, 1997), Chap. 13
10. (a) B. Krause, A.C. Dürr, K. Ritley, F. Schreiber, H. Dosch, D. Smilgies, Phys. Rev. B **66**, 235404 (2002). (b) F. Würthner, Chem. Commun. 1564 (2004)
11. M. Bendikov, F. Wudl, D.F. Perepichka, Chem. Rev. **104**, 4891 (2004)
12. D.K. James, J.M. Tour, Top. Curr. Chem. **257**, 33 (2005)
13. (a) R. Resel, Thin Solid Films **433**, 1 (2003). (b) F. Balzer, H.-G. Rubahn, Adv. Funct. Mater. **14**, 17 (2005). (c) F. Balzer, H.-G. Rubahn, PhiuZ **36**, 36 (2005)
14. (a) G. Witte, C. Wöll, J. Mater. Res. **19**, 1889 (2004). (b) D. Hertel, C.D. Müller, K. Meerholz, ChiuZ **39**, 336 (2005)
15. Some examples for depositions on alkali metal halides: (a) T. Mikami, H. Yanagi, Appl. Phys. Lett. **73**, 563 (1998). (b) E.J. Kintzel Jr., D.-M. Smilgies, J.G. Skofronick, S.A. Safron, D.H. Winkle, T.W. Trelenberg, E.A. Akhadov, F.A. Flaherty, J. Vac. Sci. Technol. A **19**, 1270 (2001). (c) D.-M. Smilgies, N. Boudet, H. Yanagi, Appl. Surf. Sci. **189**, 24 (2002). (d) E.J. Kintzel Jr., D.-M. Smilgies, J.G. Skofronick, S.A. Safron, D.H. Winkle, J. Vac. Sci. Technol. A **22**, 107 (2004). (e) T. Haber, A. Andreev, A. Thierry, H. Sitter, M. Oehzelt, R. Resel, J. Cryst. Growth **284**, 209 (2005). (f) E.J. Kintzel Jr., D. -M. Smilgies, J.G. Skofronick, S.A. Safron, D.H. Winkle, J. Cryst. Growth **289**, 345 (2006)
16. Some examples for depositions on gold (111): (a) C.B. France, B.A. Parkinson, Appl. Phys. Lett. **82**, 1194 (2003). (b) S. Müllegger, I. Salzmann, R. Resel, A. Winkler, Appl. Phys. Lett. **83**, 4536 (2003). (c) S. Müllegger, I. Salzmann, R. Resel, G. Hlawacek, C. Teichert, A. Winkler, J. Chem. Phys. **121**, 2272 (2004). (d) R. Resel, M. Oehzelt, T. Haber, G. Hlawacek, C. Teichert, S. Müllegger, A. Winkler, J. Cryst. Growth **283**, 397 (2005). (e) S. Müllegger, A. Winkler, Surf. Sci. **574**, 322 (2005). (f) S. Müllegger, A. Winkler, Surf. Sci. **600**, 1290 (2006). (g) S. Müllegger, A. Winkler, Surf. Sci. **600**, 3982 (2006)
17. Deposition on polycrystalline gold: S. Müllegger, O. Stranik, E. Zojer, A. Winkler, Appl. Surf. Sci. **221**, 184 (2004)
18. Deposition on copper (100): Y. Hosoi, N. Koch, Y. Sakurai, H. Ishii, T.U. Kampen, G. Salvan, D.R.T. Zahn, G. Leising, Y. Ouchi, K. Seki, Surf. Sci. **589**, 19 (2005)
19. Deposition on GaAs(111): R. Resel, K. Erlacher, B. Müller, A. Thierry, B. Lotz, T. Kuhlmann, K. Lischka, G. Leising, Surf. Interface Anal. **30**, 518 (2000)
20. Deposition on pyrolytic graphite and SnS_2: P.G. Schroeder, M.W. Nelson, B.A. Parkinson, R. Schlaf, Surf. Sci. **459**, 349 (2000)

21. Deposition on TiO$_2$ (110): G. Koller, S. Berkebile, J.R. Krenn, G. Tzvetkov, G. Hlawacek, O. Lengyel, F.P. Netzer, C. Teichert, R. Resel, M.G. Ramsey, Adv. Mater. **16**, 2159 (2004)
22. Deposition on glass using a rubbing technique: K. Erlacher, R. Resel, J. Keckes, F. Meghdadi, G. Leising, J. Cryst. Growth **206**, 135 (1999)
23. Examples for hot-wall epitaxy deposition on mica: (a) A. Andreev, G. Matt, C.J. Brabec, H. Sitter, D. Badt, H. Seyringer, N.S. Sariciftci, Adv. Mater. **12**, 629 (2000). (b) A. Andreev, H. Sitter, N.S. Sariciftci, C.J. Brabec, G. Springholz, P. Hinterdorfer, H. Plank, R. Resel, A. Thierry, B. Lotz, Thin Solid Films **403–404**, 444 (2002). (c) H. Plank, R. Resel, A. Andreev, N.S. Sariciftci, H. Sitter, J. Cryst. Growth **237–238**, 2076 (2002)
24. Examples for molecular beam epitaxy deposition on mica: (a) F. Balzer, H.-G. Rubahn, Appl. Phys. Lett. **79**, 2860 (2001). (b) F. Balzer, H.-G. Rubahn, Surf. Sci. **507–510**, 588 (2002)
25. F. Balzer, H.-G. Rubahn, Nano Lett. **2**, 747 (2002)
26. (a) A.Y. Andreev, C. Teichert, G. Hlawacek, H. Hoppe, R. Resel, D.-M. Smilgies, H. Sitter, N.S. Sariciftci, Org. Electron. **5**, 23 (2004). (b) C. Teichert, G. Hlawacek, A.Y. Andreev, H. Siter, P. Frank, A. Winkler, N.S. Sariciftci, Appl. Phys. A **82**, 665 (2006)
27. (a) F. Balzer, J. Beermann, S.I. Bozhevolnyi, A.C. Simonsen, H.-G. Rubahn, Nano Lett. **3**, 1311 (2003). (b) F. Balzer, L. Kankate, H. Niehus, R. Frese, C. Maibohm, H.-G. Rubahn, Nanotechnology **17**, 984 (2006)
28. H. Yanagi, T. Morikawa, Appl. Phys. Lett. **75**, 187 (1999)
29. (a) F. Balzer, V.G. Bordo, A.C. Simonsen, H-G. Rubahn, Appl. Phys. Lett. **82**, 10 (2003). (b) F. Balzer, V.G. Bordo, A.C. Simonsen, H.-G. Rubahn, Phys. Rev. B **67**, 115408 (2003)
30. (a) F. Quochi, F. Cordella, R. Orrì, J.E. Communal, P. Verzeroli, A. Mura, G. Bongiovanni, A. Andreev, H. Sitter, N.S. Sariciftci, Appl. Phys. Lett. **84**, 4454 (2004). (b) F. Quochi, F. Cordella, A. Mura, G. Bongiovanni, F. Balzer, H.-G. Rubahn, J. Phys. Chem. B **109**, 21690 (2005). (c) F. Quochi, F. Cordella, A. Mura, G. Bongiovanni, F. Balzer, H.-G. Rubahn, Appl. Phys. Lett. **88**, 041106/1 (2006)
31. J. Kjelstrup-Hansen, H.H. Henrichsen, P. Bøggild, H.-G. Rubahn, Thin Solid Films **515**, 827 (2006)
32. H.H. Henrichsen, J. Kjelstrup-Hansen, D. Engstrøm, C.H. Clausen, P. Bøggild, H.-G. Rubahn, Org. Electron. (doi:10.1016/j.orgel.2007.03.010)
33. J. Kjelstrup-Hansen, P. Bøggild, H.-G. Rubahn, J. Phys. C **61**, 565 (2007)
34. F. Balzer, H.-G. Rubahn, Surf. Sci. **548**, 170 (2004)
35. K.N. Baker, A.V. Fratini, T. Resch, H.C. Knachel, Polymer **34**, 1571 (1993)
36. M. Fanti, J. Almlöf, J. Mol. Struct. (Theochem) **388**, 305 (1996)
37. B. Champagne, E. Perpète, D. Jacquemin, S. van Gisbergen, E.-J. Baerends, C. Soubra-Ghaoui, K.A. Robins, B. Kritman, J. Phys. Chem. A **104**, 4755 (2000)
38. The second order susceptibilities β cannot be determined accurately by DFT methods because these tend to significantly overestimate these as it has been observed for other π-conjugated push–pull systems before: D. Jacquemin, J. Chem. Theory Comput. **1**, 307 (2005)
39. N. Matsuzawa, D.A. Dixon, Int. J. Quant. Chem. **44**, 497 (1992)
40. H. Schmidt, G. Schultz, Liebigs Ann. **203**, 129 (1880)

41. (a) M.L. Scheinbaum, J. Chem. Soc.: Chem. Commun. 1235 (1969). (b) T.G. Pavlopoulos, P.R. Hammond, J. Am. Chem. Soc. **96**, 6568 (1974)
42. (a) M.A. Keegstra, S. De Feyter, F.C. De Schryver, K. Müllen, Angew. Chem. **108**, 830 (1996); Angew. Chem. Int. Ed. **35**, 774 (1996). (b) V.S. Iyer, M. Wehmeier, J.D. Brand, M.A. Keegstra, K. Müllen, Angew. Chem. **109**, 1676 (1997); Angew. Chem. Int. Ed. **36**, 1604 (1997). (c) M. Müller, V.S. Iyer, C. Kübel, V. Enkelmann, K. Müllen, Angew. Chem. **109**, 1679 (1997); Angew. Chem. Int. Ed. **36**, 1607 (1997). (d) A. Fechtenkötter, K. Saalwächter, M.A. Harbison, K. Müllen, H.W. Spiess, Angew. Chem. **111**, 3224 (1999); Angew. Chem. Int. Ed. **38**, 3039 (1999). (e) S. Ito, P.T. Herwig, T. Böhme, J.P. Rabe, W. Rettig, K. Müllen, J. Am. Chem. Soc. **122**, 7698 (2000)
43. (a) J.K. Stille, R.O. Rakutis, H. Mukamal, F.W. Harris, Macromolecules **1**, 431 (1968). (b) F. Morgenroth, E. Reuther, K. Müllen, Angew. Chem. **109**, 647 (1997); Angew. Chem. Int. Ed. **36**, 631 (1997). (c) U.-M. Wiesler, K. Müllen, Chem. Commun. 2293 (1999). (d) F. Dötz, J.D. Brand, S. Ito, L. Gherghel, K. Müllen, J. Am. Chem. Soc. **122**, 7707 (2000). (e) U.-M. Wiesler, A.J. Berresheim, F. Morgenroth, G. Lieser, K. Müllen, Macromolecules **34**, 187 (2001). (f) T. Weil, U.-M. Wiesler, A. Herrmann, R. Bauer, J. Hofkens, F.C. De Schryver, K. Müllen, J. Am. Chem. Soc. **123**, 8101 (2001). (g) C.D. Simpson, J.D. Brand, A.J. Berresheim, L. Przybilla, H.J. Räder, K. Müllen, Chem. Eur. J. **8**, 1424 (2002)
44. G. Subramaniam, R.K. Gilpin, Synthesis 1232 (1992)
45. (a) H. Hart, K. Harada, Tetrahedron Lett. **26**, 29 (1985). (b) H. Hart, K. Harada, C.-J. Frank Du, J. Org. Chem. **50**, 3104 (1985). (c) K. Harada, H. Hart, C.-J. Frank Du, J. Org. Chem. **50**, 5524 (1985). (d) C.-J. Frank Du, H. Hart, K.-K.D. Ng, J. Org. Chem. **51**, 3162 (1986)
46. (a) A. Rebmann, J. Zhou, P. Schuler, H.B. Stegmann, A. Rieker, J. Chem. Res. (S) 318 (1996). (b) A. Rebmann, J. Zhou, P. Schuler, A. Rieker, H.B. Stegmann, J. Chem. Soc.: Perkin Trans. 2 1615 (1997)
47. J. Harley-Mason, F.G. Mann, J. Chem. Soc. 1379 (1940)
48. (a) A. de Meijere, F. Diederich (ed.), *Metal-Catalyzed Cross-Coupling Reactions*, 2nd edn. (Wiley-VCH, Weinheim, 2004). (b) J.J. Lie, G.W. Gribble, *Palladium in Heterocyclic Chemistry* (Pergamon/Elsevier, New York/Amsterdam, 2000). (c) N. Miyaura (Hrsg.), *Cross-Coupling Reactions* (Springer, Berlin Heidelberg New York, 2002). (d) Recent special issue on cross-coupling reactions: K. Tamao, T. Hiyama, E. Negishi (ed.), J. Organomet. Chem. **653**, 1 (2002)
49. Examples for Nickel-catalysed Kharash couplings: (a) H. Saitoh, K. Saito, Y. Yamamura, H. Matsuyama, K. Kikuchi, M. Iyoda, I. Ikemoto, Bull. Chem. Soc. Jpn **66**, 2847 (1993). (b) V.A. Ung, D.A. Bardwell, J.C. Jeffery, J.P. Maher, J.A. McCleverty, M.D. Ward, A. Williamson, Inorg. Chem. **35**, 5290 (1996). (c) A. Abdul-Rahman, A.A. Amoroso, T.N. Branston, A. Das, J.P. Maher, J.A. McCleverty, M.D. Ward, A. Wlodarczyk, Polyhedron **16**, 4353 (1997)
50. Examples for Palladium-catalysed Kharash couplings: (a) J.K. Kallitsis, F. Kakali, K.G. Gravalos, Macromolecules **27**, 4509 (1994). (b) J.K. Kallitsis, K.G. Gravalos, A. Hilberer, G. Hadziioannou, Macromolecules **30**, 2989 (1997). (c) J.M. Kauffmann, Synthesis 918 (1999). (d) R. Rathore, C.L. Burns, M.I. Deselnicu, Org. Lett. **3**, 2887 (2001)

51. Examples for Palladium-catalysed Suzuki couplings: (a) P. Liess, V. Hensel, A.-D. Schlüter, Liebigs Ann. 1037 (1996). (b) J. Fran, B. Karakaya, A. Schäfer, A.D. Schlüter, Tetrahedron **53**, 15459 (1997). (c) V. Hensel, A.-D. Schlüter, Chem. Eur. J. **5**, 421 (1999). (d) N. Sakai, K.C. Brennan, L.A. Weiss, S. Matile, J. Am. Chem. Soc. **119**, 8726 (1997). (e) B. Ghebremariam, S. Matile, Tetrahedron Lett. **39**, 5335 (1998). (f) B. Ghebremariam, V. Sidorov, S. Matile, Tetrahedron Lett. **40**, 1445 (1999). (g) J.-Y. Winum, S. Matile, J. Am. Chem. Soc. **121**, 7961 (1999). (h) F. Robert, J.-Y. Winum, N. Sakai, D. Gerard, S. Matile, Org. Lett. **2**, 37 (2000). (i) N. Sakai, D. Gerard, S. Matile, J. Am. Chem. Soc. **123**, 2517 (2001). (j) N. Sakai, S. Matile, J. Am. Chem. Soc. **124**, 1184 (2002). (k) P. Galda, M. Rehahn, Synthesis 614 (1996). (l) S. Kim, J. Jackiw, E. Robinson, K.S. Schanze, J.R. Reynolds, Macromolecules **31**, 964 (1998). (m) M.B. Goldfinger, K.B. Crawford, T.M. Swager, J. Org. Chem. **63**, 1676 (1998). (n) F.D. Konstandakopoulou, K.G. Gravalos, J.K. Kallitsis, Macromolecules **31**, 5264 (1998). (o) A. Morikawa, Macromolecules **31**, 5999 (1998). (p) B. Schlicke, P. Belser, L. De Cola, E. Sabbioni, V. Balzani, J. Am. Chem. Soc. **121**, 4207 (1999). (q) P.N. Taylor, M.J. O'Connell, L.A. McNeill, M.J. Hall, R.T. Alpin, H.L. Anderson, Angew. Chem. **112**, 3598 (2000); Angew. Chem. Int. Ed. **39**, 3456 (2000). (r) M.W. Read, J.O. Escobedo, D.M. Willis, P.A. Beck, R.M. Strongin, Org. Lett. **2**, 3201 (2000). (s) S.-W. Hwang, Y. Chen, Macromolecules **34**, 2981 (2001). (t) J.-W. Park, M.D. Ediger, M.M. Green, J. Am. Chem. Soc. **123**, 49 (2001). (u) X. Deng, A. Mayeux, C. Cai, J. Org. Chem. **67**, 5279 (2002). (v) S. Lightowler, M. Hird, Chem. Mater. **16**, 3963 (2004). (w) S. Lightowler, M. Hird, Chem. Mater. **27**, 5538 (2005)
52. (a) V. Percec, S. Okita, J. Polym. Sci.: Part A **31**, 877 (1993). (b) A. Morikawa, Macromolecules **31**, 5999 (1998). (c) Z.H. Li, M.S. Wong, Y. Tao, M. D'Iorio, J. Org. Chem. **69**, 921 (2004). (d) M. Lee, C.-J. Jang, J.-H. Ryu, J. Am. Chem. Soc. **126**, 8082 (2004). (e) J.-H. Ryu, C.-J. Jang, Y.-S. Yoo, S.-G. Lim, M. Lee, J. Org. Chem. **70**, 8956 (2005). (f) S. Welter, N. Salluce, A. Benetti, N. Rot, P. Belser, P. Sonar, A.C. Grimsdale, K. Müllen, M. Lutz, A.L. Spek, L. de Cola, Inorg. Chem. **44**, 4706 (2005)
53. M. Schiek, K. Al-Shamery, A. Lützen, Synthesis 613 (2007)
54. R.A. Knurr, S.W. Bailey, Clays Clay Miner. **34**, 7 (1986)
55. K. Müller, C.C. Chang, Surf. Sci. **14**, 39 (1972)
56. D.T. Griffen, *Silicate Crystal Chemistry* (Oxford University Press, New York, 1992)
57. First results have been published: (a) M. Schiek, A. Lützen, R. Koch, K. Al-Shamery, F. Balzer, R. Frese, H.-G. Rubahn, Appl. Phys. Lett. **86**, 153107/1 (2005). (b) M. Schiek, A. Lützen, K. Al-Shamery, F. Balzer, H.-G. Rubahn, Surf. Sci. **600**, 4030 (2006). (c) M. Schiek, A. Lützen, K. Al-Shamery, F. Balzer, H.-G. Rubahn, Cryst. Growth Des. **7**, 229 (2007)
58. M. Schiek, T. Bruhn, K. Al-Shamery, R. Koch, A. Lützen, F. Balzer, J. Brewer, H.-G. Rubahn, Unpublished results
59. F. Balzer, K. Al-Shamery, R. Neuendorf, H.-G. Rubahn. Chem. Phys. Lett. **368**, 307 (2003)
60. (a) J. Brewer, H.H. Henrichsen, F. Balzer, L. Bagatolli, A.C. Simonsen, H.-G. Rubahn, Proc. SPIE **5931**, 250 (2005). (b) J. Brewer, C. Maibohm, L. Jozefowski, L. Bagatolli, H.-G. Rubahn, Nanotechnology **16**, 2396 (2005)
61. J. Brewer, M. Schiek, A. Lützen, K. Al-Shamery, H.-G. Rubahn, Nano Lett. **6**, 2656 (2006)
62. J. Roncalli, Chem. Rev. **97**, 173 (1997)

5
Hot-Wall Epitaxial Growth of Films of Conjugated Molecules

H. Sitter

5.1 Introduction: Why Highly Ordered Organic Thin Films?

Research on conjugated organic systems is a rapidly expanding field at the turn of chemistry, condensed matter physics, materials science, and device physics due to the promising opportunities for applications of these π-electron semiconductors in electronics and photonics. Due to their interdisciplinarity, this class of materials attracted the attention of a large number of researchers and triggered the beginning of a "revolution" in organic electronics. Originating with an initial focus on the p- and n-doping of conjugated oligomers and polymers, the unique electrochemical behavior of these technological important materials enabled the development of cheap sensors. Because of the progress toward better developed materials with higher order and purity, these organic materials are now also available for organic electronic devices. More generally, organic electronics includes now diodes, photodiodes, photovoltaic cells, light-emitting diodes, lasers, field-effect transistors, electrooptical couplers, and all organic integrated circuits, and claims thereupon for key technology of the twenty-first century [1].

The area of conjugated organic semiconductors can be divided conditionally into two large parts: conjugated polymers and small conjugated organic molecules. Conjugated polymers combine properties of classical semiconductors with the inherent processing advantages of plastics and therefore play a major role in low cost, large area optoelectronic applications [2]. Unfortunately, polymer films are commonly highly disordered in the solid state and, consequently, show low charge carrier mobilities because of strong Anderson localization. Therefore, an inherent part of research in the field of organic electronics focuses on small molecule systems, in which highly ordered crystalline structures can be achieved – in contrast to the disordered, often amorphous phases of the polymers. These molecules are additionally thermally stable up to 300–400°C, can be obtained as pure materials, and processed in high-vacuum or ultrahigh-vacuum conditions. However, a technological break through of organic devices requires obligatory the use of epitaxially grown highly crystalline thin films with well-defined orientation of the molecules. Unfortunately, up to now the physical properties of molecular thin films do not compare favorable to those of single-crystal bulk materials. The significant influence of structural order on the performance of thin film devices based on small molecules will be described in detail in Chap. 11 of this monograph.

In addition to improvements in device performance, well-ordered molecular films allow the investigation of anisotropic optical and electronical properties of π-conjugated systems [3]. It should be mentioned here that such phenomena are commonly not observed in conjugated polymer thin films, which are usually disordered. Therefore, such investigations are also of considerable fundamental interest.

The challenging task for the future is to grow epitaxial layers of high crystalline quality. The main difference between organic and inorganic materials with respect to epitaxial growth is the different nature of bonds. The inorganic materials are first physisorbed and then chemisorbed on the growing surface. In the case of organic materials only physisorption occurs because no chemical bonds are formed between the molecules. As a result, the growth process is governed by very weak bonds of van der Waals type reflected by a very small sticking coefficient. That means that epitaxial growth of organic materials is performed usually at comparable low temperatures. Many attempts were made so far to grow organic materials on inert surfaces mainly by molecular beam epitaxy (MBE). An extended overview can be found in Chap. 3 of this monograph. MBE as growth method offers the advantage of in situ monitoring by reflection high energy electron diffraction (RHEED) or reflection difference spectroscopy (RDS). On the other hand growth occurs in an open system, which means that the local partial pressure of the evaporated species is comparably low in the vicinity of the growing surface. Especially in the case of van der Waals epitaxy, it is a distinct advantage to use a growth method which works as close as possible to thermodynamic equilibrium, which allows growing at relatively high vapor pressures of the organic material in the region of the substrate where the deposition occurs.

A growth method which satisfies these conditions is the so-called hot-wall beam epitaxy (HWBE) [4]. In contrast to the MBE system, HWBE uses the near-field distribution of effusing molecules at the orifice of an effusion cell [5]. The substrate can also be used to close the tube of the source like a lid forming a semiclosed growth system, which is then known as hot-wall epitaxy (HWE) [6].

5.2 Experimental Setup

5.2.1 Hot-Wall Epitaxy

A schematic illustration of a conventional HWE evaporation reactor is shown in Fig. 5.1. A heated wall, in the form of a cylindrical tube, is inserted between the source and the substrate. In this manner (a) the loss of evaporation

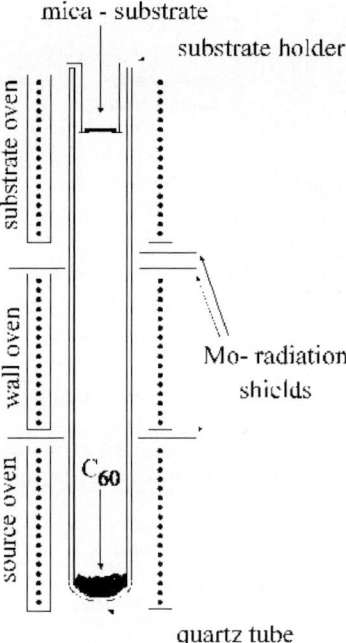

Fig. 5.1. Cross section of a hot-wall reactor with the three separated heaters to grow C_{60} films

material is strongly reduced, (b) the environment within the growth reactor is kept clean as compared to the rest of the vacuum system, and (c) a relative high partial pressure of the evaporating material can be maintained inside the reactor.

A quartz tube, with the source material at the bottom and the substrate on the top closing it tight with respect to the mean free path of the evaporated source molecules, is placed with three separated heaters into a high-vacuum chamber. The region of the growth reactor between source and substrate, called *hot wall*, guarantees a nearly uniform and isotropic flux of the molecules onto the substrate surface. The three independently heated ovens maintain the temperatures of the evaporating source material, the hot-wall tube, and the substrate, respectively. Suitable adjustment of these three temperatures enables one to achieve low supersaturation during the growth in the vicinity of the substrate. Thus an effective control over the nucleation stage and, consequently, high crystalline quality of the grown epilayers can be achieved.

To perform doping experiments, for example with Ba, a slight modification of the HWE system was necessary to add a second evaporation source for the doping material to the usual growth reactor. As shown in Fig. 5.2 the Ba-doping source is contained in a concentric quartz ampule and heated separately by oven 2. Since Ba needs a much higher source temperature than C_{60},

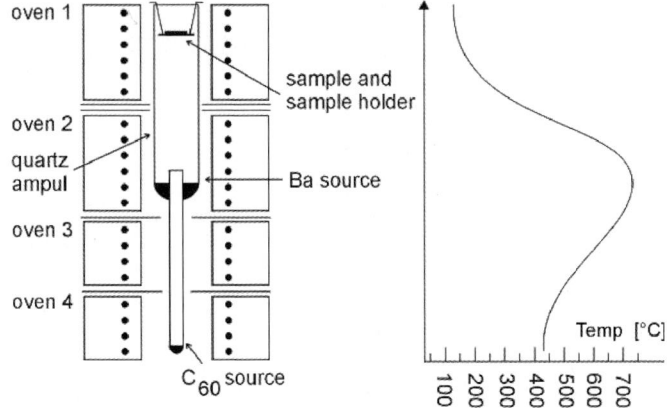

Fig. 5.2. Cross section of a hot-wall reactor to grow Ba-doped C_{60} films together with a typical temperature profile (reprinted with permission from [17])

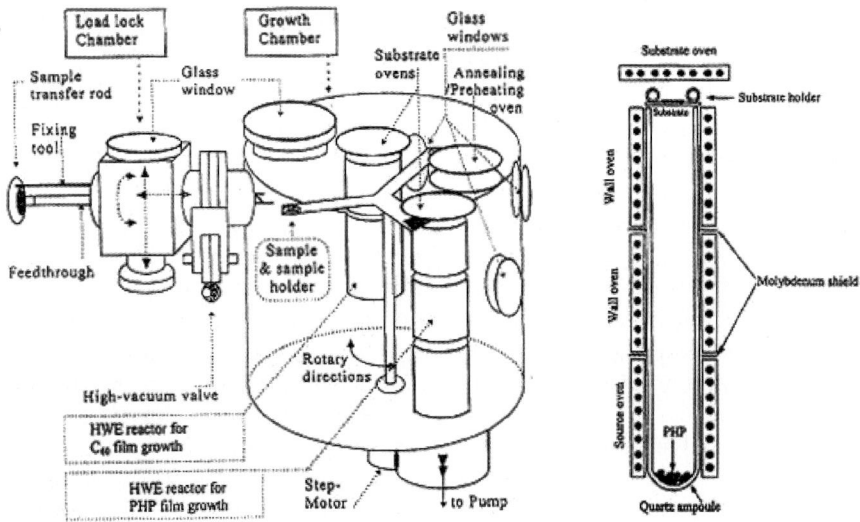

Fig. 5.3. HWBE system containing two HWE reactors for pristine C_{60} and *para*-sexiphenyl

the Ba source had to be placed above the C_{60} source. The chosen temperature profile, also depicted in Fig. 5.2, allowed controlling the partial pressure of C_{60} and Ba independently.

To increase the flexibility and to use more than one HWE reactor in one vacuum chamber, we developed a modification of HWE, called hot-wall beam epitaxy (HWBE). The HWBE reactor as shown in Fig. 5.3 differs from the conventional HWE reactor by having the substrate holder removed from the hot-wall tube to a small distance above the orifice. Accordingly, the assembly of the evaporation source together with the hot-wall tube creates a single channel extended beam source. The hot-wall tube still acts as an equilibrating element, as in conventional HWE. The substrate heater had to be redesigned as flat oven mounted on top of the substrate.

In contrast to conventional MBE systems, where the far-field distribution of the beam flux is used for epitaxial growth, in the case of HWBE the substrate is mounted very close to the source orifice. So the near-field distribution of the flux is used to obtain homogeneous layer thickness across large substrates. To grow heterostructures or multilayer systems two HWBE reactors were assembled in one growth chamber as depicted in Fig. 5.3 for the fabrication of pristine C_{60} and *para*-sexiphenyl (PSP) layers. Using a computer-controlled step motor, the sample holder could be moved to the wanted position above a growth reactor or in the position of the preheating or annealing oven. The HWBE system also contained a small load lock chamber, which allowed changing the samples without breaking the vacuum, and as a consequence increased the throughput of samples drastically.

Both modifications, the additional concentric second quartz tube as well as the change of the substrate position in the HWBE modification, did not disturb the homogeneous impingement rate of the molecules on the substrate. This fact was examined by Monte Carlo simulation [7] and could be experimentally proven by obtaining homogeneous layers over the whole cross section of the quartz tube.

5.2.2 Source Materials and Substrates

The used C_{60} source material was at first 99.4% pure. It had to be cleaned from solvents and impurities by subliming the material three times at 550°C under dynamical vacuum of 1×10^{-6} mbar and by protecting it from visible light to minimize photo-induced polymerization. About 200 mg of the cleaned material was loaded into the HWE system, which was enough to fabricate more than 50 epilayers with an area of 1 cm^2 and an average thickness of 200 nm. The Ba-doping material was loaded in a glove box under nitrogen atmosphere into the quartz ampoule to avoid any oxidation. PSP is commercially available in a higher purity grade as C_{60}, however, additional cleaning by sublimation was also necessary as described above.

As substrate material, sheets of mica were used because of the inert character of freshly cleaved surfaces, free of unsaturated bonds, which is in favor for van der Waals epitaxy. The top most layer of mica consists of K atoms forming a hexagonal grid with a periodicity which is close to the diameter of C_{60} molecules. Consequently epitaxial growth of C_{60} on mica should be initiated in the (1 1 1) direction. The mica sheets were cut into pieces (15 × 15 mm^2), cleaved in air with an adhesive tape, and immediately transferred into the vacuum chamber of the HWE or HWBE system. The substrates, before being transferred to the growth reactor, were preheated in the growth chamber for 1 h at 400°C in a separate oven to remove adhesives from the substrate surface.

NaCl and KCl substrates were cleaved with a razor blade in the laboratory atmosphere, transferred into the growth chamber, and heat treated before layer growth. A clear dependence of the layer quality on the heat treatment could be observed as described below.

5.2.3 Characterization Methods

Immediately after growth a first inspection of the surface morphology was performed by Nomarski interference microscopy. A more detailed analysis of the surface structure was obtained by atomic force microscopy (AFM), using a Digital Instruments Nanoscope III in the tapping mode. A detailed description of the surface structure analysis by AFM can be found in Chap. 1 of this book.

The growth rate of the pristine and doped C_{60} layers was determined by measuring the layer thickness of the individual layers after growth using a surface profilometer.

High-resolution X-ray diffraction (HRXD) was selected as the most appropriate method to evaluate the overall crystalline quality of the films, because it is nondestructive and takes a large region of the epilayer into account. The measured rocking curves give information about crystalline defects like dimension and tilt of mosaic blocks, density of misfit dislocations, and existence of lateral structures and surface corrugation [8]. Our HRXD measurements were performed with $Cu_{K\alpha}$ radiation filtered by a Ge four crystal monochromator. Since the intensity of the C_{60} reflex is very low, the scans were typically carried out with a 0.003° step width and an integration time up to 1 min per step. A much more sophisticated structural analysis by XRD was performed by pole-figure measurements as described in Chap. 6.

To determine the electrical properties of the Ba-doped C_{60} layers, which are very sensitive to air because of oxidation processes, an in situ arrangement had to be used. The electrical resistance and the Hall voltage had to be measured in the growth chamber without breaking the vacuum. This was achieved by evaporating four gold contacts on a mica substrate, which were soldered by silver paste to very thin gold wires fixed to the sample holder of the HWE system. A schematic drawing of our special sample holder with the used wiring is depicted in Fig. 5.4.

This wiring in van der Pauw geometry in connection with current feedthroughs in the vacuum chamber allowed a continuous in situ measurement of the electrical resistance of the growing layer. After the growth the sample could be transferred on top of a permanent magnet, which was also mounted in the growth chamber and delivered a magnetic field of 0.3 T on its surface, where the sample rested during the Hall effect measurements. In that way, carrier concentration and carrier mobility could be measured inside the growth chamber.

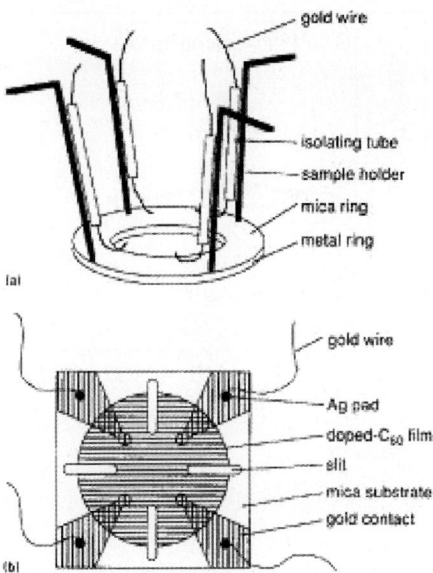

Fig. 5.4. (a) Sample holder with the attached wiring to measure in situ during growth the resistance of the deposited layer. (b) Sample configuration for in situ resistivity and Hall effect measurements in van der Pauw geometry

5.3 Pristine and Ba-Doped C_{60} Layers

5.3.1 C_{60} Films on Mica Substrates

In a first set of experiments the source temperature was varied from 360 to 440°C at a fixed substrate temperature of 140°C and a wall temperature of 400°C. The results show that the growth rate is proportional to the effusion rate, indicating that the growth mechanism in our system is source limited [9].

The influence of variations of the wall temperature on the growth rate and crystalline structure is of no great importance. Only for wall temperatures below 340°C the growth rate decreases significantly due to augmenting condensation of the source material in the upper part of the growth reactor near the substrate. At wall temperatures around 440°C the contrary is true: The growth rate is higher with respect to the experiments performed before. The reason for this behavior is reevaporation of condensed source material in the wall and substrate region, which was used in earlier experiments, where the wall temperature was lower than 440°C. So for the following experiments and all samples described and presented below, a source temperature of 400°C and a wall temperature in the same range (380–400°C) were applied [10].

The influence of the substrate temperature (T_{sub}) on the growth and crystalline quality of C_{60} films grown in our HWE system was studied by varying T_{sub} in the range from 100 to 200°C at a fixed source and wall temperature of

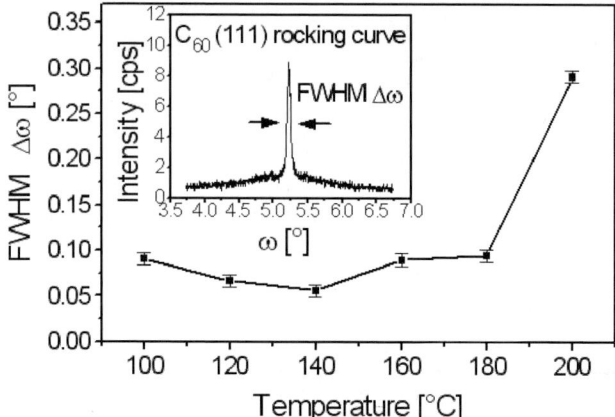

Fig. 5.5. FWHM of the C_{60} (1 1 1) rocking curve vs. the applied substrate temperature (reprinted with permission from [17])

400°C. In Fig. 5.5 the full width at half maximum (FWHM) of the (1 1 1) rocking curve of C_{60} films, which are about 120 nm thick, is plotted as a function of T_{sub}. Regarding temperatures between 100 and 200°C, the C_{60} film grown at 140°C has a minimal FWHM of $0.057° \pm 0.006°$ (200 arcsec $\pm$ 20 arcsec), indicating a nearly perfect monocrystalline growth. For lower temperatures there is a gentle increase of the FWHM, at higher temperatures a significant jump can be observed between 180 and 200°C resulting from a change in the shape of the rocking curve. The 120 nm thick films grown at substrate temperatures between 100 and 180°C exhibit a narrow Gaussian shape rocking curve as shown in the insert of Fig. 5.5.

The influence of the film thickness on the more complex shape of the rocking curve is demonstrated in Fig. 5.6a–d. For the thicker epilayers, an additional very broad peak (FWHM = 0.9°) appears with the narrow one sitting on top. The broad peak can be observed for films thicker than 160 nm and becomes dominant with increasing layer thickness [11].

Studying the line shape of the rocking curves in detail (see inset of Fig. 5.7), the intensity of the broad peak (I_d), which was evaluated by calculating the area under the curve, is increasing proportional with the thickness of the film, whereas the intensity of the narrow peak (I_c) stays approximately constant. The ratio of the intensity of the narrow peak I_c to the intensity of the broad peak I_d is plotted in Fig. 5.7 vs. the film thickness for two different substrate temperatures.

This behavior can be explained by a very simple and phenomenological model. We assume that after the initial island formation a perfect C_{60} film grows in a layer-by-layer mode with extended monomolecular terraces.

Exceeding a certain critical thickness t_c, the growth mode changes significantly to an island growth. In this growth regime, many defects are formed in the layers, like dislocations and discontinuities due to slightly tilted and

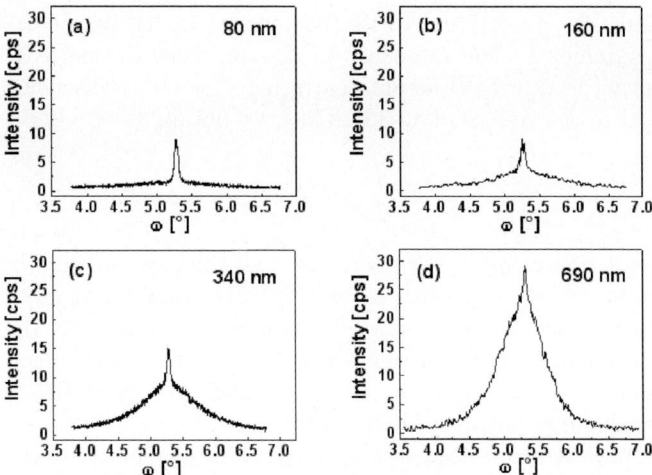

Fig. 5.6. Change of the shape of the C_{60} (1 1 1) rocking curve with increasing film thickness

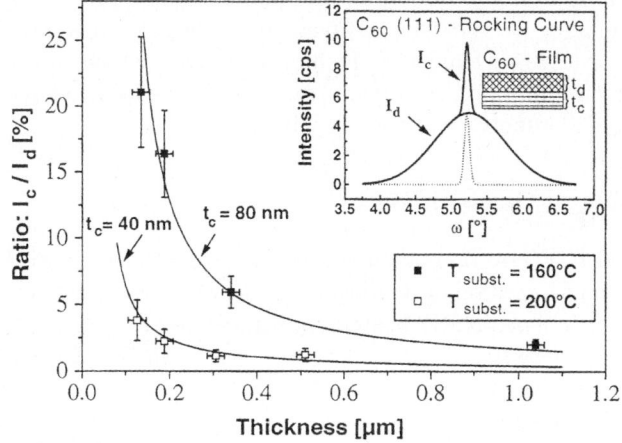

Fig. 5.7. Ratio of the intensities I_c and I_d as a function of the total film thickness. The *full squares* are experimental values for $T_{sub} = 160°C$ and the *hollow ones* for $T_{sub} = 200°C$. The *lines* represent a fit assuming that the intensity I_d is increasing linearly with the thickness of the disturbed part of the C_{60} film (thickness t_d), whereas I_c is a result of coherent scattering from the perfectly grown part of the C_{60} layer (thickness t_c)

rotated grains, which are responsible for diffuse scattering and, therefore, for the enormous broadening of the rocking curve. However, the narrow peak, which is the contribution of coherent scattering of the perfect C_{60} layer, is almost undisturbed. We assume that the intensity I_d, resulting from incoherent scattering from small mosaic blocks in the part of the layer with the imperfections, is proportional to its thickness t_d, as it was observed by our X-ray measurements. The intensity I_c, coming from the perfect layer with a thickness t_c, is due to coherent scattering. Therefore, the intensity I_c is proportional to the square of t_c [12]. Thus, we obtain

$$\frac{I_c}{I_d} = \frac{\alpha_1 t_c^2}{\alpha_2 t_d}, \qquad (5.1)$$

with $\alpha_1, \alpha_2 = $ const. and $t_c + t_d = t$, the total thickness of the film.

Mean values of the constants α_1 and α_2 were evaluated from the measure of intensities I_c and I_d. The only fitting parameter used was the thickness of the perfect C_{60} layer t_c. As shown in Fig. 5.7, the experimental findings could be fitted with our model by assuming different values of t_c for the data obtained at different substrate temperatures.

The results indicate that the critical thickness t_c decreases with increasing substrate temperature [t_c ($T_{sub} = 160°C$) = 80 nm, t_c ($T_{sub} = 200°C$) = 40 nm].

The additionally installed annealing oven allowed performing postgrowth annealing processes without breaking the vacuum. The improvement in crystalline quality of the C_{60} epilayers by annealing was investigated by HRXD [13]. The improvement of the annealed layer, which was baked 20 min at a temperature of 130°C, is documented by the decrease of the FWHM to 140 arcsec and the increase of the peak intensity.

5.3.2 Doping of C_{60} with Ba

In a modified HWE system, as described above, we have grown Ba_xC_{60} layers [14]. The incorporation of Ba during growth was obtained by coevaporation of Ba from an independently heated Ba source (see Fig. 5.2). As soon as Ba was used in the growth reactor, the growth rate of the C_{60} layers was mainly controlled by the Ba source temperature (T_{Ba}) and only slightly dependent on the C_{60} vapor pressure. Figure 5.8 shows the total thickness of Ba-doped C_{60} layers after a growth time of 7 h. For $T_{Ba} \leq 600°C$ the results follow an exponential function, for $T_{Ba} \geq 600°C$ saturation can be observed. This behavior can be interpreted by the assumption that the incorporated Ba makes a charge transfer to C_{60} [15] and controls the sticking coefficient for the C_{60} molecules which are always presented in a surplus in the vapor. However, when the surface coverage of Ba is saturated, also the growth rate cannot be increased further.

The resistivity of the Ba-doped layers was measured in situ by a four-point probe during growth (see Fig. 5.4). We found that the resistivity depends only on the applied T_{Ba}. For $T_{Ba} \leq 600°C$, an exponential decrease of the resistivity can be observed; however, for $T_{Ba} \geq 600°C$, saturation is reached.

We assume that for $T_{Ba} \geq 600°C$ the growing surface is covered with a saturation value of Ba atoms and no further incorporation and, consequently,

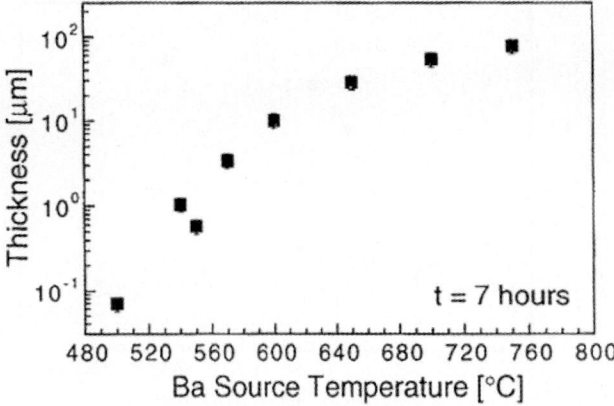

Fig. 5.8. Thickness of Ba-doped C_{60} layers as a function of Ba source temperature after a growth time of 7 h

no further decrease in resistivity can be achieved. It is worthwhile to note that the growth rate and the resistivity of the Ba-doped layers show the same trends. For $T_{Ba} = 520°C$ the growth rate and the resistivity are comparable to pristine C_{60} layers, which means that the Ba partial pressure is too low to give an effective Ba deposition and, consequently, no charge transfer occurs influencing the sticking coefficient and the resistivity. The saturation of the Ba incorporation is observed in the growth rate as well as in the resistivity for $T_{Ba} \geq 600°C$, which can be interpreted in both cases by a limitation in the Ba incorporation rate at electrical active sites.

The permanent magnet, installed in the vacuum chamber of the HWE system, allowed to measure in situ Hall effect on Ba_xC_{60} epilayers. The sample holder together with the wiring and the contacted Ba_xC_{60} epilayer could be transferred from the HWE system to the permanent magnet. The evaluation of the data gave n-type conductivity with a carrier concentration between 10^{13} and 10^{18} cm^{-3}. The results are shown in Fig. 5.9 as a function of the used Ba source temperature, which means that each data point represents a different growth experiment. For lower Ba source temperatures, the C_{60} layers were semi-insulating. For Ba source temperature larger than 600°C, the carrier concentration saturates around 2×10^{18} cm^{-3}.

Even more surprising were the data obtained for the mobility of the carriers that are much higher than expected [16]. The mobility that was measured for each sample is plotted as a function of the obtained carrier concentration in Fig. 5.10. The increase in mobility with increasing carrier concentration together with the maximum value of 6×10^3 cm^2 V^{-1} s^{-1} caused some doubts on our experimental findings. So, we performed in addition to the Hall effect measurement in situ magnetoresistance measurements, which gave in all experiments the same results with a maximum deviation of 5% in comparison to the data obtained from Hall effect.

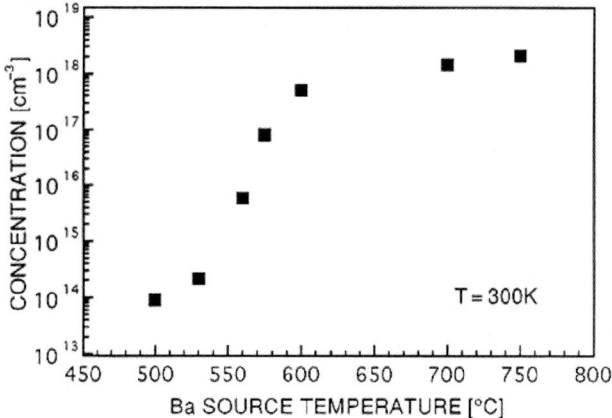

Fig. 5.9. n-type carrier concentration as a function of Ba source temperature

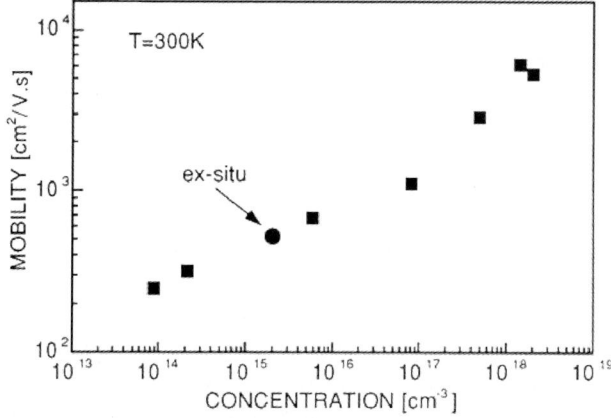

Fig. 5.10. In situ measured carrier mobility as a function of n-type carrier concentration. For comparison the ex situ measured data point is given by a *full circle*

To test the in situ electrical characterization, some Ba-doped C_{60} layers were overgrown with pristine C_{60} to protect them from oxidation when exposed to air [17]. In that way the Ba-doped C_{60} layers could be transferred to a standard ex situ Hall effect measurement apparatus. The obtained result was in excellent agreement with the in situ data and is inserted in Fig. 5.10 labeled "ex situ."

5.4 Highly Ordered Films of *Para*-Sexiphenyl

In comparison to the spherical symmetric C_{60} molecule, an elongated small molecule with a pronounced axis like *para*-sexiphenyl (p-6P) should be described. This molecule consists of six carbon rings aligned in a straight direction and interconnected at the corners of the carbon rings forming a so-called *para*-linkage. Each dangling bond of the carbon atoms is saturated by hydrogen. p-6P crystallizes monoclinic with the lattice constants $a = 8.09$ Å, $b = 5.56$ Å, $c = 26.24$ Å, $\beta = 98.17°$ [18]. The long molecular axes of all molecules within the unit cell are oriented parallel to each other, whereas adjacent molecular planes are tilted about 66°, which form the typical herringbone structure of p-6P.

First attempts to grow thin films of p-6P were made by physical vapor deposition on isotropic inorganic substrates like glass and GaAs. Even in those layers a high degree of order was found by HRXD [19]. Further improvement in structural order of p-6P films was achieved by evaporation of p-6P on rubbed layers of previously deposited p-6P, serving as an orientation-inducing layer. The results reveal that the orientation of the p-6P molecules in the top layer is effectively influenced by the rubbed layer. The long axes of the p-6P molecules are oriented parallel to the rubbing direction. However, the pole-figure measurements document that the rubbing procedure does not determine the direction of all p-6P molecules, a certain amount of crystallites possesses a different orientation [20].

The next step in increasing the structural order was made by MBE [21]. The p-6P molecules were deposited on GaAs substrates with a miscut of 2° relative to the (0 0 1) surface. After deoxidizing the substrates in As atmosphere, p-6P was evaporated from an effusion cell heated to 230°C. Epitaxial growth was found up to a maximum substrate temperature of 150°C. Changing the substrate temperature from 90 to 170°C, the island density increases following an Arrhenius law and the mean thickness of the p-6P islands changes from 120 to 400 nm. The structural analysis of these islands was made by electron diffraction in a TEM. The diffraction patterns indicate that the growth of the sexiphenyl islands on the substrate is well defined. The vertical orientation of the sexiphenyl molecules corresponds to the (10 0 1) or to the (11 0 −2) contact plane. The molecular chains form an angle of about 40° to the surface of the substrate.

Needle-like structures of p-6P were obtained by the mask-shadowing vapor deposition technique, as shown in Fig. 5.11 [22]. p-6P molecules were vapor deposited onto a KCl (0 0 1) surface kept at 150°C through a mesh mask having round holes of 0.6 mm in diameter, which faced the KCl substrate with an intervening space of 0.5 mm. A p-6P film was formed under the holes of the mask and in addition needle-like crystals were formed in the shadowed region of the KCl surface. The orientation was orthogonal along the KCl [1 1 0] directions. The length of the needles becomes longer away from the edge of the hole and reaches more than 100 μm. The growth of the needles is explained by a manifold reflection of the p-6P molecules between the substrate surface and the mask as schematically depicted in Fig. 5.11.

Some portion of p-6P molecules further intrudes and migrates into the shadowed region. These molecules probably can desorb from the surface because the temperature of the shadowed region is heated slightly higher than

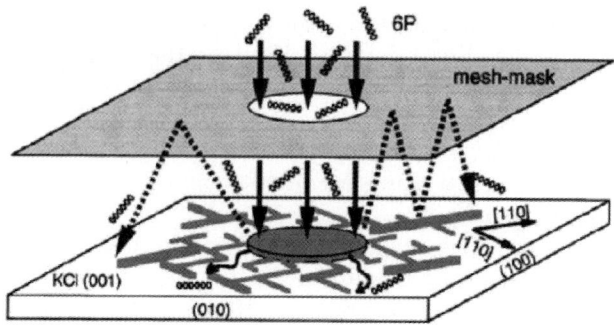

Fig. 5.11. p-6P needle-like crystals grown on KCl substrate by the mask-shadowing vapor deposition technique (reprinted with permission from [22])

the exposed hole. Due to this temperature gradient, the molecules in the shadowed space would be repeatedly reflected onto the KCl surface and the mask wall, intrude into the deep, and then settle at the growing edge of the needles. Under UV light excitation at $\lambda = 365$ nm using a conventional inverse fluorescence microscope, the needles show a blue light emission with bright spots at the tips of the needles. When the excitation was focused on a local region with a round aperture, such a spotty radiation still occurred at the tips of the needles extending outside the excited region. The distance between the emitting tip and the excited edge reaches almost 50 μm. Since it is not believable that the excitons travel so long a distance in organic crystals, this can be attributed to the self-waveguided emission in the needle-like crystals. The light emitted at the excited region is confined inside the crystal and propagated along the needle axis, then radiated from the tip. This self-waveguided effect is based on the uniaxial molecular orientation, and affected by the size and morphology of the needle-like crystal. The electron diffraction pattern taken from a single needle reveals that the molecular axis of p-6P lies parallel to the KCl surface and is aligned perpendicular to the needle axis.

Obviously the manifold impingement of the p-6P molecules caused by the shadowing of the mask described above is the important growth environment to obtain such self-organized structures. This growth condition is automatically given in an HWE system by the local equilibrium of the p-6P vapor with the growing surface. Therefore, it could be expected that similar needle-like structures can be obtained for p-6P grown by HWE.

5.4.1 Needles and Islands of p-6P on KCl Substrates

KCl crystals with its well-defined (0 0 1)-oriented surface are known as versatile model substrates for organic thin film deposition [23]. The p-6P films were grown by HWE in a vacuum of about 6×10^{-6} mbar after preheating of the KCl substrates in the growth chamber. The substrate temperature was

80–90°C and the p-6P source temperature was fixed at 240°C. The growth time was varied between 45 s and 40 min. The film morphology was imaged by AFM operated in air in contact mode. AFM test measurements done both in contact and tapping modes reveal that p-6P films were not damaged if the contact mode was used. XRD studies were performed at the Cornell High Energy Synchrotron Source (CHESS, Ithaca, USA) using monochromatic radiation with $\lambda = 1.25\,\text{Å}$.

Since KCl substrates are quite hygroscopic, in a first set of experiments the influence of the substrate preparation on the surface morphology was investigated [24]. The KCl substrates were freshly cleaved in air and immediately transferred to the growth chamber. Besides that it was necessary to perform a preheat treatment of the substrates in the vacuum of the growth chamber to get rid of surface contamination mainly caused by water molecules. A clear influence of the preheating temperature (T_{pre}) and the preheating time (t_{pre}) could be observed. At $T_{\text{pre}} = 150°\text{C}$ we applied for comparison a preheating process for $t_{\text{pre}} = 30\,\text{min}$ and $t_{\text{pre}} = 10\,\text{h}$. With increasing preheating time the density of the obtained needles decreased drastically. At a lower $T_{\text{pre}} = 90°\text{C}$ the influence of the preheating time was much less pronounced. The reason for this effect is probably the longer diffusion length of p-6P molecules on the better preheated KCl surface. Because very long preheating times are not practical, preheating conditions were fixed for the following growth experiments at $T_{\text{pre}} = 150°\text{C}$ and $t_{\text{pre}} = 30\,\text{min}$.

In a next set of experiments, the influence of the growth time on the surface structure was investigated. Figure 5.12 shows the AFM images of the surface after a growth time of 45 s, 10 min, and 40 min, respectively. It is evident that under HWE conditions p-6P can form needles (see Fig. 5.12a) and plate-like crystallites (see Fig. 5.12b,c) at the same time. As shown below, this is in contrast to HWE growth of p-6P on mica, where only needles are formed. As shown in Fig. 5.12a, the needles are clearly the initial growth stage of the films. They start to grow most probably direct on the KCl surface (the background of Fig. 5.12a is very similar to typical AFM images of a bare KCl substrate after cleavage), generating a rectangular network in accordance with

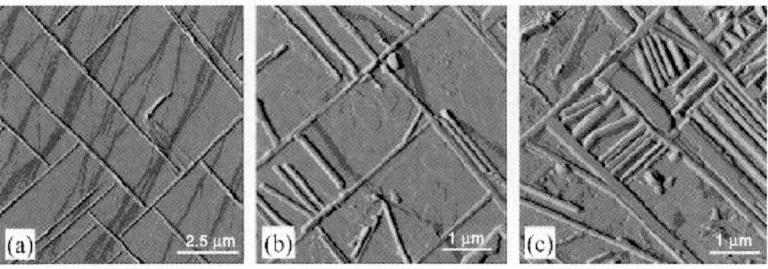

Fig. 5.12. AFM images of the p-6P films grown for (**a**) 45 s, (**b**) 10 min, and (**c**) 40 min (reprinted with permission from [24])

the surface symmetry of the substrate. That means a real epitaxial growth in the sense of the definition given in the introduction occurs.

In this stage the needles are in average 10 nm high, 100 nm wide, and 3–4 µm long. After a few minutes of deposition time, terrace-shaped islands develop between the needles, as depicted in Fig. 5.12b. The terraces are in average 2.6 nm high, which corresponds to one monolayer of standing molecules of p-6P. Further growth (Fig. 5.12b,c) is characterized by a coexistence of the constantly growing needles and islands, whereby the last ones are clearly limited by the rectangular network of the needles. The number of terraces on layered islands tends to increase with increasing surface coverage: at most 2, 5, and 9 terraces were observed in the films deposited for 5, 10, and 40 min, respectively. After 40 min of deposition almost the whole surface between needles is covered with layered terraced-shaped islands (Fig. 5.12c). Nevertheless, it was never observed that the needles grow across the islands, or vice versa.

In accordance with [25], the AFM images show that the needles mainly grow along two orthogonal KCl [1 1 0] directions, referring to the KCl [1 0 0] edge. This bidirectional epitaxial orientation seems to be originating from an interaction between the linear p-6P molecules and the ionic rows along KCl [1 1 0] and [1 −1 0] directions. The epitaxial orientation of the needles is almost perfect for thin p-6P layers (Fig. 5.12a), but the amount of misaligned needles clearly increased with increasing growth time (see Fig. 5.12b,c). The additionally appearing needles form an angel of 17° to the originally grown needles and represent another growth direction, due to a different crystallographic orientation of the needles. A detailed description of this phenomenon can be found in Chap. 6 of this monograph and in [26, 27].

It is worth mentioning that, independent from growth time, no needle crossing is observed, in agreement with previous optical investigations. The needles always terminate when a needle comes to another orthogonal oriented one, as shown for example in Fig. 5.13. This T-end shaped growth was explained in [28] by assuming that the needles nucleate at the step edges of KCl and then extend along these edges. In contrast, our AFM investigations do not indicate that surface steps act as nucleation centers for the formation of the needles. Moreover, as also clearly shown in Fig. 5.13, needles can grow over the surface steps, at least if the step height is below 100 nm.

The crystallographic orientation of the needles and therewith the orientation of the p-6P molecules inside the needles was investigated by XRD measurements using a synchrotron source, which allows one to measure even weak intensities as those from thin organic layers. To obtain spectra, where the diffraction peak positions of the organic material do not depend on the equipment and the energy of the synchrotron beam, a so-called L-scan technique was used [29]. In principle, an L-scan is a $\theta/2\theta$-scan, where the diffraction intensity is measured not as usual as a function of 2θ but as multiples of the lattice constant of the substrate $L_{KCl} = 6.2917$ Å. The L_{hkl} value can be calculated as

$$L_{hkl}/L_{KCl} = d_{hkl}, \tag{5.2}$$

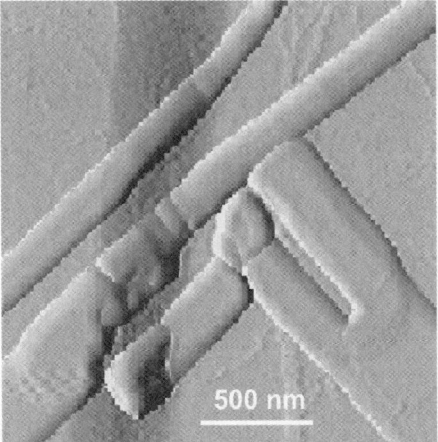

Fig. 5.13. AFM image of a T-shaped contact region of p-6P needles (reprinted with permission from [24])

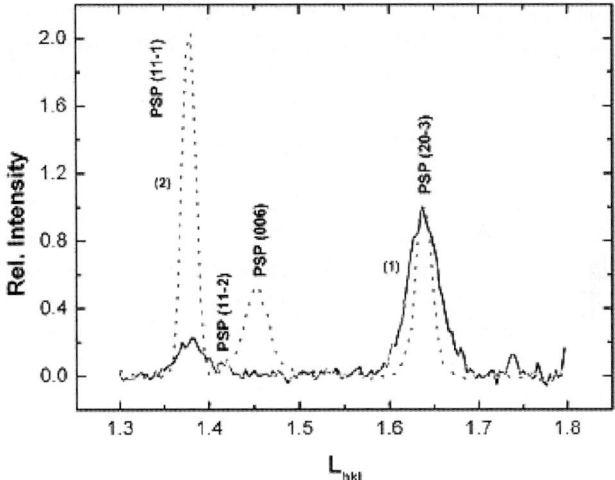

Fig. 5.14. L-scans for p-6P layers grown on KCl. The p-6P layers were grown for (**1**) 45 s and (**2**) 40 min (reprinted with permission from [24])

where h, k, and l are the corresponding Miller indices and d is the net plane distance of the p-6P planes.

Figure 5.14 shows typical L-scans for p-6P films grown within 45 s and 40 min at 80°C. Note that the AFM morphologies of these films are depicted in Fig. 5.12a,c. The scans reveal at least four different growth orientations parallel to the surface of the substrate: (1 1 1), (1 1 2), and (2 0 3) reflections represent crystal structures containing molecules lying on the substrate surface, whereas the (0 0 6) reflection is characteristic for an orientation with standing p-6P molecules [30]. In full agreement with the results obtained from AFM investigations described above, the (0 0 6) reflection appears only for the thicker film grown for 40 min. Correspondingly, the needles comprise three epitaxial orientations, where in each case the long molecular axis is aligned approximately parallel to the KCl (0 0 1) surface (edge-on), while in the terraced-shaped islands the molecules are approximately perpendicular to the substrate surface (head-on).

5.4.2 Islands and Nanofibers of p-6P on Mica Substrates

Also for this set of experiments, the p-6P source material was carefully cleaned by a threefold sublimation process in dynamic vacuum. As substrate, freshly cleaved muscovite mica (0 0 1) was used. After insertion into the growth chamber the mica substrates were preheated at 90°C for 30 min to remove surface contaminations like water and CO_2. In contrast to the KCl substrates, no clear influence of the preheating time or temperature on the surface morphology of the p-6P layers could be observed, which can be explained by the more inert surface of muscovite mica.

For the growth the HWBE system, as described earlier, was used. The vacuum during growth was 6×10^{-6} mbar. The source temperature was 240°C for all experiments presented in this chapter. The substrate temperature was changed in the range between the lowest possible values of 90 and 150°C. The wall temperature was in the range of 240–260°C.

The surface morphology of the p-6P layers was investigated ex situ by AFM using a Digital Instruments Multimode IIIa scanning probe microscope with an AS-130(J) scanner. All measurements were done under ambient conditions using tapping mode to eliminate the risk of surface damage due to lateral forces between tip and surface. In contrast to the p-6P layers on KCl, the layers on mica were very sensitive to AFM investigations in contact mode. Due to the low interaction potential between the inert substrate surface and the p-6P molecules, the surface structure was very easily disturbed by the AFM tip in contact mode. Most of the presented images have been obtained with Pt-coated Si-tips. Some of the high-resolution images were recorded using high density carbon tips. These tips consist of a carbon whisker attached to a conventional Si-tip. These probes have a tip radius of less than 5 nm and an opening angel of less than 10°. The resulting high aspect ratio allows to precisely image crystallites with nearly vertical side walls. To draw statistical information out of the AFM pictures, at least 100 objects per growth stage were analyzed.

Thermal desorption spectroscopy (TDS) was performed in an UHV chamber with a background pressure of 1×10^{-10} mbar. For TDS the sample was attached to a resistivity heated steel substrate holder, which was heated at a rate of $1\,\mathrm{K\,s^{-1}}$. The desorbing species were detected with a quadrupole mass spectrometer [31].

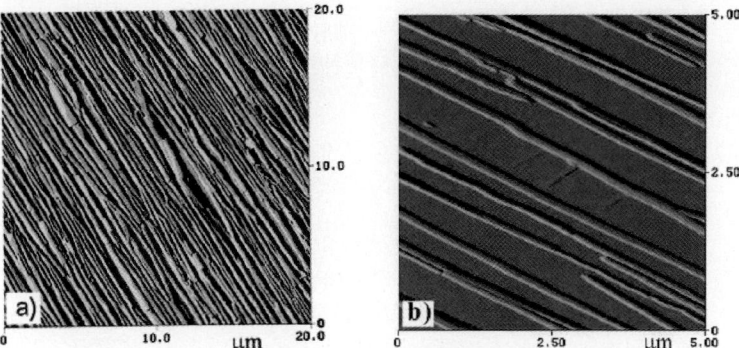

Fig. 5.15. AFM images characterizing the self-organized growth of p-6P on mica substrates after a deposition time of (**a**) 60 min and (**b**) 60 s

In a first trial, we have set the substrate temperature to 90°C and selected a growth time of 60 min. Figure 5.15a shows the surface morphology of such a p-6P layer with a characteristic long-range ordered structure with a strongly expressed preferential direction [32]. This structure consists of large oriented crystallites looking like very long nanofibers, which are parallel to each other and separated by flat areas in between. These flat areas between the nanofibers are much more pronounced in an earlier growth stage, as depicted in Fig. 5.15b. In this case the growth was interrupted after 1 min. Again a clear structure with nanofibers was obtained with a length to width ratio of 500. On perfectly cleaved mica substrates, this preferential direction does not change over the entire film surface. However, on mica substrates containing cleavage steps, the preferential direction of the nanofibers can change by 120° from terrace to terrace, while the morphology of the nanofibers remains identical. The turn around of the nanofibers by 120° is based on an interaction between the p-6P molecules and the mica substrate, which can exhibit two different surface planes. A detailed explanation of this unique phenomenon can be found in Chap. 6 of this book and in [33]. The surface roughness between the nanofibers is 3–4 nm. On the other hand, freshly cleaved mica shows a surface roughness of 0.2–0.3 nm. This can be taken as a first hint that the areas between the nanofibers are also covered with p-6P.

To investigate this clear self-organized growth in more detail, the surface morphology was studied in dependence of the substrate temperature used during the growth [34]. For this series of experiments, the source temperature and therewith the impingement rate of the molecules on the growing surface as well as the growth time (45 s) were kept constant. The obtained surface structures for substrate temperatures of 80, 90, 110, and 150°C are depicted in the AFM images in Fig. 5.16. At lower temperatures mainly islands are formed, while at higher temperatures the typical nanofibers start to appear. At medium temperatures islands and elongated chains of islands coexist. Due to the larger surface mobility of the impinging molecules at higher substrate temperatures,

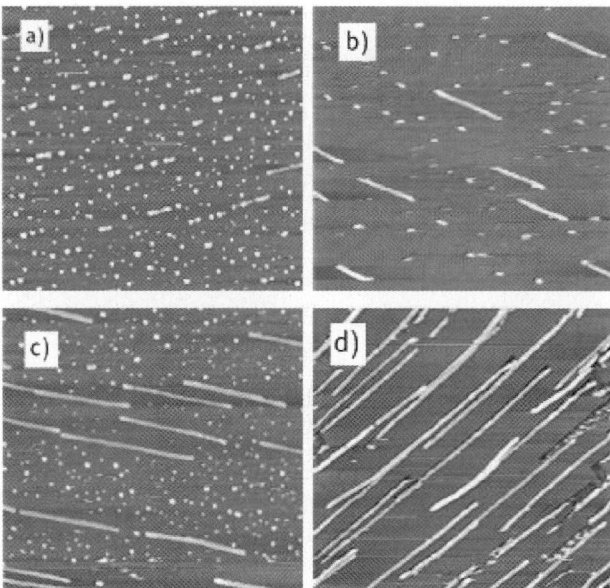

Fig. 5.16. AFM images characterizing the surface morphology at different substrate temperatures: (**a**) 80°C, (**b**) 90°C, (**c**) 110°C, and (**d**) 150°C (reprinted with permission from [34])

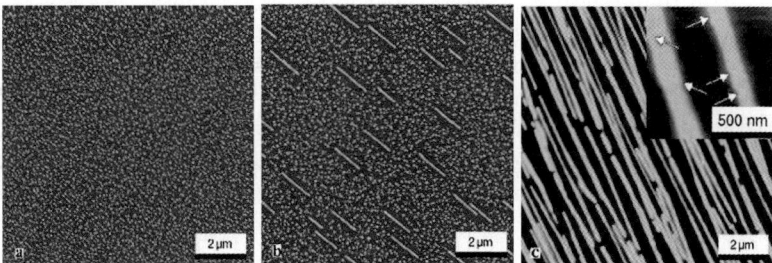

Fig. 5.17. Different growth stages of p-6P films on mica: (**a**) pure island growth after 30 s, (**b**) islands and chains after 45 s, and (**c**) long-term growth morphology after 9 min, consisting only of chains; z-scale for all pictures is 50 nm

the self-assembling process becomes more pronounced. However, out of this experiments it was not clear if the single molecules are the only mobile species on the surface, or if the whole islands still can move on the growing surface to undergo this drastical shape transition from islands to elongated chains and finally into extended nanofibers.

Only a detailed time-dependent study of the surface morphology could help to obtain a clearer picture of the growth process [35]. Figure 5.17 shows representative images of the three different stages of p-6P growth on mica (0 0 1).

Fig. 5.18. Detailed 3D image of the single 800 nm long chain after 35 s of p-6P deposition (intermediate stage); $1 \times 1\,\mu m$ scan; z-scale is 30 nm (reprinted with permission from [36])

The first stage is shown in Fig. 5.17a and is characterized by the formation of randomly distributed islands with uniform size of about $100 \times 50 \times 10\,nm^3$ (length × width × height). The surface morphology changes drastically as soon as a critical density of islands is reached between 30 and 35 s of growth: a rearrangement of islands occurs resulting in self-organized parallel chains coexisting with isolated islands (intermediate growth stage as shown in Fig. 5.17b). With increasing time these chains become progressively longer and closer to each other, whereas the isolated crystallites disappear. Figure 5.17c shows the corresponding advanced growth stage after 9 min. The surface structure in this stage is made up only of parallel chains with a very high aspect ratio in length.

Figure 5.18 shows a detailed AFM scan of a single chain in an intermediate stage. This chain is 850 nm long and has a height and width of 15 and 75 nm, respectively. The size of the features within the chain and that of the islands that surround the chain are very similar. It seems that the displayed chain is made up from approximately 14 islands. It is important to mention here that much longer chains show also the same internal structure as such a shorter one shown in Fig. 5.18.

Figure 5.19 presents the island size and density evolution over the first 60 s of p-6P deposition [36]. One can see that the islands reach their final height of 18 nm after 10 s and their final width and length, of 50 and 100 nm, after approximately 30–35 s (first stage). Simultaneously the island density increases continuously and reaches a saturation value of $35\,\mu m^{-2}$. The first chains appear after approximately 35 s of the p-6P deposition marking the start of the intermediate growth stage, in which individual islands and chains coexist. From this time on, the size and the density of the islands do not increase anymore and saturate at the values given above. In contrast, the chain length strongly increases from 800 nm to $\approx 2\,\mu m$ between 35 and 55 s of deposition. However, their width and height increase much slower.

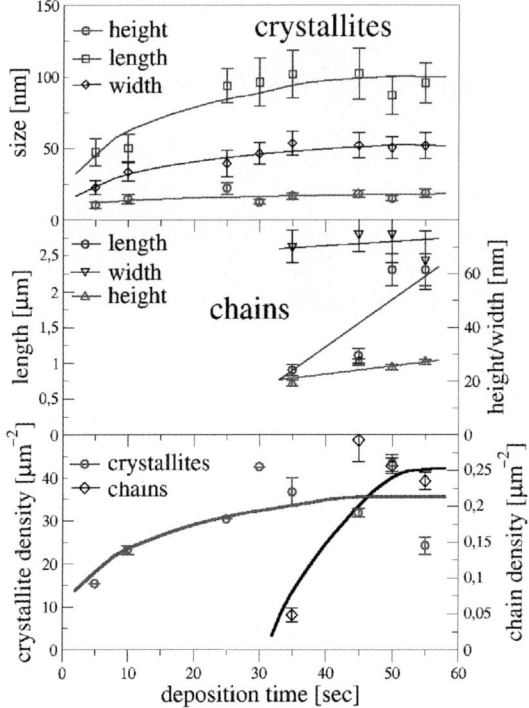

Fig. 5.19. Evolution of length, width, and height as well as density of islands and chains. The longer axis of the entities is called length and the shorter one width. *Continuous lines* are just guides to the eye (reprinted with permission from [36])

Figure 5.20 shows the evolution of the chain length distributions in the intermediate stage. One can see that at the beginning (35 s) the chains exhibit a very narrow length distribution with an average length of 800 nm (see also Fig. 5.17). The minimum and maximum chain lengths are here around 500 and 1300 nm, respectively. If more material is deposited, the average chain length increases and the chain length distribution slightly broadens. Most important, however, is the existence of a threshold length, below which chains seem to be energetically unstable. This threshold length depends on surface coverage, namely it increases from 500 nm (about 10 islands) after 35 s of deposition to 1200 nm (about 24 islands) after 55 s (see Fig. 5.20).

As can be clearly shown from Fig. 5.21, in the intermediate stage the chains are surrounded by a denuded zone that is free of islands. This denuded zone is found at all four sides of the chains. In Fig. 5.21 two rectangles of the same size are drawn. The number of islands in the lower rectangle is approximately equal to the number of islands needed to form the chain in the upper rectangle. From this observation and from the interior chain structure, we conclude that the chains are spontaneously formed by rearrangement of the individual

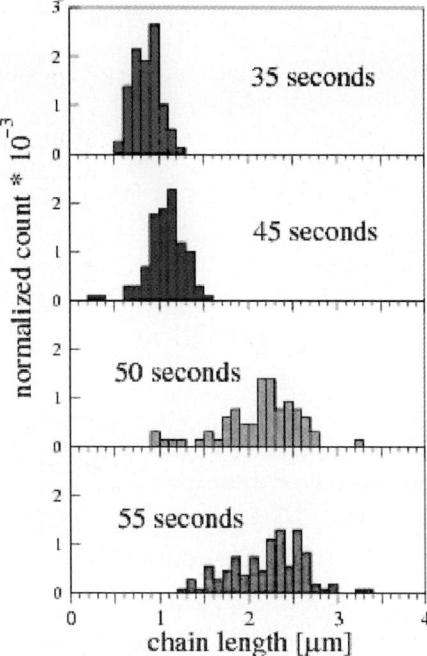

Fig. 5.20. Evolution of chain length distribution as a function of deposition time from 35 to 55 s (reprinted with permission from [36])

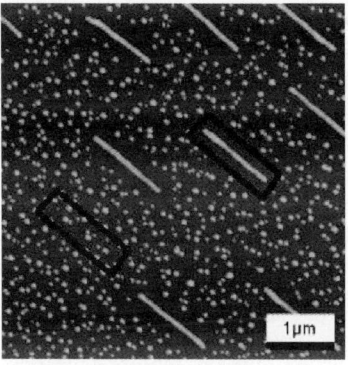

Fig. 5.21. AFM image revealing denuded zones around chains. The amount of material in the lower rectangle is approximately equal to the amount of material needed to build the chain in the upper rectangle; z-scale is 50 nm (reprinted with permission from [36])

islands as entities. This idea is also supported by recent selected area electron diffraction and TEM measurements, which revealed the existence of highly ordered different oriented p-6P domains within the chains (see Chap. 6 in this book).

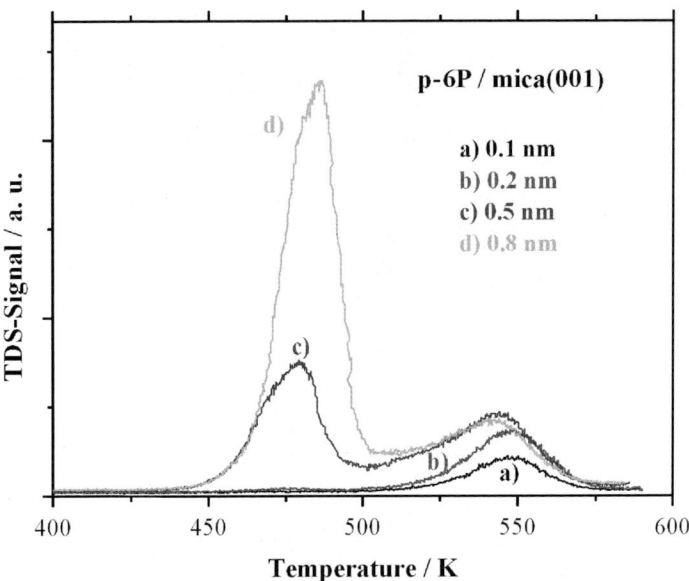

Fig. 5.22. Thermal desorption spectra for p-6P on mica (0 0 1). Adsorption temperature is 360 K, heating rate is $1\,\mathrm{K\,s^{-1}}$. The mean thickness of the individual films as measured with the quartz microbalance is indicated (reprinted with permission from [36])

Before the evolution of the growth morphology is discussed in detail, another experimental result should be reported concerning the question whether the islands and chains grow directly on the mica substrate or on a wetting layer of lying p-6P molecules. As described above the increased surface roughness between the nanofibers gave a first hint for the existence of such a wetting layer. The direct proof for the existence of such a wetting layer arises from thermal desorption spectra presented in Fig. 5.22. For small coverage one observes a desorption peak at around 550 K, which saturates at a mean coverage of 0.2–0.3 nm. Considering the van der Waals dimensions of p-6P ($0.35 \times 0.67 \times 2.85\,\mathrm{nm^3}$), this is a clear indication that a rather strongly bonded wetting layer of flat lying molecules exists. For higher coverage, a second peak arises at around 480 K which does not saturate. The common initial slope and the quite sharp decrease of the trailing edge of this peak are a clear indication of a zeroth-order desorption. Thus, we can attribute the second peak to desorption from a multilayer, which corresponds in our case, to the islands sitting on the one monolayer thick wetting layer of lying p-6P molecules.

The formation of the final surface morphology, dominated by chains, can be divided in three steps. First, at low p-6P coverage, only 3D islands are formed (Fig. 5.17a). The size evolution of the islands as depicted in Fig. 5.19 is typical for a strain-controlled growth mechanism [37]. When more material

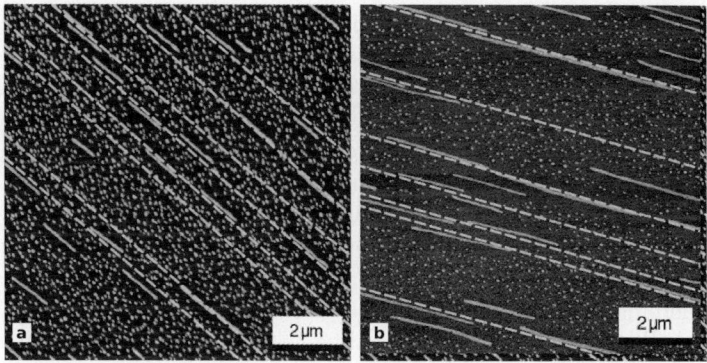

Fig. 5.23. AFM images of as-grown (*left*) and annealed (*right*) p-6P films on mica (0 0 1). The *dashed green lines* indicate the proposed linear defect array; z-scale is 50 nm (reprinted with permission from [36])

is deposited the lateral and vertical size of the islands saturates quickly and a new entity is formed instead of adding molecules to already existing structures. As a result the density of the islands per unit area increases almost linearly in the first 30 s.

In a second step, islands and chains of islands coexist (Fig. 5.17b). This growth morphology starts to evolve as soon as a critical density of islands is exceeded locally. For the spontaneous rearrangement of the islands we propose the following scenario. The strain that is induced locally by the islands into the wetting layer leads to the formation of a linear defect in the wetting layer (see Fig. 5.23a) whose orientation is related to the substrate geometry [38]. This defect stimulates the mobile islands to touch each other in a specific direction and is so acting as a nucleation center for the chain. The rearrangement process is performed by the islands as entities, since we can still recognize the individual islands. Once the islands are incorporated into the chain they are trapped there, forming a stable nucleus (see threshold length in Fig. 5.20).

The threshold length of the formed chain is determined by the amount of the excess density. If – in a next step – a new island is formed, the amount of excess density added by a single island is higher as in the preceding step, because the area covered by islands is now smaller than before (parts of the original area are now covered by the already built chains and their denuded zones). This leads to the formation of a longer chain because more material has to be removed from the area covered with islands. This can be done by either extending an existing chain or creating a new chain. Such a new chain is formed by either using the extended linear defect induced by an already existing chain in the vicinity, or inducing a new line defect finally resulting in a one-dimensional defect array. The former case is vividly shown in Fig. 5.23a where frequently two or more chains line up along the same linear defect. When a certain number of chains are formed the density increases less since there is always a chain close to the newly formed island that can be extended.

If the growth is interrupted at this point and the sample is kept at the deposition temperature for a certain time (annealing), a further reduction of the strain can be achieved by incorporation of surrounding islands into the already existing chains. Usually, no more new chains are formed during this annealing process. This can be understood, if one takes into account that no more new islands will be formed during annealing. As the critical density that is necessary to induce the linear defect cannot be exceeded, no new nucleation centers are formed. However, if a new chain is formed during the annealing process, it has to use one of the already existing nucleation centers and is therefore always aligned with a previously formed longer chain. The experimental result of such an annealing process is displayed in Fig. 5.23b. Indeed, we observe an increase in the average length of the chains, but no increase in the density of the defect array. Furthermore, this experiment is an additional hint that the islands are mobile.

In a third stage, no more islands are formed on the wetting layer because the surface is completely covered with chains and their capture zones (Fig. 5.17c) start to overlap. Either new material is included directly at the ends of existing chains or new islands are formed on top of them (insert in Fig. 5.17c).

5.5 Conclusion

It could be shown that HWE is a powerful method for growing thin films of small conjugated molecules. In particular high crystallographic order was obtained for C_{60}, Ba-doped C_{60}, and p-6P layers on mica as well as on KCl substrates.

In the case of C_{60} layers even further improvement of the crystalline quality could be achieved by in situ annealing, resulting in an FWHM of the rocking curve of 140 arcsec, which is still the best value ever reported for epitaxially grown C_{60} layers.

The incorporation of Ba in the C_{60} layers caused a significant charge transfer from Ba atoms to the C_{60} molecules, resulting in an increase of the growth rate and at the same time an effective n-type doping was obtained. Surprisingly also a very high mobility of the electrons in the Ba-doped C_{60} was observed, which is very unusual for metallic-like conductivity.

In the case of p-6P, highly ordered self-organized nanostructures were obtained by HWE on mica and KCl substrates. On KCl substrates different kinds of morphologies were obtained, namely wider and smaller needle-like structures and terraced-shaped islands between them. One kind of the needles is always perfectly aligned along [0 1 1] directions of the KCl substrate, other needles are shorter and wider and not always parallel to the [0 1 1] direction but also epitaxially oriented. All needles, no matter which orientation, consist of herringbone layers parallel to the needle axis. The growth of the needles can be described by a typical VW growth mode, whereas the terraced-shaped islands inhibited FM growth. Moreover there was no wetting layer found between the needles.

For the growth of p-6P on mica substrates, three different growth morphologies depending on the amount of deposited material have been observed. The first growth stage is dominated by the formation of strain-controlled randomly arranged 3D islands, which turned out as a typical example of SK growth mode. The second stage is characterized by the coexistence of these mobile islands and parallel aligned chains, which are surrounded by depletion zones, due to the rearrangement of the islands. The driving force for the spontaneous formation of the chains is a critical density of islands that locally induce a linear defect in a wetting layer. When this density is reached no more islands can be formed without interfering with already existing entities. Further deposition of material leads to the third growth stage, where only p-6P chains are observed.

The possibility to dope C_{60} layers by Ba and on the other hand the particular growth mode of p-6P on mica opens the field for a variety of applications. In the first case, organic field-effect transistors with high field-effect mobility and reasonable on/off ratios seem to be possible. In the case of the p-6P nanofibers, one can speculate with graphoepitaxial methods to control the places where the nanofibers should be formed on the substrate. The possibility of transferring the grown nanofibers to another substrate would open up an even wider range of applications.

Acknowledgments

This book chapter is based mainly on the careful and patient work of many diploma and Ph.D. students. The C_{60} layers were grown by D. Stifter and G. Matt. The Ba-doping experiments were performed by T. Nguyen Manh; A. Andreev and A. Montaigne Ramil used the HWE system to grow the numerous layers of p-6P nanofibers. The surface morphology was investigated in collaboration with the institute of physical chemistry, lead by N.S. Sariciftci and the institute of physics at the Montan-University Leoben. In these institutions A. Andreev, H. Hope, Ch. Teichert, and G. Hlavacek spent a lot of time to obtain clear and informative AFM pictures. The crystallographic structure was investigated by R. Resel, T. Haber, and H. Plank at the institute of solid state physics, Graz University of Technology. The technical assistance by O. Fuchs and E. Rund is also acknowledged.

The research was mainly financed by the Austrian Science Foundation within the project cluster "Highly Ordered Organic Epilayers" (especially, FWF projects P-15155, P-15625, P-15626, and P-15627) and within the National Research Network "Interface Controlled and Functionalised Organic Films" (especially, FWF projects NFN-S9706, NFN-S9707, and NFN-S9708). Part of this work was performed within the Christian Doppler society dedicated laboratory on plastic solar cells funded by the Austrian ministry of economic affairs and Konarka Austria GmbH.

References

1. A. Heeger, Curr. Appl. Phys. **1**, 247 (2001)
2. M. Granström, M. Harrison, R. Friend, *Handbook of Oligo- and Polythiophenes* (Wiley-VCH, Weinheim, 1999), p. 405 ff
3. A. Andreev, G. Matt, C. Brabec, H. Sitter, D. Badt, H. Seyringer, N. Sariciftci, Adv. Mater. **12**, 629 (2000)
4. J. Humenberger, K. Gresslehner, W. Schirz, H. Sitter, K. Lischka, MRS Proc. **216**, 53 (1991)
5. J. Humenberger, H. Sitter, Thin Solid Films **163**, 241 (1988)
6. A. Lopez-Otero, Thin Solid Films **49**, 3 (1978)
7. J. Humenberger, H. Sitter, Thin Solid Films **163**, 679 (1989)
8. V. Holy, J. Kubena, E. Abramof, K. Lischka, A. Pesek, E. Koppensteiner, J. Appl. Phys. **74**, 1736 (1993)
9. D. Stifter, H. Sitter, Fullerene Sci. Technol. **4**, 277 (1996)
10. D. Stifter, H. Sitter, J. Cryst. Growth **156**, 79 (1995)
11. D. Stifter, H. Sitter, Thin Solid Films **280**, 83 (1996)
12. M. Krivoglaz, *X-Ray and Neutron Diffraction in Nonideal Crystals* (Springer, Berlin Heidelberg New York, 1996)
13. H. Sitter, D. Stifter, T. Nguyen Manh, Thin Solid Films **306**, 313 (1997)
14. H. Sitter, T. Nguyen Manh, Cryst. Res. Technol. **34**, 605 (1999)
15. A. Oshiyama, S. Saito, N. Hamada, Y. Miyamoto, J. Phys. Chem. Solids **53**, 1457 (1992)
16. H. Sitter, A. Andreev, G. Matt, N. Sariciftci, Mol. Cryst. Liq. Cryst. **385**, 51 (2002)
17. H. Sitter, A. Andreev, G. Matt, N. Sariciftci, Synth. Met. **138**, 9 (2003)
18. K. Baker, A. Fratini, T. Resch, H. Knachel, W. Adams, E. Socci, B. Farmer, Polymer **34**, 1571 (1993)
19. R. Resel, G. Leising, Surf. Sci. **409**, 302 (1998)
20. K. Erlacher, R. Resel, J. Keckes, F. Meghdadi, G. Leising, J. Cryst. Growth **206**, 135 (1999)
21. K. Erlacher, R. Resel, S. Hampel, T. Kuhlmann, K. Lischka, B. Müller, A. Thierry, B. Lotz, G. Leising, Surf. Sci. **437**, 191 (1999)
22. H. Yanagi, T. Morikawa, Appl. Phys. Lett. **75**, 187 (1999)
23. T. Mikami, H. Yanagi, Appl. Phys. Lett. **73**, 563 (1998)
24. A. Andreev, T. Haber, D.M. Smilgies, R. Resel, H. Sitter, N. Sariciftc, L. Valek, J. Cryst. Growth **275**, e2037 (2005)
25. H. Yanagi, T. Ohara, T. Morikawa, Adv. Mater. **13**, 1452 (2001)
26. T. Haber, A. Andreev, A. Thierry, H. Sitter, M. Oehzelt, R. Resel, J. Cryst. Growth **284**, 209 (2005)
27. T. Haber, M. Oehzelt, R. Resel, A. Andreev, A. Thierry, H. Sitter, D.M. Smilgies, B. Schaffer, W. Grogger, R. Resel, J. Nanosci. Nanotechnol. **6**, 698 (2006)
28. H. Yanagi, T. Morikawa, S. Hotta, K. Yase, Adv. Mater. **13**, 313 (2001)
29. D. Smilgies, N. Boudet, H. Yanagi, Appl. Surf. Sci. **189**, 24 (2002)
30. H. Plank, R. Resel, S. Purger, J. Keckes, A. Thierry, B. Lotz, A. Andreev, N. Sariciftci, H. Sitter, Phys. Rev. B **64**, 235423 (2001)
31. S. Müllegger, O. Stranik, E. Zojer, A. Winkler, Appl. Surf. Sci. **221**, 184 (2004)
32. H. Sitter, G. Matt, A. Andreev, C. Brabec, D. Badt, H. Neugebauer, N. Sariciftci, MRS Proc. **598**, BB3.39 (2000)

33. H. Plank, R. Resel, A. Andreev, N. Sariciftci, H. Sitter, J. Cryst. Growth **237–239**, 2076 (2002)
34. A. Andreev, H. Sitter, C. Brabec, P. Hinterdorfer, G. Springholz, N. Sariciftci, Synth. Met. **121**, 1379 (2001)
35. A. Andreev, C. Teichert, G. Hlawacek, H. Hope, R. Resel, D.M. Smilgies, H. Sitter, N. Sariciftci, Org. Electron. **5**, 23 (2004)
36. C. Teichert, G. Hlawacek, A. Andreev, H. Sitter, P. Frank, A. Winkler, N. Sariciftci, Appl. Phys. A **82**, 665 (2006)
37. C. Teichert, Phys. Rep. **365**, 335 (2002)
38. H. Plank, R. Resel, H. Sitter, A. Andreev, N. Sariciftci, G. Hlawacek, C. Teichert, A. Thierry, B. Lotz, Thin Solid Films **443**, 108 (2003)

6

Crystallography of Ultrathin Organic Films and Nanoaggregates

T. Haber and R. Resel

Currently, technical applications of organic nanoaggregates are discussed for various purposes like displays, electronic circuits, sensors and optical waveguides. In all these cases the optical and electronic properties (and also the combination of both) are decisive parameters. Why are the structural properties of the nanoaggregates so important? The reason is the strong relationship between the structure (arrangement of the molecules within the bulk state) of the nanoaggregates and the application-relevant properties: optical emission and electronic charge transport.

The optical absorption and emission of organic molecules is highly directed. Therefore, a strong optical anisotropy is observed for ordered bulk materials [1, 2]. For example in the case of nanoneedles of the molecule sexiphenyl (or hexaphenyl) light with polarization along the long molecular axes is emitted [3, 4]. Also the electronic charge transport shows anisotropic behaviour within an ordered bulk state [5]. It is generally accepted that the intermolecular charge transport happens along overlapping π-conjugated segments of neighbouring molecules [6, 7]. Therefore, specific directions within an ordered bulk state show enhanced ability for charge transport.

Applications based on nanoaggregates which are prepared on surfaces are always associated with specific directions of light absorption/emission and charge transport. The orientation of the molecules as well as the directions of dense π-packing of neighbouring molecules are defined by the crystalline properties of the nanoaggregates. The orientation of the crystallites on the substrate surface is determined by the orientation of the molecules relative to the substrate surface. The alignment of the crystallites along specific surface directions is determined by the in-plane alignment of the molecules relative to the substrate.

One further important parameter for application is the size and shape of the crystalline domains. Generally, large single crystalline domains with regular shape are preferred. Grain boundaries disturb the regular lattice and interrupt the periodic overlapping of neighbouring molecules, this considerably influence the charge transport [8, 9].

While crystal structure investigations of inorganic nanoaggregates are frequently performed, organic nanoaggregates are much less studied. The experimental difficulties are small diffraction signals in case of X-ray diffraction and neutron diffraction. Surprisingly, neutron diffraction works for thin organic films [10]. The advantage of electron diffraction studies is the good sensitivity, however, the degradation of the organic material due to the electron bombardment is a real problem.

X-ray diffraction studies of thin organic films also show some difficulties. The low scattering probability of the involved atoms (mainly carbon and hydrogen) together with the small scattering volume results in diffraction patterns with small intensities. However, thin films of molecular crystals show acceptable intensity which can be observed by standard laboratory equipment. In case of films with low crystallographic order and also for ultrathin films (with a thickness of few monolayers) synchrotron radiation has to be used. The low crystal symmetry (triclinic, monoclinic or orthorhombic) together with large lattice constants imply a huge number of diffraction peaks which are frequently overlapping.

Good results for structural characterization of organic nanoaggregates are obtained by combining different diffraction techniques. X-ray diffraction has low sensitivity but high resolution regarding the reciprocal space. It can be combined quite nicely with transmission electron microscopy (TEM) techniques which have high sensitivity with high spacial resolution. Supplementary information to the crystalline order is the morphology of the nanoaggregates; it can be obtained by diverse microscopy methods. In case of transmission electron techniques both, diffraction and microscopy, can be combined so that the morphology and the crystal structure properties can be related.

The main goals of crystallographic investigations are the determination of basic crystalline properties like crystalline phases, preferred orientations (or epitaxial orders) of crystallites and crystal sizes. The correlation of crystallographic order with the film morphology reveals basic information on the growth mechanisms of organic nanoaggregates on surfaces.

6.1 Overview

The organic nanoaggregates of this chapter are built up by conjugated molecules, which are composed from aromatic units like phenyl or thiophene rings. Due to the π-conjugation these conjugated segments are rigid units of the molecules. Basically two types of molecular shape exist. On the one hand rod-like molecules which have a long molecular axis and a comparable small width; the conjugation is extended along the long molecular axis. On the other hand there are disc like molecules with an extended conjugation within the molecule. Basically, the crystal structures of aromatic based molecules are classified in three different types [11, 12]: herringbone type, sandwich herringbone type and the stacked type. Schematic pictures of the three different types of crystal structures are shown in Fig. 6.1.

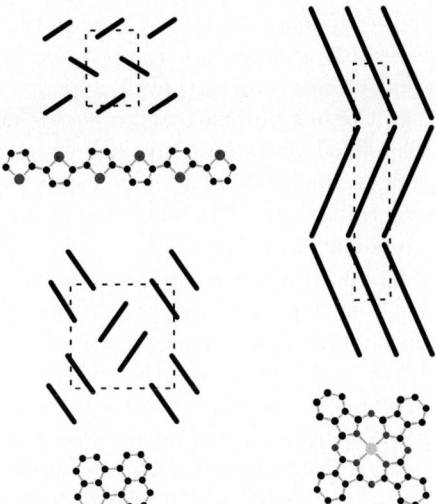

Fig. 6.1. The three different types of crystal structures of conjugated molecules: the herringbone structure (*top left*), the sandwich herringbone structure (*bottom left*) and the stacked structure (*right*). The molecules are drawn in projection of their conjugated segments and the basal plane of the crystallographic unit cells is drawn by *dashed lines*

Disc-shaped molecules pack in a stacking of the discs above each other. A large overlapping of the π-conjugated segments appears along the stacking direction. Columns of stacked molecules pack as close as possible to each other. Examples for molecules which crystallize in the stacked structure are coronene, hexabenzocoronene and phthalocyanines (example in Fig. 6.1).

Rodlike molecules pack in a layered herringbone structure: the aromatic planes of neighbouring molecules are tilted relative to each other, while the long molecular axes are parallel to each other. Molecules like sexiphenyl (example in Fig. 6.1), sexithiophene and pentacene form this type of crystal structure.

The third type of crystal structure is the sandwich herringbone which is observed for molecules which have an elongated shape in between disc and rodlike shape. These molecules form dimers in the crystal structure and these dimers are arranged in a herringbone type. The molecules perylene (example in Fig. 6.1) and pyrene form a sandwich herringbone structure.

Basically, two possible orientations of neighbouring molecules are observed: a stacking and a strong tilting of the conjugated segments. The strong tilting of the conjugated segments is a consequence of quadrupolar electrostatic forces of neighbouring molecules which overcome the van der Waals type interactions [13, 14]. It is obvious that van der Waals interactions are responsible for the stacking of neighbouring molecules which results in a structure according to the close packing principle [15].

The appearance of different crystal structures of a single molecular material (polymorphism) is a frequent phenomenon regarding molecular crystals [16, 17]. In most cases polymorphism appears within one type of crystal structure. E.g. Copper–phthalocyanine shows eight different polymorph phases but all refined phases are stacked structures [18]. Few cases are known where a single molecular material shows two different (of the above mentioned) structure types. The molecules perylene and pyrene can crystallize in the herringbone structure as well as in the sandwich herringbone structure. The molecule fluorene shows a transition from a herringbone type to a stacking type structure under high pressure of about 2 GPa [19].

The typical shapes of crystalline nanoaggregates on surfaces show a large variety. The type of crystal structure together with the preparation conditions plays an important role for size, shape and crystalline order of the nanoaggregates. Nanoaggregates consisting of disc-shaped molecules form mainly island type structures. Islands with non-regular shape extended in both lateral dimensions are observed [20]. Surface corrugations and the crystal symmetry of the underlying substrate are affecting the film morphology quite often [21–23].

Molecules that crystalize in the herringbone structure show two distinct morphologies of nanoaggregates. Islands with lateral expansion in two dimensions with partly dendritic like structures are observed [24–26]. These islands show a terraced morphology with defined height variations between the individual terraces [27,28]. Therefore, this morphology is denoted as "terraced islands". The second morphology is islands with an extension in one single direction, these are elongated islands [29–31]. In case of an extreme extension of the single direction, this morphology is called nanoneedles [32–34]. The widths and heights of the nanoneedles range from few nm up to 100 nm, but the length could be up to several mm [35]. Sometimes a single nanoneedle shows considerably height variation along the needle direction, therefore, they are sometimes called nanochains [36]. Nanoaggregates grown from molecules which crystallize in the sandwich herringbone structure form basically the same morphologies: terraced islands and nanoneedles [37].

6.2 Crystal Structure of Rodlike Conjugated Molecules

Figure 6.2 shows a schematic drawing of the herringbone structure of sexiphenyl. The drawing is based on the β-structure of sexiphenyl which is stable at room temperature and regularly observed [38]. The top part of the figure shows two layers of the crystal structure. The thickness of one layer is approximately the length of the molecule which is 2.6 nm, the molecules are tilted by an angle of 73° to the layer. The bottom part of the figure shows a single layer projected along the long molecular axes. All aromatic units of a single molecule are arranged in one plane, the aromatic planes of neighbouring molecules are side tilted by an angle of 66° (herringbone angle). The plane between two

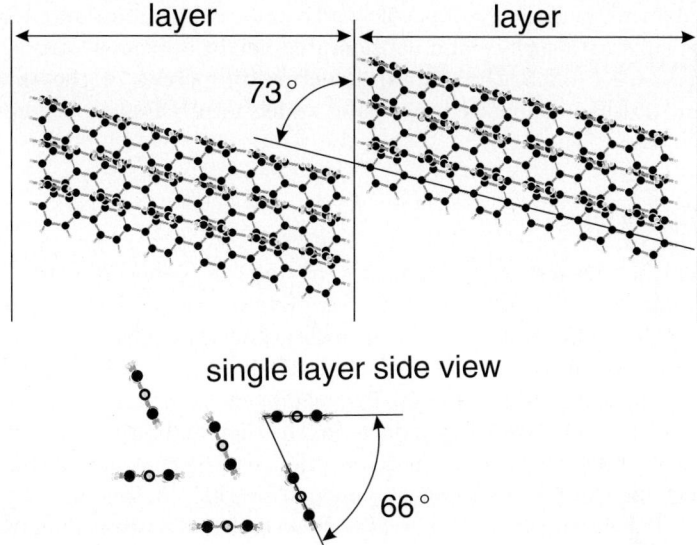

Fig. 6.2. The layered herringbone structure shown on the example sexiphenyl. Two layers (*top part*) and the herringbone arrangement within a single layer (*bottom part*)

layers shows the typical characteristics of a cleavage plane: the layers consist of densely packed molecules with the feature that the intralayer interactions are considerably higher than the interlayer interactions [39, 40].

Single crystals show a morphology of thin plates [41]. Energetic considerations of the crystal growth explain this morphology. For a single molecule the gain of energy is much larger, by attaching itself to an existing herringbone layer than by starting a new layer. As a consequence an anisotropic growth rate appears. The lateral dimensions of the plate-like crystals increase rapidly while the thickness of the plate-like crystals grows with a considerably smaller rate. In some cases, the crystals are more sheets than plates with a thickness only in the µm range [38].

The morphology of the nanoaggregates on surfaces is a consequence of the anisotropic crystal growth characteristics. If the plane of a herringbone layer is formed parallel to the surface of the substrate, the resulting growth morphology of crystals is terraced islands. The layers of the crystal structure are extended along the substrate surface in the two lateral dimensions, the growth in the vertical dimension happens much slower in a terrace like manner. The step heights of the single terraces are approximately the length of a single molecule. Another type of growth morphology appears if the crystal growth starts from molecules oriented with their long molecular axes parallel to the surface of the substrate. The growth of the crystallites happens in the two dimensions of the herringbone layers: one growth direction is along the substrate surface and the second growth direction is perpendicular to the substrate surface. Nanoneedles with small width and large elongation along the substrate surface and considerable height appear.

In Sects.6.2.1–6.2.3, the crystal structure properties of the three most prominent rod-like conjugated molecules which form nanoneedles on surfaces are introduced.

6.2.1 Oligoacenes

The scientific interest on oligoacenes as organic semiconductors already started many decades ago; e.g. the first observation of electroluminescence of an organic semiconductor was on anthracene [42]. The crystal structure of anthracene is quite intensively studied [43–49]. Less data are available for larger oligomers like tetracene [50–52], pentacene [50, 51, 53, 54] and hexacene [50]. The oligoacenes crystallize in the herringbone structure. These molecules are fully conjugated, therefore, they can be considered as fully rigid with no change in the molecular conformation due to packing in a crystal structure. The symmetry of the crystal structures are centrosymmetric as the symmetry of the molecules themselves. Anthracene crystallizes monoclinic within spacegroup $P2_1/c$; longer oligomers crystallize mainly triclinic within the spacegroup $P\bar{1}$. Interestingly, the long molecular axes of neighbouring molecules are not fully parallel to each other, a distortion of several degrees appears. High pressure experiments reveal that the herringbone angle of neighbouring molecules decreases with increasing pressure, a reduction of about 10° appears at an external pressure of 10 GPa [55]. Several polymorph phases are observed for oligoacenes, however, only the herringbone structure is observed. Anthracene shows weak tendency to form polymorph phases [56], while two polymorph phases are observed for tetracene. The molecule pentacene shows a huge variety of different polymorphs which are formed especially in thin films. It seems that all of them are of herringbone type; triclinic, monoclinic and orthorhombic crystal structures are suggested [57–59]. Frequently the so-called "thin film phase" is observed which is formed in the first monolayer on surfaces as a herringbone structure [60].

6.2.2 Oligophenylenes

The outstanding physical properties of oligophenylenes are the optical properties in the visible blue range with high photoluminescence quantum yield and excellent chemical stability so that this class of molecules are one of the best choice for blue electroluminescence [61]. Since many decades the crystal structures of oligophenylenes (terphenyl [62–69], quaterphenyl [70, 71], quinquephenyl [38], hexaphenyl or sexiphenyl [38, 72–75], septiphenyl [38]) are a focus of interest, since these molecules show clear differences in the molecular conformation between the isolated molecule and the molecule packed in the solid state [76,77]. Within an isolated molecule in the gas phase the repulsion of ortho-hydrogen atoms force neighbouring phenyl rings to be tilted relative to each other. Tilt angles up to 40° are determined. Within the crystallized state packing forces tend to reduce these tilt angles down to few degrees.

The refined crystal structures of oligophenylenes are herringbone type which are centrosymmetric within the spacegroup $P2_1/c$. In difference to the oligoacenes, the long molecular axes of all molecules within the crystal structure are parallel to each other.

Three distinct herringbone type crystal structures are known for sexiphenyl: One phase is detected at low temperatures (below 110 K), but could not be fully solved. However, important details could be obtained. The conformation of a single molecule shows large tilt angles of adjacent phenyl rings relative to each other. Within the low temperature structure this tilting is extended periodically over the whole crystal lattice [38]. At ambient temperature sexiphenyl shows a crystal structure with another molecular conformation: the molecule appears as average flat with all phenyl rings in one plane. However, it seems that also in this crystalline state a tilt angle of adjacent phenyl rings is present, but reduced to some degrees. Thermal librations of the phenyl rings together with statistical disorder of the tilt angles over the whole crystal induce displacements which are not fully solved up to now. A third crystal structure is also known for sexiphenyl. The presence of a substrate surface during the crystal growth can induce a third – up to know unsolved – structure [74, 78]. Recently, an additional phase transition was detected around room temperature [79]. Despite the fact that sexiphenyl can take different molecular conformations, the tendency to form polymorph phases is considerably weaker than for pentacene.

6.2.3 Oligothiophenes

The scientific interest on oligothiophenes appeared 20 years ago [80, 81]. The molecules are characterized by a low band gap. Therefore it can be used for electronic applications. The crystal structure of terthiophene [82, 83], quaterthiophene [84, 85], quinquethiophene [83], sexithiophene [86–88], septithiophene [83] and octathiophene [89] is known.

The molecular conformation of the oligothiophenes is flat which means that all thiophene rings are in one plane. Due to the repulsion of orthohydrogen atoms the thiophene rings are arranged in trans conformation (compare Fig. 6.1) which means that in adjacent thiophene rings the sulphur atoms are located at the opposite sides. The pentagon shaped thiophene rings and the resulting molecular binding geometry induce a drastic change in conformation by tilting neighbouring thiophene rings: A tilt induces a small kink in the long molecular axis. Therefore, in comparison with oligophenylenes librational movements are strongly reduced in the crystallized state.

The crystal structures of oligothiophenes are of herringbone type which are centrosymmetric within the spacegroup $P2_1/c$. In difference to the oligoacenes, the long molecular axes of all molecules within the crystal structure are parallel to each other.

Oligothiophenes have a weak tendency to form polymorph phases, although three polymorph phases are known for sexithiophene. However, only two types are observed within nanoaggregates [90, 91]. One reason for the low tendency of polymorph formation could be the molecular shape. The individual thiophene rings determine a stronger structured van der Waals surface, in comparison to oligoacenes. According to the closed packing principle the packing of strongly structured rigid van der Waals surfaces to a crystal structure results in much deeper energy minima as it is the case for weakly structured rigid van der Waals surfaces.

6.3 Experimental Methods

This section introduces experimental methods which are used for the crystallographic characterization of organic nanoaggregates. The techniques are in principle known from textbooks, however, experimental difficulties have to be addressed for a successful application to organic nanostructures. Furthermore specific terms are introduced. Only the combination of the different techniques makes a complete structural characterization of organic nanostructures possible.

6.3.1 Fundamentals

The fundamental background of diffraction will not be discussed in this book since there are numerous textbooks that deal with this topic. Familiarity with keywords such as the reciprocal space and the diffraction conditions, e.g. the *Laue-Equations*, *Bragg's law* and the Ewald construction, respectively, is essential to understand the following investigation methods. Due to its importance *Bragg's law* be posted once more:

$$2d_{hkl} \sin\left(\frac{2\theta}{2}\right) = \lambda. \tag{6.1}$$

The net plane distance of the diffracting net plane is denoted as d_{hkl}, λ is the used wavelength and 2θ is the diffraction angle and defined as the angle of the deflection of the beam. Besides the Bragg's law another condition has to be fulfilled, namely that the incident angle has to equal the outgoing angle concerning the scattering lattice plane. In other words the scattering vector $\boldsymbol{q}$, which is defined as the difference between the diffracted and the incoming wave vectors (compare Fig. 6.3), has to be parallel to the normal vector $\boldsymbol{n}$ of the scattering lattice plane. Since the angle θ is often used as the angle between the incident beam and the surface of the specimen it is important to note that θ is not necessarily defined as $\frac{2\theta}{2}$ and that vice versa the *diffraction angle* 2θ is a self-contained angle which is to use in *Bragg's law* and not necessarily defined as twice the angle θ.

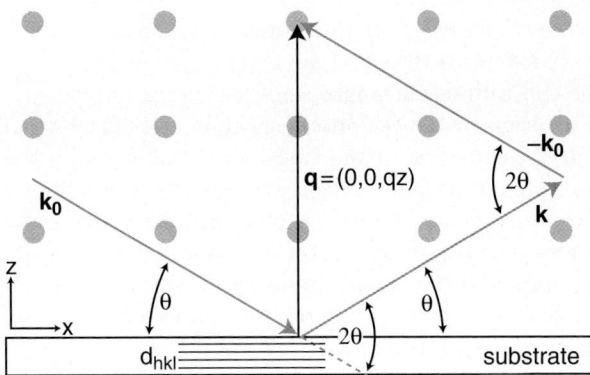

Fig. 6.3. Principle of a specular scan: by simultaneously increasing the incident and outgoing angle θ planes parallel to the surface are probed dependent on their net plane distance d_{hkl}. In the reciprocal space this is a q_z scan and intensity is detected when the vector $\boldsymbol{q}$ hits a reciprocal lattice point (*grey points*)

The so far mentioned aspects are exclusively based on geometric conditions for constructive interference. Furthermore the intensities of these reflections give numerous information about the structure of the solid and are therefore of large interest. In principle the intensity is dependent on numerous factors such as the *structure factor*, the *polarization factor*, the *Lorenz factor*, the *absorption factor*, the *plane-multiplicity factor* and the *temperature factor*. A detailed description of these factors as well as their influence on the intensity of a diffracted peak is described in numerous textbooks [92].

Concerning organic nanoaggregates it is to emphasize that an important result of the *structure factor* is that usually there are only few net planes that yield sufficient diffracted intensity. Due to the low order number of the involved atoms the scattering amplitudes are quite small. The crystal symmetries of organic nanoaggregates are low (triclinic or monoclinic) and therefore a huge number of diffraction peaks appear. The high intensity peaks are usually anisotropically distributed within a small angular range and often overlapping each other. The large lattice constants result in large interplanar distances and thus the diffraction peaks are observed at small $|\boldsymbol{q}|$ values and small diffraction angles.

6.3.2 Specular Scans

A schematic drawing of the principle of a specular scan (also known as a $\theta/2\theta$ scan, L scan or q_z scan) is shown in Fig. 6.3. The incoming beam hits the surface under the angle θ and the diffracted beam is detected under the same outgoing angle. By increasing the angle of incidence and the angle of detection simultaneously, the scattering vector is always normal to the surface, and the varied parameter is the *diffraction angle* 2θ. Due to this geometry only net planes parallel to the surface of the sample can be detected. This arrangement probes according to Bragg's Law (6.1) different net plane distances by a variation of the diffraction angle since the used wavelength is constant. Considering this scan in the reciprocal space it is a variation of the scattering vector q in the direction of the surface normal. Usually this direction is defined as the z-direction (the surface itself as the x–y plane) and therefore the scan corresponds to an elongation of q in z-direction and is thus called q_z scan. Whenever the scattering vector equals a reciprocal space vector (and hence hits a reciprocal lattice point, indicated as small grey circles in Fig. 6.3) scattered intensity is to expect. The relation between the reciprocal and real space concerning this scan type is:

$$q_z = \frac{4\pi}{\lambda} \sin\left(\frac{2\theta}{2}\right). \tag{6.2}$$

For the sake of completeness it is to emphasize that each reciprocal space point corresponds to a certain net plane and that they are equally denoted with (hkl).

The evaluation of specular scans, in the context of thin organic films, is basically based on calculated powder patterns of crystals with known crystal structure. As shown below, in most of the investigated specimens the structures of the organic nanostructures are identical to the known single crystal structures. The powder patterns can be simulated with the software *POWDERCELL 2.3* [93]. The position of the experimentally observed diffraction angles (or $|q|$ values) can be compared with the simulations and thus crystal phases can be identified. If the measurement shows a superimposed pattern of different crystal structures, polymorph phases are present in the thin organic film. Unfortunately the intensities are not really assignable to experimentally observed intensities, since the calculation is always based on randomly oriented crystallites (three-dimensional powder). In case of a deviation of the measured intensities from the simulated ones, a preferred alignment is most probably present. Quantitative phase analyses of organic nanostructures are extremely difficult to reasonably perform since the integral intensity of the three-dimensional reciprocal lattice point has to be determined which is dependent on the in-plane mosaicity, the out-of-plane mosaicity and the crystal size.

6.3.3 Rocking Curves

As shown in Fig. 6.4 organic crystallites are not always perfectly aligned. In the depicted case there are crystallites with the same out-of-plane orientation but small deviations from the perfect alignment. This is also denoted as out-of-plane mosaicity whereby a smaller mocaicity means a better alignment. The rocking curve measurement is the method to investigate the mosaicity. By a simultaneous variation of the incident angle (θ_{in}) and the outgoing angle (θ_{out}) and keeping the diffraction angle constant ($2\theta = \theta_{in} + \theta_{out}$) the

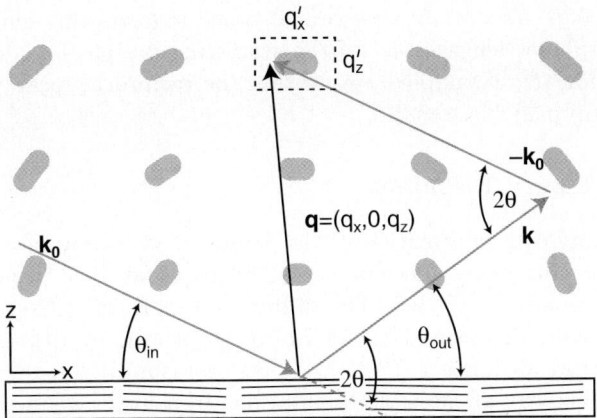

Fig. 6.4. The rocking is performed by increasing the incident angle and decreasing the outgoing angle so that the diffraction angle is constant. The alignment of a certain net plane is probed. In the reciprocal space the measurement corresponds to a q_x scan at an approximately fixed q_z. The *dotted square* indicates a reciprocal space map that can be obtained by performing several rocking curves at slightly different q_z values

alignment of a certain net plane is probed. The better the alignment of the crystallites the smaller is the width of the rocking curve. Please note that randomly distributed crystals (three-dimensional powder) would not give a peak in a rocking curve. A possible way to illustrate rocking curves is to use the deviation ($\Delta\theta$) from the mean value as x-axis, which is defined as

$$\Delta\theta = \theta_{\text{in}} - \frac{2\theta}{2}. \tag{6.3}$$

The mosaicity in an organic thin film leads to broadening of the reciprocal lattice points on a sphere around the origin as shown in Fig. 6.4. The distance of the reciprocal lattice point to the center, which is dependent on the net plane distance, is not affected. In case of randomly oriented crystallites a reciprocal lattice point would be spread over a whole sphere. A rocking curve in reciprocal space represents a tilt of the vector $\boldsymbol{q}$ in the x-direction. Since the rocking curve is performed in a small angular range the q_z component stays rather constant and thus the scan is called q_x scan. The correlation between the measured angles and the reciprocal space in this context is given by:

$$q_x = \frac{4\pi}{\lambda} \sin\left(\frac{2\theta}{2}\right) \sin(\Delta\theta). \tag{6.4}$$

By combining the specular scan technique with the rocking curve technique a reciprocal space map can be obtained. This is a two-dimensional map with the axes q_x and q_z and therefore gives information about both, the crystalline properties of single crystallites and the mosaicity of those crystallites within the organic film. To obtain a reciprocal space map rocking curves are performed at different lengths (q_z) of the scattering vector. The dotted square in Fig. 6.4 indicates a sampled area within the reciprocal space which would lead to a reciprocal space map.

6.3.4 Pole Figure Technique

To get the complete information of the spatial distribution of crystallites in organic films pole figure measurements are required. The principle of this technique is shown in Fig. 6.5. The diffraction angle 2θ is fixed to detect a certain net plane distance. The orientation is probed by tilting the sample stepwise around an angle ψ (0 to 90°) and rotating it for $\varphi = 0\text{--}360°$ for each step. This way all possible orientations of a lattice plane are checked. Whenever the normal vector $\boldsymbol{n}$ of the specific net plane is parallel to the scattering vector diffracted intensity will be detected. In the reciprocal space a pole figure measurement corresponds to a movement of the scattering vector $\boldsymbol{q}$ over the whole upper hemisphere with constant radius. The hemisphere is scanned by tilting $\boldsymbol{q}$ stepwise and performing a complete conoidal rotation at

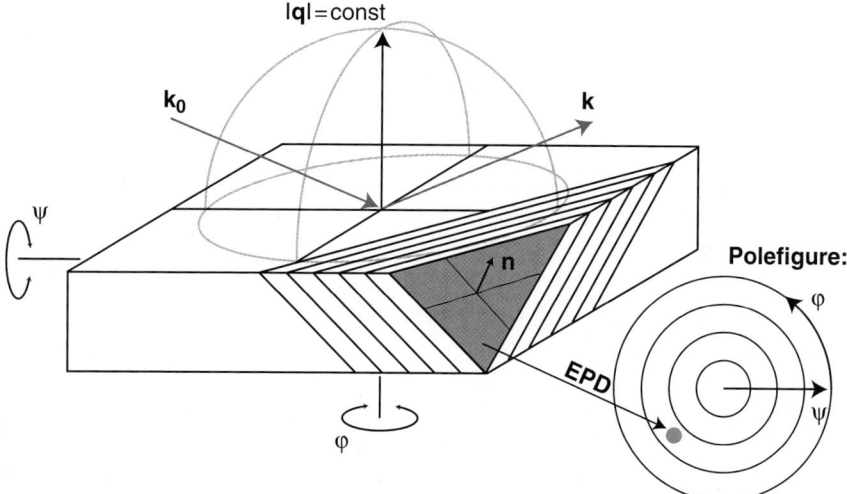

Fig. 6.5. Principle of a pole figure measurement: The scattering angle 2θ and thus the absolute value of the scattering vector $|\boldsymbol{q}|$ are kept constant at a value of a certain net plane. The spatial distribution of this particular net plane is probed by tilting the sample stepwise around ψ from 0 to 90° with a complete (360°) rotation of φ for each step. When the normal vector ($\boldsymbol{n}$) of the investigated plane is parallel to $\boldsymbol{q}$ diffracted intensity will be detected and this is recorded in a pole figure as enhanced pole density (EPD). In the reciprocal space this measurement corresponds to a stepwise tilt of $\boldsymbol{q}$ with a conoidal rotation at each step. This way the reciprocal space is probed for one hemisphere for a certain $|\boldsymbol{q}|$ (as indicated by *grey lines*) and thus for a certain net plane distance d_{hkl}

each step. Whenever it hits a reciprocal space point intensity is detected and recorded in a pole figure. The orientation of this plane is defined by the tuple (φ/ψ). Since the net plane normals are called poles, areas in the pole figure where diffracted intensity is measured are denoted as enhanced pole densities (EPD).

Commonly the use of pole figure measurements is restricted to already known crystal structures. In this case the net plane distances of strongly diffracting planes are well known and the according diffraction angles can be set. Recording a couple of pole figures for different net planes usually enables an unambiguous determination of the orientation of the crystallites in the solid and therefore in case of two or more involved crystal orientations their relative alignment. This is one of the most powerful feature of this technique and of particular importance when investigating organic nanostructures. Based on the relative alignment of the thin film crystallites to the substrate orientation the epitaxial relationships can be determined.

The evaluation of pole figures is performed using the software *STEREOPOLE* [94]. This program allows a simulation of the expected peaks based on the crystal structure and orientation. The simulations and the measured pole figures are superimposed to check whether they fit together or not. Of course they have to match in every single pole figure measured for different net planes. If congruency is achieved the spatial distribution of the crystallites is solved and the epitaxial relationships can be directly gained.

6.3.5 Surface Diffraction

When investigating very thin films, with thicknesses of few monolayers, the scattered intensity in the q_z direction becomes negligible small. Regarding the reciprocal space the reciprocal lattice points become elongated in the z-direction (reciprocal lattice rods) which is due to the low range of periodicity [95]. In the extreme case of a single monolayer the periodicity in the z-direction is completely lost and the reciprocal space points are homogeneous rods normal to the surface as illustrated in Fig. 6.6. This makes the above mentioned techniques useless. Since the loss of periodicity due to the thickness affects only the component normal to the surface, information parallel to the surface can still be obtained by performing surface diffraction. In this case the components of the scattering vector are often denoted differently. The component q_z is called the normal component $q_\perp$. Since this component is very small regarding surface diffraction, the absolute value is often approximated by the component parallel to the surface $q_\parallel$ which corresponds to $(q_x^2+q_y^2)^{1/2}$. An important fact, when performing surface diffraction, is that the incident angle should be smaller than the critical angle of total reflection which is usually below half a degree. That way a wave that propagates parallel to the surface, while being exponentially damped with penetration depth, is created [96]. This results in a very low penetration depth, raising energy in the first layers and thus increasing scattering signals from the surface.

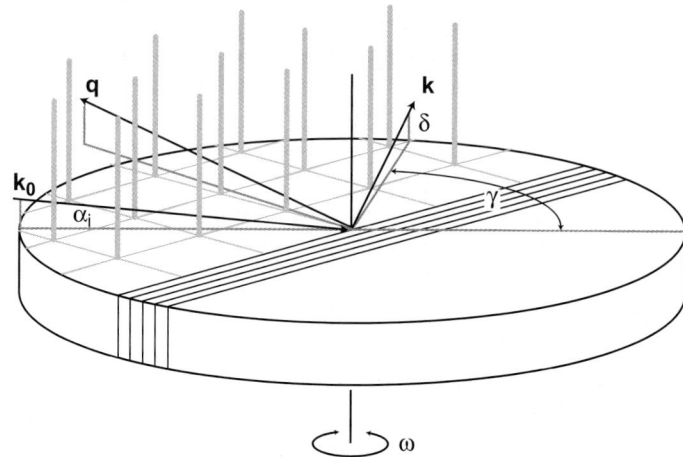

Fig. 6.6. Principle of surface diffraction: The incident angle is chosen below the critical angle of total reflection to create an intense surface wave. The diffraction angle γ is parallel to the surface, δ is kept very small, and therefore planes that are perpendicular to the surface can be detected. The reciprocal space of a single monolayer is illustrated by light grey rods. A γ/ω scan and a pure ω scan correspond to a $\theta/2\theta$ scan and a rocking curve, respectively, in specular arrangement

The principle of a surface diffraction experiment is shown in Fig. 6.6. The incident angle is smaller than the critical angle of total reflection and the diffracted beam is measured under the angle δ to the surface. The diffraction angle in this arrangement is the angle γ and the in-plane measurement γ/ω ($q_\parallel$ scan) is the equivalent to the specular scan explained in Sect. 6.3.2. To probe the in-plane alignment of the nanoaggregates the diffraction angle γ is fixed at a value corresponding to a specific plane and a pure ω scan is performed (in-plane rocking curve). This is the pendant to the rocking curve concerning out of plane arrangement as explained in Sect. 6.3.3. Additional information about the thickness and roughness can be obtained from quite homogeneous films and layers by probing the rod ($q_\perp$ by a variation of δ) [96].

6.3.6 Line Profile Analysis

The shape of a diffraction peak concerning a $|q|$ scan (i.e. $\theta/2\theta$ scan or γ/ω scan) includes numerous important information, particularly about crystallite size and microstrain. Usually line profile analysis (LPA) is performed by fitting the peaks using model functions that are composed by a mixture of a Gauss and a Lorentz (Cauchy) function. The most widely used model functions are the *Pearson VII*, the *Voigt* and the *Pseudo Voigt* functions. All these functions depend on a set of parameters such as the position of the peak maximum ($2\theta_0$), the maximum intensity (I_0), the width of the peak and a weight factor between the Lorentz and the Gauss contributions. A very important parameter is the integral breadth (β) of the resulting peak. This is defined as the width of a rectangular box having the height I_0 and the same integral intensity (area) as the peak which can be expressed as

$$\beta = \frac{1}{I_o} \int_{-\infty}^{\infty} I(2\theta)\, d2\theta. \tag{6.5}$$

In order to be able to fit a peak reasonably there are some components that have to be considered. First of all the peaks have to be background corrected. Furthermore the equipment itself, including all the settings, the geometry and the fact that an X-ray tube usually emits not solely one wavelength, causes a certain peak broadening. How to consider the different aspects in detail is explained in [97]. When all those effects are taken into account a peak can be properly fitted and the broadening of a diffraction peak can be ascribed to two reasons, the crystallite size and the microstrain. The integral breadth is thus composed by

$$\beta = \beta_{\text{size}} + \beta_{\text{strain}}, \tag{6.6}$$

where β_{size} is the peak broadening according to the crystallite size and β_{strain} represents the influence of lattice distortions. The broadening of size effects can be derived from the slit interference function using small numbers of periodicity which gives

$$\beta_{\text{size}} = \frac{\lambda}{D \cos \theta}, \tag{6.7}$$

that is known as the *Scherrer equation* with the average size parameter D. Please note that β has to be used in radians. The broadening is dependent on the diffraction angle via $1/\cos \theta$. The peak broadening of strain effects is obtained by the differentiation of Bragg's law with respect to 2θ and gives [98]

$$\beta_{\text{strain}} = 2\xi \tan \theta, \tag{6.8}$$

where ξ is the integral line breath of the strain distribution. Obviously the broadening of the peak due to microstrain is dependent on $\tan \theta$. Since the two effects of broadening show different dependencies on θ the actual contributions can be separated by analysing several peaks. Insertion of (6.7) and (6.8) in (6.6) and multiplying by $\cos \theta / \lambda$ gives

$$\beta \frac{\cos \theta}{\lambda} = \frac{1}{D} + \xi \frac{2 \sin \theta}{\lambda}, \tag{6.9a}$$

$$\beta^* = \frac{1}{D} + \xi s. \tag{6.9b}$$

The equation shows that a plot of β^* ($= \beta \cos \theta / \lambda$, using β in radians) versus s ($= 2 \sin \theta / \lambda$) should give a straight line with the slope ξ and the intercept $1/D$. Usually the abscissa of this plot is given by $s = \frac{2 \sin \theta}{\lambda} = \frac{|q|}{2\pi}$ which differs from the scattering vector only by the constant 2π. The reciprocal breadth

β^* at $s = 0$ gives the reciprocal value of the average size of the crystallites. This way to separate the influences of strain and size on the peak broadening is called Williamson–Hall analysis [97, 99].

Performing a Williamson–Hall analysis requires several peaks to analyse to be able to find the line of best fit. This makes the LPA very difficult in case of anisotropic nanoaggregates. The dimensions of the crystallites vary with the direction and thus conclusions drawn from LPA are only valid in the actual direction of the scattering vector q. Therefore only series of higher order diffraction peaks can be reasonably taken into account when applying a Williamson–Hall analysis. This again is commonly restricted to a single diffraction series concerning organic nanoaggregates, namely the poles normal to the herringbone layers (typically $h00$ or $00l$ diffractions). The size and strain of the nanoaggregates can thus be determined only in the direction perpendicular to the herringbone layers. This direction is typically normal to the surface in case of terraced island growth where the molecules are standing on the surface. The diffraction peaks are detected in a specular scan and q is thus pointing out of the sample. The obtained information is thus the vertical size and strain normal to the surface. In contrast to this arrangement needle like growth is typically formed by crystallites with the herringbone layer parallel to the needle direction intersecting the surface plane inclining a rather high angle. Therefore the peak series will be detected by an in-plane scan or under a rather small out-of-plane angle δ (compare Fig. 6.6). This means that the obtained information is about lateral size and strain, more precisely normal to the needle direction and almost parallel to the surface. Finally it is to say that the mentioned method is a good method to compare the crystallite sizes to observed morphologies and thus be able to decide whether the aggregates are polycrystalline or rather single crystalline in the particular direction. However, the theory is based on a lot of assumptions (especially of the appearance of the crystallites) and thus certainly not the most reliable way to determine absolute values of crystallite sizes. For further methods of line profile analysis the reader is referred to [97].

6.3.7 Transmission Electron Microscopy

As shown X-ray diffraction is a powerful tool for structural analyses of thin films with a high angular precision regarding both, the diffraction angles and the alignment angles. Furthermore it allows almost destruction free investigations and a simultaneous characterization of both, the substrate and the thin film. Since X-ray diffraction is an integral method over a large surface area it fails when investigating individual crystallites of the thin film. These small structures are not individually accessible by X-ray diffraction. Therefore, TEM and diffraction are a perfect complementary method to X-ray investigations. Although the crystal structure and its alignment cannot be determined with high precision TEM allows a simultaneous investigation of the morphology

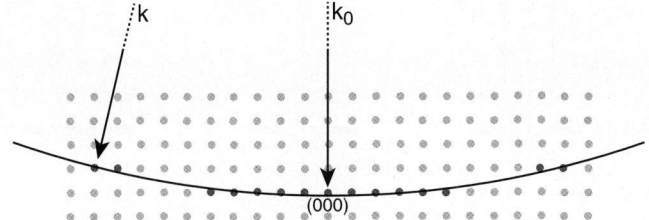

Fig. 6.7. The principle of electron diffraction: since the wave vector k is very long with respect to typical distances in the reciprocal space the Ewald sphere hits numerous reciprocal space points with different scattering vectors simultaneously. The recorded diffractions are all deriving from reciprocal lattice points that correspond to planes almost perpendicular to the surface

and the according crystal order with high spatial resolution. This means that the present alignments can be determined with X-ray diffraction with high precision and can be assigned to different morphological aggregates by the use of TEM.

The theoretical thoughts about diffraction are initially associated to the interaction of X-rays with a crystal but due to the wave-particle-dualism of electrons most of the results can directly be applied unchanged to electron diffraction. In a transmission electron microscope typical electron energies are 100–200 keV which corresponds to wavelengths of several picometers. This means that the wave vector k is very long in comparison to typical distances in the reciprocal space. As depicted in Fig. 6.7 the Ewald sphere hits numerous reciprocal space points and therefore all those diffractions (with different scattering vectors q) can be detected simultaneously. Since the hit reciprocal space points correspond to diffracting planes that are almost perpendicular to the surface, the diffraction pattern gives also information about the in-plane alignment. As the name of the instrument already includes, this technique works with transmitted electrons and thus the sample has to be sufficiently thin (a few nanometers) to work reasonably. Therefore, the organic films have to be separated from the substrate whereat a destruction of the nanoaggregates cannot be excluded. Furthermore, it is observed that the organic crystals are instable under the electron beam, even if properly prepared to avoid charging, some of the crystallites change or lose their crystallographic order within seconds. This makes TEM investigations of organic nanostructures challenging.

A schematic view of the electron ray path in the TEM is given in Fig. 6.8 for three different operation modes. The first one is the bright field mode (BF) which simply gives an enlarged image of the specimen and thus a morphological information. The enlargement is based on simple optical considerations. The diffraction pattern of the same illuminated part of the specimen is obtained by changing the strength of the intermediate lens so that the

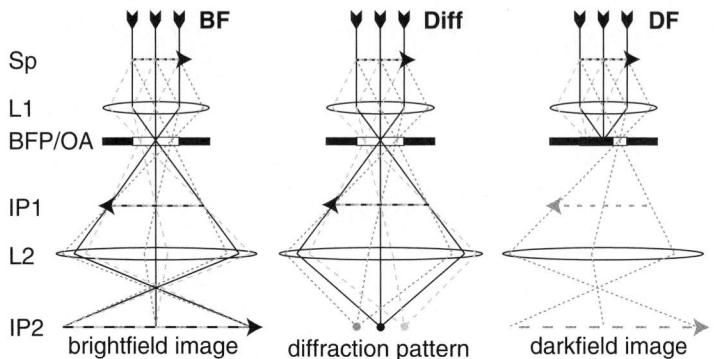

Fig. 6.8. The path of the electron beam in a transmission electron microscope for three different operation modes: bright field (BF), diffraction (Diff) and dark field (DF) mode. The bright field mode gives an enlarged image of the specimen's (Sp) morphology whereas the diffraction pattern, and thus the crystal order, is observed in diffraction mode. To switch between these two modes the strength of the intermediate lens (L2) has to be changed. By selecting a certain diffraction spot in the back focal plane (BFP) with the objective aperture (OA) a dark field image is obtained which is a real image of the specimen but only parts that are contributing to the specific diffraction spot and thus equally ordered parts are observed (L1: objective lens, IP1: image plane 1, IP2: image plane 2)

back focal plane is represented on the image plane two. Since the illuminated part can be chosen very small the orientation of individual crystallites can be measured. The specimen in Fig. 6.8 is composed by differently bright parts allegorizing differently scattering parts of the sample leading to individual diffraction spots. By selecting a diffraction spot in the back focal plane by the use of an aperture, the primary beam is masked and the image is dark. Only parts that contribute to the selected spot are detected in this mode. The result is a dark field image where all crystallites that are equally aligned and whose alignment contributes to the diffraction spot will be bright and everything else will be dark and thus changes in the crystallite order will be seen as an abrupt change in intensity in the image. A combination of the presented techniques allows an assignment of the crystal structure to different morphological aggregates and gives information about the quality of the crystallites, e.g. whether it is about small single crystals or not.

The evaluation of the diffraction patterns is performed using the software *cerius*^{2} which is a commercially available program developed by *ACCELRYS*. The diffraction patterns are simulated based on the crystal structure and the preliminarily characterized orientations by X-ray studies. This emphasizes the importance of using both complementary methods to get a more detailed view of the structural behaviour of organic nanoaggregates.

6.4 Crystallographic Order within Nanoaggregates

When investigating the crystal order within nanoaggregates first of all it is to clarify whether the thin film is actually crystalline, simply amorphous or even a mixture of both. If crystallites are present, their crystal structure is of interest, whether they show known single crystal phases or they form individual thin film phases. Having worked out the crystal structure of the crystallites, their alignment is to find out. In a first step it is to check if the crystallites show a preferred contact plane, which is a crystallographic net plane that is parallel to the surface of the substrate. Thin films that are composed by crystals with a certain contact plane while having them randomly oriented regarding their *in-plane* order are also denoted as two-dimensional powders. As a result, both the specular scan and the rocking curve, lead to a sharp peak, whereas the EPDs in the pole figures are smeared out over the whole φ-range so that they create rings in the pole figure. In some cases the crystallite size can be obtained in specific directions.

Furthermore, films are prepared that show both, an *out-of-plane* alignment as well as an *in-plane* alignment (azimuthal alignment). In this case the film is called epitaxially ordered and the epitaxial relationships can be determined. To clarify what is epitaxy we go along with the definition of [100]: *Any structure-dependent intergrowth (overgrowth) of two chemically and/or structurally different crystalline or subcrystalline phases is called epitaxy.* It is to say that the driving forces of epitaxial growth in case of organic materials can differ from those of inorganic materials. Commonly inorganic epitaxy is understood in terms of a lattice mismatch of the two involved materials. In the case of organic materials where van der Waals interactions play a major role this lattice matching condition is drastically relaxed and other mechanisms are found to be responsible for the alignment [101, 102]. Three different types of epitaxial alignment are regularly observed. The first one is, in consistency with the inorganic counterpart, the commensurable type where the two involved surface lattices align regarding to a minimization of the lattice mismatch. The second type of alignment is an alignment where distinct directions of the contact plane align parallel to distinct directions of the substrate surface plane. Concerning rodlike molecules an additionally observed mechanism is the alignment of initially adsorbed molecules. They align with their long molecule axes parallel to distinct surface directions, do not change this alignment during crystal growth and thus determine the alignment of the crystallites.

The knowledge about the crystalline order is not enough to understand the film growth and be able to control it. The main topic of structural investigations is to determine the above-mentioned cases of alignment dependent on the used materials and the applied growth conditions. Any kind of change regarding the crystallographic order (type of epitaxy, crystal alignment, ratios of differently aligned crystallites, crystal sizes, polymorphism, crystallinity) associated with a change of the growth conditions (such as growth temperature, growth time, surface preparation, equipment) or a change of either the substrate material or even the film material can give important information to understand and control the film growth. In the following some different kinds of organic thin films are presented and accompanied by illustrative examples to give an idea of the varieties of crystalline order according to different systems and of the possibility of controlling the thin film formation.

6.4.1 Out-of-Plane Order

The out-of-plane order gives information about preferred contact planes and the quality of the alignment of the associated crystallites. This information is obtained by specular scans and rocking curves, respectively, where the scattering vector is perpendicular to the surface and thus it is called out-of-plane order. Specular scans yield the planes parallel to the surface. The alignment of the individual net planes is checked by rocking curves and if a peak in the rocking curve is observed a preferred alignment is found. Since there are only few net planes with sufficient scattering intensity pole figure measurements are necessary to solve the complete spatial distribution unambiguously. Nevertheless it is empirically found that in general the contact planes are strongly diffracting planes, mostly even cleavage planes.[1]

As an example the growth of sexiphenyl on KCl(100) is chosen [103,104]. The specular scan of a thin film is shown in Fig. 6.9a. Four peaks are detected

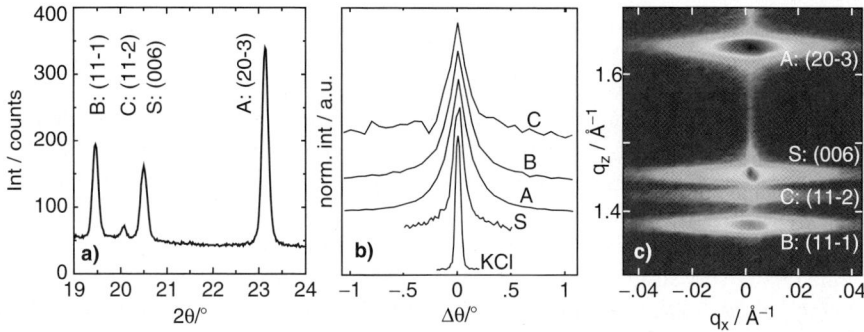

Fig. 6.9. Out-of-plane X-ray studies of sexiphenyl nanoaggregates grown by hot wall epitaxy. The specular scan (**a**) gives the crystallographic planes parallel to the surface; four orientations are found and denoted by capital letters. The rocking curves (**b**) reveal that each single orientation is well aligned, actually they are not much broader than the KCl substrate. A combination of a series of rocking curves at different q_z values yields a space map (**c**) which includes the information of (**a**) and (**b**)

---

[1] A cleavage plane is a densely packed plane in the crystal that seems to be suited for cleavage or which is likely to form a side during crystal growth.

that can be referred to the diffractions 11–1, 11–2, 006 and 20–3. The 200 diffraction of KCl is not shown explicitly. The according rocking curves, including the rocking curve of KCl(200), are shown in Fig. 6.9b. It is seen that the substrate shows a very sharp peak, as expected from a single crystal. Furthermore, also the rocking curves of the organic material are rather sharp (only four times the width of the substrate), especially the width of the 006 peak is only twice the substrate ones. The results of specular scans and rocking curves yield that the observed crystallites show different contact planes and that those differently aligned crystallites are well oriented. One can say that there are different orientations present each with a certain net plane well aligned parallel to the substrate surface. To avoid mixing up orientations and diffraction peaks the orientations are usually denoted arbitrarily. In the present case crystallites with the (006) plane parallel to the surface are denoted as orientation S which is related to the molecular order with standing molecules. All other observed orientations are composed by lying molecules and are simply denoted as orientations A, B and C according to the planes (20–3), (11–1) and (11–2) parallel to the surface, respectively. A simultaneous view of all orientations is shown in Fig. 6.9c by a reciprocal space map. The intensity is scaled logarithmically. The four orientations lead to four different peaks at certain q_z values while the q_x broadening represents the quality of the alignment according to the individual orientations. It is to emphasize that these experiments give exclusively information about the out-of-plane order which means, about the planes parallel to the surface and their quality of alignment. So far no information about a possible in-plane alignment (or azimuthal alignment) is obtained.

A schematic drawing of the observed orientations is given in Fig. 6.10. Orientation S is composed by almost upright standing molecules (end-on) with the herringbone layer parallel to the surface. The molecules are represented by the three lowest carbon atoms of the terminal phenyl ring. Orientations A, B and C are composed by differently aligned edge-on lying molecules. In the case of orientation A the molecule long axes are completely parallel to the surface while molecules within one molecular row represent one herringbone layer that points out of the surface. Concerning this orientations all molecules within one row are equally tilted with respect to the surface while every other row of molecules is tiled mirror wise (the other edge of the molecules is down). This explains the large lattice constant of the surface unit cell regarding this orientation. In the cases of orientations B and C which are very similar to each other the molecule long axes are slightly tilted to the surface. Again a row of molecules represents a herringbone layer that points out of the surface. In contrast to orientation A adjacent rows of molecules are equal to each other but there are alternate tilt angles of the molecules within one row. The resulting surface unit cells of the individual orientations are indicated by black quadrangles.

As mentioned above the results from such experiments do not yield a complete determination of the existing orientations. Additional pole figure

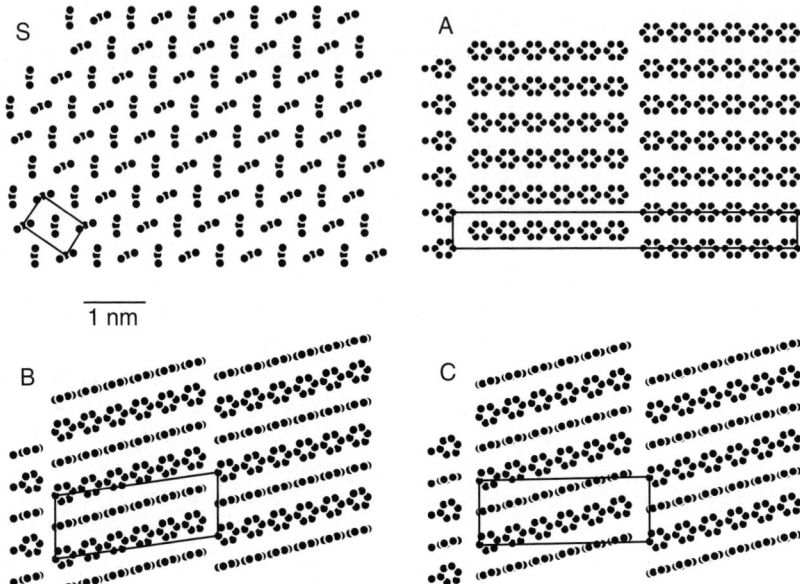

Fig. 6.10. Top views of the observed orientations of sexiphenyl on KCl(100). Crystallites with the contact plane (00l) are denoted as orientation S. They are formed by almost standing molecules which are represented by the lowest three carbon atoms of the lowest phenyl ring. Orientations A, B and C ((20–3), (11–1) and (11–2) parallel to the surface) are formed by differently lying molecules. Different dimensions of the surface unit cells of the individual orientations are indicated by *solid lines*

measurements confirm these results and show that there are no more other orientations present concerning these specimens. For the sake of completeness it is to say that another contact plane is regularly observed concerning sexiphenyl which is not observed on KCl. This orientation is formed by flat-on molecules with the contact plane (21–3) and denoted as orientation D.

6.4.2 In-Plane Order (Organic Epitaxy)

To determine the complete orientation of thin film crystallites on a substrate, a set of pole figure measurements is necessary. This gives the complete spatial distribution of the involved crystallites. If there is a crystal order of the thin film crystallites, the relative alignment to the substrate is called epitaxial relationship. Before starting the illustration of epitaxial order, the denotation of epitaxial relationships should be clear which is explained in the following.

Grammar of Epitaxy

To finally write down an observed epitaxial relationship there are mainly two different types of denotation in use. Both of them describe the relative crystallographic order unambiguously by giving two statements, the out-of-plane order and the in-plane order. In both cases the first statement contents the substrate surface plane and the contact plane which are parallel to each other and thus this statement defines the out-of-plane order. Consequently the second statement contents the in-plane alignment but there are different ways of its description. The first way is to give two parallel crystal directions of the substrate and the film which are situated in the substrate surface plane and the contact plane, respectively.[2] The combination of parallel planes and two parallel directions within these planes determines the relative alignment of the two involved crystals unambiguously. The denotations look like

$$\text{orientation X:} (hkl)_{\text{sub}} || (hkl)_{\text{film}} \text{ and } (uvw)_{\text{sub}} || (uvw)_{\text{film}}. \quad (6.10)$$

By convention low indexed directions that are parallel to each other are chosen but as one immediately sees the choice is not mandatorily unique. This means that there are maybe different parallel directions which describe the same alignment which makes a direct comparison of orientations denoted this way sometimes difficult.

The second description of the in-plane order is based on the knowledge of the surface unit cells of the substrate surface plane (a_1, a_2, α) as well as of the contact plane (b_1, b_2, β). If these surface unit cells are known the declaration of the angle ρ between the two base vectors a_1 and b_1 defines the relative alignment unambiguously. Based on these numbers the epitaxial matrix C can be determined which is a useful instrument to directly evaluate the type of commensurability of the alignment [40]. This matrix is defined by

$$\begin{bmatrix} b_1 \\ b_2 \end{bmatrix} = C \cdot \begin{bmatrix} a_1 \\ a_2 \end{bmatrix} = \begin{bmatrix} C_{11} & C_{12} \\ C_{21} & C_{22} \end{bmatrix} \cdot \begin{bmatrix} a_1 \\ a_2 \end{bmatrix}, \quad (6.11a)$$

$$C_{11} = \frac{b_1 \sin(\alpha - \rho)}{a_1 \sin(\alpha)}, \quad (6.11b)$$

$$C_{12} = \frac{b_1 \sin(\rho)}{a_2 \sin(\alpha)}, \quad (6.11c)$$

$$C_{21} = \frac{b_2 \sin(\alpha - \rho - \beta)}{a_1 \sin(\alpha)}, \quad (6.11d)$$

$$C_{22} = \frac{b_2 \sin(\beta + \rho)}{a_2 \sin(\alpha)}. \quad (6.11e)$$

Concerning the epitaxial matrix C there are some rules which give direct information about the lattice commensurability of a certain alignment. If all elements of C are integer values the alignment is a *point-on-point-commensurism* (POP) which means that each lattice point of the overlying contact plane hits a lattice point of the substrate surface. Another good lattice

---

[2] A certain crystal direction $[uvw]$ is situated in a certain crystal plane (hkl) if the inner product is zero: $uh + vk + wl = 0$.

match is present in the case that one whole column of C consists of integers.[3] This means that it is about a *point-on-line-coincidence-I* (POL-I) where each surface lattice point of the film hits a primitive lattice line of the subjacent surface lattice. Furthermore there is another defined coincidence where all elements of C are rational numbers which is called *point-on-line-coincidence-II* (POL-II) where only some lattice points of the film surface lattice hit a primitive substrate surface lattice line. In this case a supercell, which is an integer multiple of the initial film surface unit cell, can be constructed to get POL-I. A fourth type of lattice match which cannot be directly seen by the matrix C is described in [105]. This describes a lattice match where a certain lattice line of the overlying lattice coincides with a certain lattice line of the subjacent lattice and thus each overlying lattice point hits a certain surface lattice line. This type of lattice match is called *line-on-line-coincidence* (LOL) and the main difference to POL is that each lattice point of the contact plane hits a certain substrate surface lattice line that differs from a primitive one. However, the first two cases (POP and POL-I) give definitely two easily understandable explanations for epitaxial growth and can be directly seen by the epitaxial matrix.

In the case of POL-II the evaluation has to be handled with care. Since all numbers are naturally afflicted by errors it is very easy to interpret all numbers as rational numbers and declare an observed alignment as POL-II. It is to emphasize that due to the fact that the lattice match condition is relaxed in case of organic epitaxy the epitaxial matrix need not lead to the explanation for a certain alignment. As already mentioned the epitaxy might also be driven by an alignment of certain crystal directions or even by the alignment of single adsorbed rodlike molecules.

Organic Epitaxy of Sexiphenyl on TiO$_2$(110)

To give an illustrative example of organic epitaxy the material combination sexiphenyl on TiO$_2$(110) (rutile) is chosen for illustration [106]. The TiO$_2$ surface is twofold symmetric consisting of rows of oxygen atoms leading to an epitaxial alignment of the rodlike molecules. The epitaxial relationships are determined by a set of pole figures. Two of these pole figures are shown in Fig. 6.11 where the rare measurements are depicted on the left side and the superposition with the simulations on the right side. The pole figure A is measured at $2\theta = 42.4°$ to detect 21-3 diffractions of sexiphenyl. The observed enhanced pole densities (EPDs) are assigned to the 21-3 diffractions of crystallites with their (20-3) and (-203) planes parallel to the surface. The determined in-plane alignment is cross checked by the second pole figure measured at $2\theta = 30°$. The numerous EPDs fit perfectly with the simulated data, furthermore all of them can be explained by the found orientations and the

---

[3] In case of hexagonal substrate surface lattices ($\alpha = 60°$) coincidence is also present if the sum of both rows ($C_{11} + C_{12}$ and $C_{21} + C_{22}$) are integers. In case of the choice $\alpha = 120°$ the differences ($C_{11} - C_{12}$ and $C_{21} - C_{22}$) have to be considered.

TiO$_2$(110) single crystal. Therefore the found alignments are determined relatively to the substrate which gives the epitaxial relationships. It is to say that the orientations (20–3) and (–203) are equivalent to each other with an in-plane rotation of 180°. Therefore the two observed orientations can be summarized in one orientation with a twofold symmetry which is induced by the twofold symmetric surface. In this case the orientation (20–3) is chosen arbitrarily and denoted as orientation A. Two parallel low-indexed directions of the substrate and the film that are situated parallel to the surface are shown in Fig. 6.11 by a grey arrow. Thus the epitaxial relationship is determined to:

$$A : (110)_{\text{TiO2}} || (20-3)_{6P} \text{ and } [1-10]_{\text{TiO2}} || [0 \pm 10]_{6P}. \qquad (6.12)$$

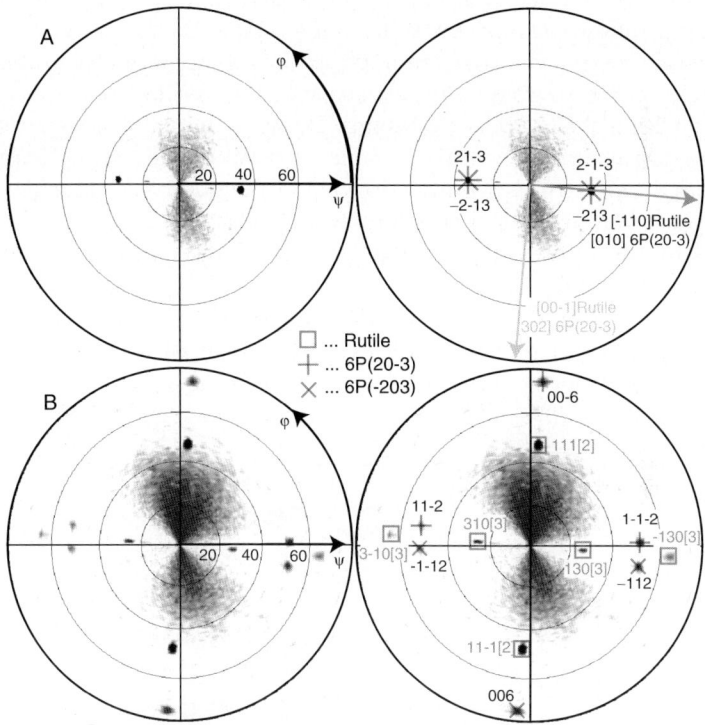

Fig. 6.11. Set of pole figures of sexiphenyl on TiO$_2$(110) measured at (**a**) $2\theta = 42.4°$ and (**b**) $2\theta = 30°$ to probe {21–3} and {11–2}, respectively. The experimental results are shown on the left side while the simulations are superimposed to them on the right side. As seen the indexation of the first pole figure is not unambiguous because the found EPDs are composed by a superposition of the orientations (20–3) and (−203). The indexation of the second pole figure is unambiguous and reveals additionally the orientation of the TiO$_2$ substrate and thus the epitaxial relationship. This is indicated by two parallel directions which are represented by the *grey arrow* in the upper right pole figure

The twofold symmetry is included by the ± in the direction of sexiphenyl. To demonstrate the problem concerning this way of description an equivalent denotation is given representing the same epitaxial relationship. This occurs because another two directions are parallel to each other as indicated by the light grey arrow in Fig. 6.11. In this case the twofold symmetry is included by the ± in the crystal direction of TiO_2:

$$A: (110)_{TiO2} \parallel (20-3)_{6P} \,\&\, [00 \pm 1]_{TiO2} \parallel [302]_{6P} \,. \qquad (6.13)$$

Based on these results a schematic drawing of both, the film orientation and the substrate surface can be realized. At this point it is to say that the drawing represents the relative alignment of the two involved bulk structures and not necessarily the interfacial structure. This is due to the fact that with out-of-plane X-ray diffraction only crystallites of a certain thickness can be detected and thus the interfacial structure might differ due to an undetectable monolayer order or due to strain and relaxation in the first few layers close to the interface within the organic crystallites. However, in many cases a reason for a certain alignment can be directly concluded out of the schematic drawing.

Figure 6.12 shows a schematic drawing of the observed epitaxial relationship between the sexiphenyl crystals and the substrate. The substrate is represented by the top most layer that consists of oxygen rows parallel to the $[011]TiO_2$ crystal direction. The organic crystals are represented by one molecular layer regarding the contact plane (20–3). The two directions that are found parallel to each other are indicated by vertical arrows (compare (6.12)). The according surface unit cells are listed in Table 6.1.

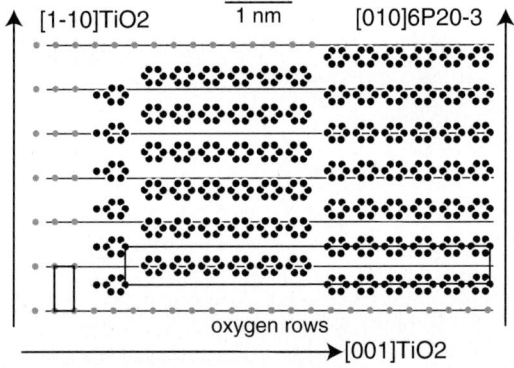

Fig. 6.12. The molecular alignment of the observed epitaxial order of sexiphenyl on TiO_2. The parallel planes are $(110)TiO_2$ and (20–3)6P, the parallel crystal directions are $[1–10]TiO_2$ and [010]6P. The surface unit cells are represented by *black squares* whereby the surface unit vectors of the particular surface unit cell are parallel to each other. In this case no lattice match is observed but the long molecule axes are aligned parallel to the oxygen rows of the TiO_2 surface

Table 6.1. Surface unit cells of the observed orientation and the substrate based on the bulk structures

	TiO$_2$(110)	6P(20-3)
a_1 (Å)	2.959	54.60
a_2 (Å)	6.497	5.568
α (deg)	90	90

Based on these surface unit cells and the knowledge of the angle between the two surface unit cell vectors $a_{1,\text{TiO2}}$ and $b_{1,6\text{P}(20-3)}$ which is in this case $\rho = 0°$ the epitaxial matrix can be calculated (6.11):

$$C = \begin{bmatrix} 18.45 & 0 \\ 0 & 0.86 \end{bmatrix}. \tag{6.14}$$

According to this epitaxial matrix no lattice match is observed but looking at the schematic drawing there are two outstanding parallelisms that could explain the alignment. On the one hand it is seen that the herringbone layers that are represented by molecular rows along the [010]6P direction are parallel to the [1−10]TiO$_2$ direction. This could mean that the epitaxy is driven by an alignment of distinct crystal directions parallel to each other. On the other hand the molecule long axes are found parallel to the oxygen rows on the TiO$_2$ surface (Fig. 6.12). This seems to be a more reasonable explanation for the found alignment, which means the epitaxy is driven by the alignment of initially adsorbed and aligned molecules. Especially in this case strain and relaxation is the first layers of the organic material cannot be excluded. However, concerning these investigation methods neither the existence of strain and relaxation nor their absence can be proven, but anyway, the reason for the found alignment can be explained.

6.4.3 Relation Between Crystal Structure and Film Morphology

In general not only the crystal structure and the alignment are of interest but also the thin film morphology. This is typically standardly probed by atomic force microscopy (AFM). However, an assignment of the crystalline order to the nanoaggregates is not always possible by a simple combination of XRD and AFM. This is related to the fact that usually there are differently oriented crystallites present side by side. Sometimes, as shown in Sect. 6.4.1 considering the example sexiphenyl on KCl(100), even different contact planes are detected. In the latter case four different contact planes were detected by X-ray investigations and additionally AFM images reveal that there are four different types of morphologies [103]. These are shown in Fig. 6.13 where three different types of needles are clearly visible, namely long needles (A), thin needles (B) and thick needles (C). While the long needles are always

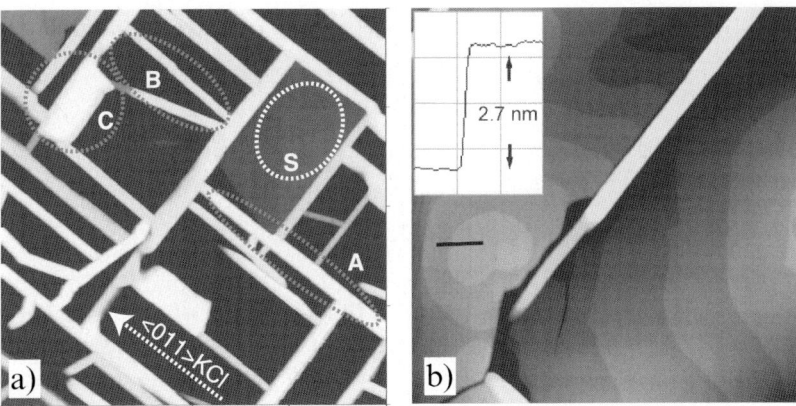

Fig. 6.13. Atomic force microscopy images of a sexiphenyl thin film on KCl(100) grown by hot wall epitaxy [103]. Four different morphologies are observed in (**a**) which is a $10 \times 10\,\mu m^2$ image (height scale 500 nm): long needles (A), thin needles (B), thick needles (C) and terraced islands (S). A more detailed view of the terraced islands is shown in the $5 \times 5\,\mu m^2$ image (**b**) with a height scale of 100 nm. The step height of one single terrace is determined to 2.7 nm which corresponds to one layer of standing molecules. Reprinted with permission from [103]

aligned parallel to the $[011]_{KCl}$ direction the thin and thick needles are found parallel as well as tilted to it (17° tiltangle). Additionally terraced islands are found as a fourth type of morphology. This suggests itself that the different contact planes form differently shaped nanoaggregates. As shown in Fig. 6.13b the step height of the terraced islands is 2.7 nm which equals the length of standing molecules which corresponds to the contact plane (00l). Therefore in this particular case a reasonable assignment of the orientation S to the terraced island morphology can be made without any other proof. To be sure about the assignment, and moreover to be able to assign the different crystal orientations to the different types of needles, transmission electron microscopy and diffraction are required.

As shown in Fig. 6.14 transmission electron microscopy enables a simultaneous investigation of both, the morphology and the crystal order, with high spatial resolution. Thus separate nanoaggregates can be investigated in terms of their crystal order. As assumed above the diffraction pattern of terraced islands corresponds to crystals with the contact plane (00l) which correspond to standing molecules and the orientation is denoted as orientation S. Furthermore the other three crystal orientations A (20–3), B (11–1) and C (11–2) can be assigned to differently shaped needles. The diffraction pattern deriving from long needles corresponds to orientation A with the contact plane (20–3). The thin needles show a diffraction pattern that derives from the contact plane (11–1) and are thus assigned to orientation B. Finally orientation C, contact plane (11–2), leads to a diffraction pattern as observed in the case of thick short needles and are thus assigned to each other.

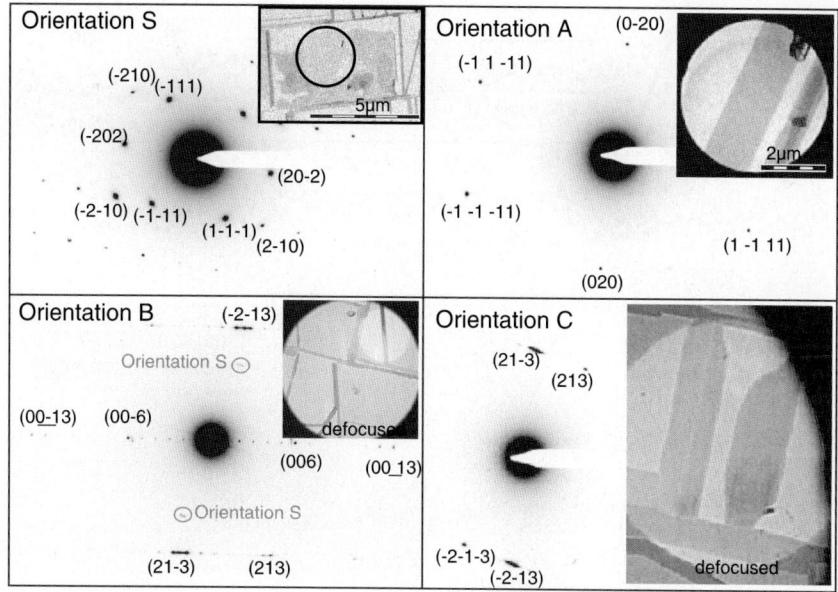

Fig. 6.14. Transmission electron microscopy images and diffraction patterns of different morphologies of sexiphenyl on KCl(100) [103]. Terraced islands lead to the diffraction pattern of orientation S which corresponds to the contact plane (00l). Orientation A, which corresponds to contact plane (20–3) is assigned to long needles. Contact plane (11–1) (orientation B) leads to thin needles and (11–2) (orientation C) yields thick short needles. Reprinted with permission from [103]

This method gives complementary information to the X-ray investigation and cannot substitute the latter. The interpretation of the electron diffraction pattern is much easier if they are based on X-ray results and moreover the high angular precision and the epitaxial relationships cannot be attained with TEM.

As already mentioned above the TEM yields an unambiguous assignment of morphology and crystal structure. However, this proof is not always necessary to be able to assign a certain crystal order to a morphology. As already shown a perfect hint for crystallites of standing molecules is a terraced like growth with single step heights in the order of the length of a molecule. Moreover, there are also material combinations where a comparison of the morphology to the found crystal orientations makes a further investigation unessential. This is illustrated considering the example sexithiophene on TiO_2. In this case AFM investigations yield a needle like growth with two different needle directions symmetrically tiled by 21.5° to the $[1-10]TiO_2$ direction. Structural investigations reveal that the (010) plane of sexithiophene is parallel to the surface and that the in-plane order contains two alignments symmetrical around $[1-10]TiO_2$. A comparison of the determined molecular orientation and the AFM images directly allows an assignment

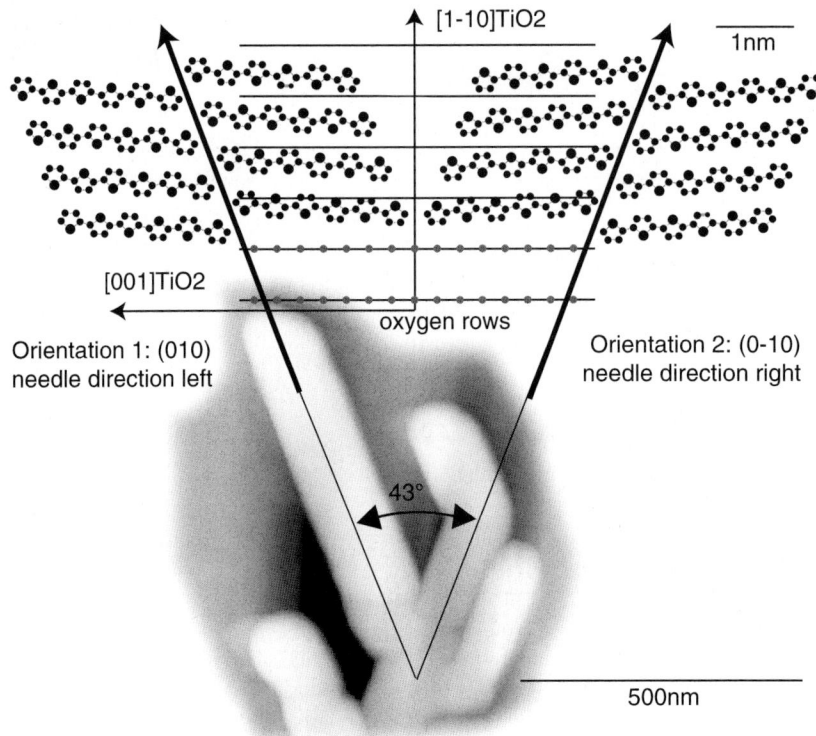

Fig. 6.15. A comparison of the observed molecular alignments of sexithiophene on TiO$_2$(110) with the morphology imaged by atomic force microscopy. An assignment of the morphology to the crystal orientation is possible

of the two crystal alignments to the two found needle directions. This is illustrated in Fig. 6.15. The horizontal lines represent the oxygen rows of the TiO$_2$(110) surface in the [001]TiO$_2$ direction, while the vertical arrow points in the direction [1–10]TiO$_2$. The molecular alignment which is determined by X-ray investigations reveals molecular layers symmetrically tilted around the [1–10]TiO$_2$ direction. These layers determine the needle growth direction which is indicated by two bold black arrows. This is a reasonable conclusion when comparing this alignments with the according AFM image on the bottom of Fig. 6.15.

6.4.4 Crystallite Size

Concerning crystallite size it is to remind that the molecular crystals of rodlike molecules are formed by herringbone layers. Single molecules achieve the highest energy gain when attaching at the edge of a herringbone layer. This way a single crystal is typically enlarged in the two dimensions of the herringbone layer while the elongation perpendicular to it is comparably small. This explains the highly anisotropic appearance of the nanoaggregates. Furthermore this behaviour leads to the typically observed morphologies of terraced islands and needles. The islands are formed by standing molecules where the herringbone layer is parallel to the surface and thus the aggregates are enlarged in the two surface dimensions (lateral size) while the height of islands is typically rather small. Needles are formed by lying molecules where the herringbone layer intersects the surface approximately perpendicular to it. Since the molecules on the surface are most likely attached at the edge of a herringbone layer the aggregate grows predominantly in one direction. This results in lateral sizes that are large in one surface dimension while small in the other surface dimension. The vertical size of needle like nanoaggregates is typically larger than this of simultaneously grown terraced islands.

Terraced Islands

An illustrative method to determine single crystalline domain sizes is dark field imaging by TEM. With this technique a particular crystal orientation can be selected and directly imaged microscopically. In this image crystals of the selected orientation appear bright while anything else is dark. A sudden change of the crystal order will be detected as a sudden loss of intensity of the signal in the image. As an example for terraced islands again the nanoaggregates of sexiphenyl on KCl(100) is chosen [104]. Figure 6.16a shows a bright field TEM image of simultaneously grown sexiphenyl needles and islands. The diffraction pattern (not shown explicitly) shows a combination of the actual diffraction patterns of the observed crystal orientations (compare Fig. 6.14). Recording a dark field image by selecting a strong diffraction spot assigned to orientation S leads to the image presented in Fig. 6.16b. The island is bright while all needles are dark. Since the island is uniformly bright and does not show any

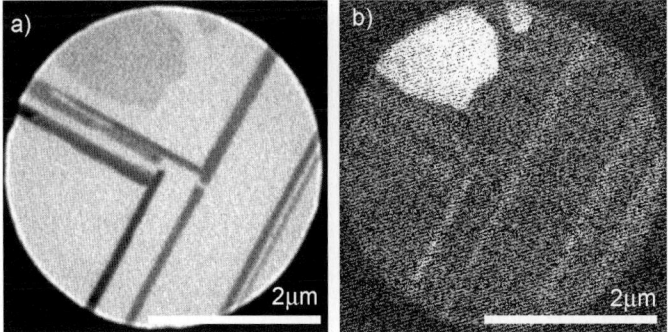

Fig. 6.16. A bright field transmission electron microscopy image (**a**) and a dark field image of the same sample area (**b**). The bright area in (**b**) reflects the single crystalline nature of the terraced island. Reprinted with permission from [104]

internal boarders, except at the edge of the island itself, it can be concluded that the whole island is a sexiphenyl single crystal. The lateral dimensions of this single crystal can be directly obtained by the dark field image.

Needle Like Growth

Concerning needles both, the length of nanoneedles such as the size of the crystallites forming needles, is of large interest since grain boundaries show a large influence on physical properties. Nanoneedles of a certain length are not necessarily single crystalline and thus the information of the crystallite size in comparison to the morphology is extremely important. Figure 6.17 shows a comparison of two sexiphenyl needles grown on mica(001) and KCl(100). It is to say that the image of the needle grown on KCl(100) corresponds to the same part of the sample presented in Fig. 6.16. Furthermore both needles are equally oriented having the contact plane (11–1).

The needle grown on mica (Fig. 6.17a) shows dark and bright parts and sharp boarders between them. This means that this needle is composed by small crystallites where the lateral dimensions are rather small and were determined to a few tens of nm [107]. In contrast to this, Fig. 6.17b shows a dark field image of sexiphenyl on KCl [104]. As seen the needle grown on KCl(100) is completely bright while all the other aggregates (differently oriented needles and terraced islands) are dark. No grain boundaries are detected within this needle which suggests a single crystalline nature. The length of this needle exceeds the imaging area but single crystalline needles up to several microns are observed [104].

Furthermore the crystallite size can be probed by high resolution TEM. Figure 6.18 shows a high resolution image of a sexiphenyl needle grown on KCl(100) [104]. The observed lines within the needle parallel to the needle direction are identified as single molecular layers. The distance between the

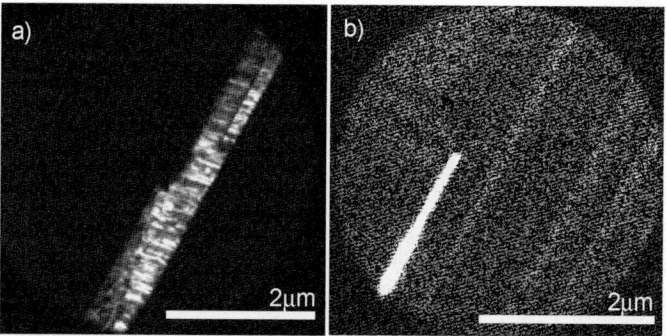

Fig. 6.17. Two TEM images recorded in dark field mode. The needle grown on mica(001) (**a**) is obviously composed by small crystallites while the particular needle grown on KCl(100) (**b**) which appears completely bright is of single crystalline nature

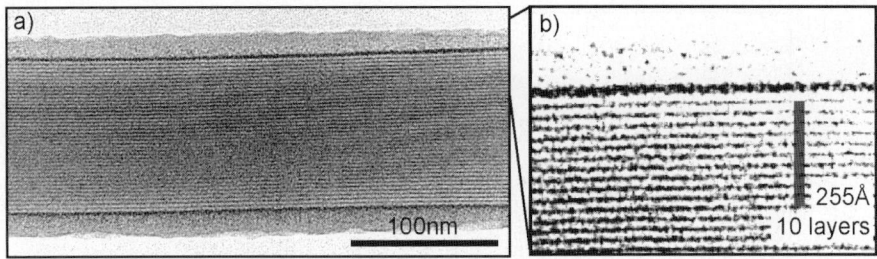

Fig. 6.18. A high resolution TEM image of a sexiphenyl needle grown on KCl (**a**) with a zoom in (**b**). The stripes parallel to the needle are identified as herringbone layers. This proves the single crystalline nature of the needle

layers fits perfectly to the length of the sexiphenyl molecules and thus the layers are identified as the actual herringbone layers parallel to the needle direction. The continuity of the herringbone layers concerning the two lateral directions, parallel to the needle direction and perpendicular to it, prove the single crystalline nature of this needle. It is to emphasize that the single crystalline character of needles is proven by two different TEM methods.

Line Profile Analysis

Besides the TEM there is another possibility of crystallite size determination, namely line profile analysis (LPA) of X-ray diffraction peaks (compare Sect. 6.3.6). As already mentioned a whole series of diffraction peaks having the same direction of q is required for reasonable performance of LPA. In the case of small organic crystals there is typically only one series of higher order diffraction peaks that can be detected, namely the peaks deriving from the planes parallel to the herringbone layer. As an example for LPA the material sexithiophene grown on Cu(110) is chosen [108]. AFM images reveal lateral sizes of the nanoaggregates which are smaller than 100 nm. Specular scans yield that sexithiophene grows with the (010) plane parallel to the surface and a distinct in-plane alignment is observed by surface diffraction. An in-plane $|q|$ scan (compare Sect. 6.3.5) to detect the (h00) series is presented in Fig. 6.19a. The measurement was performed using synchrotron radiation and peaks up to (32.00) are detected. The reciprocal breadths β^* are calculated and plotted over s (compare (6.9)). The line of best fit (Fig. 6.19) finally reveals the average crystallite size in the direction perpendicular to the herringbone layers. In this case it is determined to 50 nm which is in good agreement with the AFM results. The absolute value is of course just an estimation of the real crystallite size but since the AFM images show nanoaggregates of similar sizes it is concluded that the nanoaggregates are small single crystallites in this particular direction.

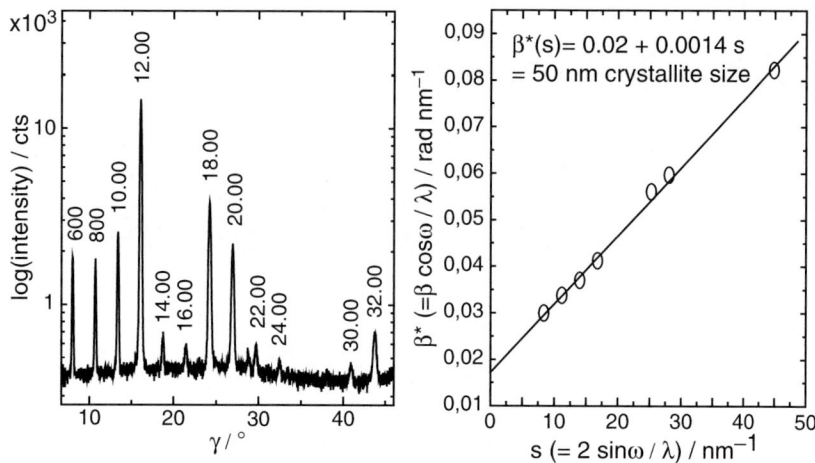

Fig. 6.19. An in-plane $|q|$ scan of the $(h00)$ series of (010) oriented sexithiophene crystallites grown on a Cu(110) surface (**a**). The line profile analysis of the scan reveals the average size of the crystallites (**b**)

6.4.5 Polymorphism

Most of the structural analyses are based on the knowledge of the single crystal structure. As long as the thin organic layers show the known crystal structure the analyses are straight forward. However, it is regularly observed that thin films of organic molecular crystals show several polymorph phases. Mostly these polymorph phases are very similar to the known crystal structure and thus the polymorphism is only detectable by X-ray diffraction due to the high angular resolution.

As an example sexiphenyl thin films prepared on isotropic substrates at different temperatures are shown [109]. The growth of sexiphenyl at a substrate temperature of 160°C leads to smaller diffraction angles regarding the $(00l)$ series, as shown in Fig. 6.20 [74, 110]. This again means that the herringbone layers are slightly thicker compared to the well known β-phase of sexiphenyl. Also a coexistence of these polymorph phases is reported [78, 111, 112]. A schematic drawing of the estimated crystal structure of the γ-phase is shown in the inset of Fig. 6.20a. It is to add that not only the position of the diffraction patterns differ from each other but also the intensities of the peaks. As seen the diffraction peaks 00$\underline{12}$ and 00$\underline{13}$ are clearly detectable in case of the β-phase while there is no signal measured at the γ-phase.

Unfortunately the polymorph phases are often very similar to each other so that they cannot be easily separated from each other. The determination of the existence of polymorph phases is often restricted to the diffraction peaks deriving from crystal planes parallel to the herringbone layers because in this orientation the largest variation of the interplanar spacing is observed.

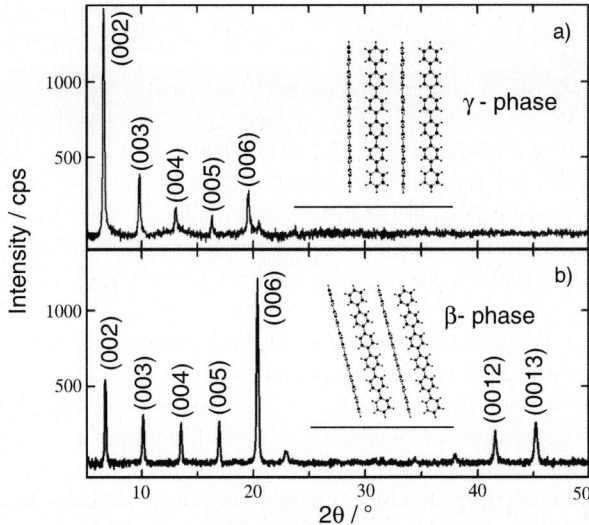

Fig. 6.20. Specular scans of two different phases of sexiphenyl on native oxide silicon substrates. The γ-phase shows slightly smaller 2θ angles regarding the (00l) series. This means that the interplanar spacing is slightly enlarged compared to the β-phase (**b**). Reprinted with permission from [109]

6.5 Early Stage Growth

The formation of specific crystallographic orientations of crystalline nanoaggregates on surfaces is determined by the film growth kinetics and growth energetics. The film growth is energetically determined by a competition of two different interactions: first, the interaction between individual molecules and second, the interaction of the individual molecules with the substrate surface [40]. If the molecule/molecule interaction dominates over the molecule/substrate interaction, the molecules pack as densely as possible at the substrate/molecule interface. As a consequence cleavage planes of the molecular crystal are formed parallel to the surface of the substrate. If the molecule/substrate interaction is dominant, the molecules are specifically oriented and aligned relative to the substrate surface. The film growth is determined by this specific orientation of the molecules in the first monolayers [113].

Besides the growth energetics the migration of the molecules on the surface is important for the formation of crystalline nanoaggregates. Experimental parameters like surface purity, substrate temperature and deposition rate affect the mobility of molecules on surfaces an thus have a large influence to the film morphology and crystalline order [30, 114]. In most cases it is observed that the order in the first monolayers determines the orientation of the crystallites in the films [3, 115–117]. Only in case of restricted molecular migration (very low substrate temperature, high deposition rates) the order in the first monolayer is not decisive for the molecular order within the whole film [118].

Low interaction of the molecules with substrates is observed in case of oxygen, water or carbon terminated surfaces [119], glass and polymer surfaces [111, 120], on single crystalline surfaces of dielectric materials like mica(001) [121], alkali halides [104] and TiO_2(110) [106]. As a consequence of the low interaction cleavage planes of the crystalline nanoaggregates are formed in the first monolayer(s) parallel to the substrate surface. In the following paragraphs some examples are given for the film formation with low molecule/substrate interaction.

Figure 6.21 shows two types of growth morphologies of sexiphenyl thin films which are induced by the orientation of the molecules in the first molecular layers at the interface. The left part of Fig. 6.21 shows terraced islands with a defined step height between the individual terraces which are typically formed by standing molecules from the first monolayer. The packing of the molecules within the first upright standing monolayer appears in a similar matter than in the three-dimensional crystal [60]. The film growth on top of upright standing molecules continues with the molecular orientation of the first layers so that crystallites are formed. Typical crystallite sizes (in direction perpendicular to the substrate surface) of 90 nm and larger are observed [122, 123]. Please note that polymorph crystal structures can appear, which are mediated by the molecular orientation and packing of the molecules in the first monolayers (surface mediated polymorphs) [74, 124].

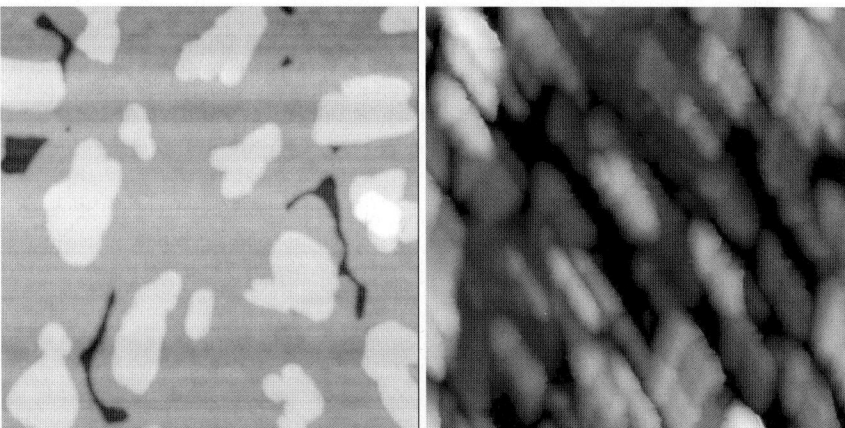

Fig. 6.21. Morphology of sexiphenyl thin films taken by atomic force microscopy. The *left part* shows a $1 \times 1\,\mu m^2$ micrograph of a film with a nominal thickness of few monolayers prepared on a native oxide surface, the z-range is 8 nm. Reprinted with permission from [111]. The *right part* shows a $4 \times 4\,\mu m^2$ micrograph of a film with a nominal thickness of 4,000 nm prepared on a uniaxially rubbed surface, the z-range is 300 nm. Reprinted with permission from [122]

The right part of Fig. 6.21 shows the morphology of a sexiphenyl film grown on a rubbed surface. Elongated islands with a defined alignment are observed. Crystallographic investigations reveal uniaxially aligned crystallites with the long molecular axes perpendicular to the rubbing direction [125,126]. The fact that both, the morphology as well as the crystal structure properties, are uniaxially aligned reveal that the orientation of the molecules in the first monolayers define the film growth up to the top layers.

The morphology of sexiphenyl nanoaggregates on mica(001) develops from an early growth morphology with small rectangular islands to an advanced growth morphology with a needle (or chain) type character [127]. Although the first monolayer covers the complete mica surface by lying molecules an island type morphology appears as an early growth morphology [128,129]. The left side of Fig. 6.22 shows a TEM image (inset) of the islands together with a transmission electron diffraction pattern of the islands. Three different types of crystalline alignments are found in the early growth stage which are in good agreement with the crystalline orientations determined by X-ray diffraction on films with the needle like morphology [107,121]. The right part of Fig. 6.22 shows a scanning electron diffraction pattern of a film which was grown at a substrate temperature of 435 K. The advanced growth stage with needle (or chain) like character is clearly visible, but, additional islands of the early growth stage appear between the needles. It could be shown by TEM dark field methods that the needles are composed of crystallites of different orientations

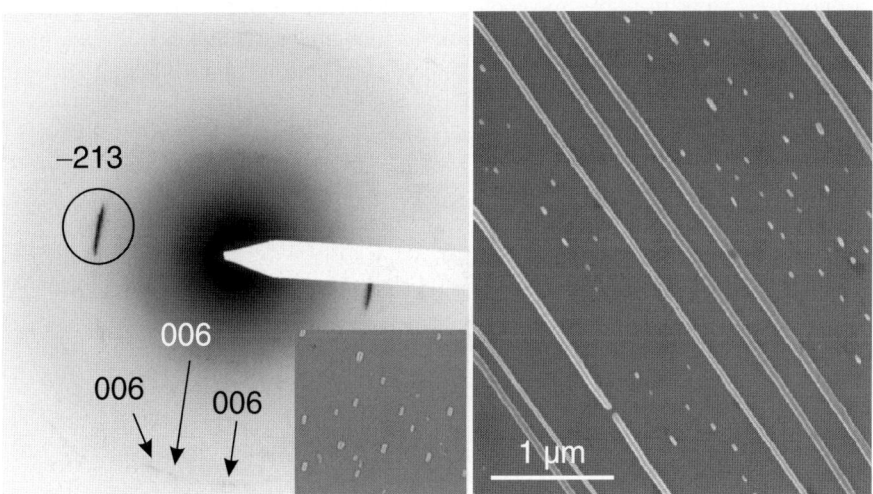

Fig. 6.22. Sexiphenyl on mica(001) grown by hot wall epitaxy. The *left part* shows a transmission electron diffraction and the microscopy picture of the early growth stage. Reprinted with permission from [107]. Copyright 2003, Elsevier. The *right part* shows a scanning electron microscopy image of a film in the advanced growth state

and alignments [107]. In general needle like growth of sexiphenyl on mica is observed independently of the growth methods and growth conditions. However, it is proposed that the microscopic structure of the nanoneedles varies from a composition of small crystallites to elongated single crystallites [35].

The growth of sexiphenyl on KCl(100) surfaces shows a combined needle and terraced island-like morphology which appears from lying and standing molecules, respectively [110, 130]. The morphology of the initial growth stage is built by needles which form an extended rectangular network reflecting the fourfold symmetry of the substrate. These needles are formed from crystallites having their (20–3) plane parallel to the substrate surface. The long axes of the molecules are aligned parallel to the surface of the substrate and furthermore parallel to distinct surface directions, and the aromatic planes are oriented edge-on. In advanced growth stages other crystal orientations appear with different cleavage planes parallel to the surface of the substrates. The later formed crystal orientations are built partly by lying molecules (low growth temperature) and partly by upright standing molecules (high growth temperature) [131, 132].

In case of sexiphenyl grown on TiO_2 surfaces the orientation of the crystallites relative to the substrate surface can be adjusted by the substrate temperature during the film growth procedure. At elevated substrate temperatures the crystallites grow as terraced islands with upright standing molecules [133]. It is observed that the crystal structure of sexiphenyl is already present in the first monolayer and the alignment of the crystallites is extremely pronounced (low in-plane mosaicity with FWHM of 0.2°). At low substrate temperatures crystallites with lying molecules and a typical needle like morphology can be prepared [106]. Figure 6.23 shows an in-plane rocking curve of

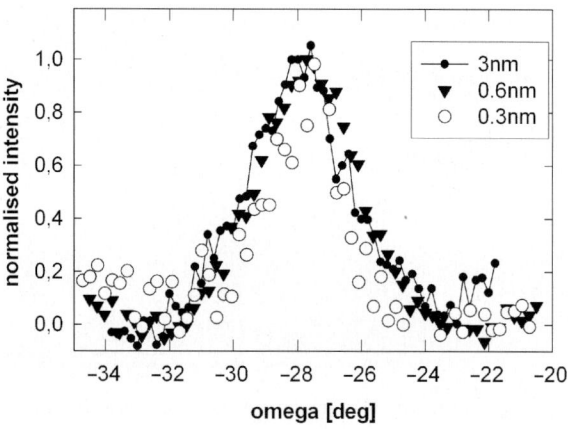

Fig. 6.23. X-ray diffraction of sexiphenyl films with nominal thickness of 0.3 (1 monolayer of lying molecules), 0.6 and 3 nm grown on a reconstructed $TiO_2(110)$ surface. The scan is an in-plane rocking curve on the 026 reflection taken in grazing incidence geometry

the 026 reflection taken from films with average thicknesses of one, two and ten monolayers of laying molecules. It clearly reveals that the large in-plane disorder of the crystallites (high in-plane mosaicity with FWHM 4°) develops already in the first monolayers.

The above mentioned examples are chosen as representatives for a low molecule/substrate interaction. Large interaction of the molecules with the substrates is observed for metal surfaces [134] and on surfaces of inorganic semiconductors [135]. In these cases the molecules are adsorbed with defined orientation and alignment at specific sites of the surface. The out-of-plane orientations and in-plane alignments of the crystallites are determined by the geometry of the firstly adsorbed molecules.

Conjugated molecules tend to orient flat-on on metal surfaces, i.e. the conjugated segments of the molecules are oriented parallel to the surface of the substrate [134]. This is observed also for sexiphenyl on the Al(111) surface [116, 136, 137]; and especially in the detailed study of quater- and sexiphenyl on Au(111) surfaces [117, 132, 138, 139]. On Al(111) surfaces it is found that the sexiphenyl crystals show specific alignments so that the long molecular axes are aligned parallel to the $\langle 1$–$10 \rangle$ directions of the Al(111) surface with the phenyl rings parallel to the surface. Assuming that the crystal growth starts from an initially adsorbed molecule (oriented and aligned as described above) four different alignments of crystallites should result, which is indeed experimentally observed. Therefore, the molecular orientation and alignment of sexiphenyl on Al(111) could be concluded from crystallographic investigations on 30 nm thick films (Fig. 6.24).

Thermal desorption spectroscopy reveals that the first monolayer of sexiphenyl completely covers the Au(111) surface. However, the first half

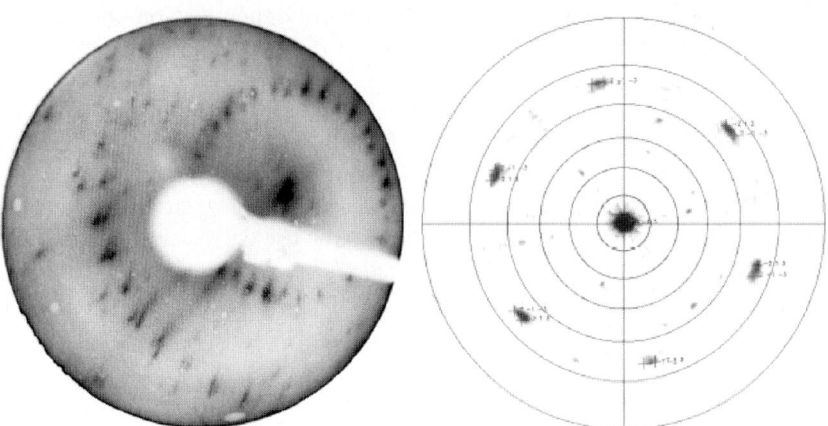

Fig. 6.24. Diffraction pattern of sexiphenyl thin films on Au(111) prepared at 300 K substrate temperature. Low energy electron diffraction of a single monolayer taken at 20 V (*left part*); pole figure of the 21–3 reflection of a film with the thickness of 30 nm, enhanced pole densities of sexiphenyl are marked by crosses (*right part*)

monolayer shows another energetical characteristic than the second half of the monolayer [139, 140]. In the first half of the monolayer the molecules are flat-on but in the second monolayer the molecules are edge-on located in between the flat-on molecules. This arrangement within the first complete monolayer represents an pre-stage of the crystal structure. These results are obtained by the combination of low energy electron diffraction and near edge X-ray absorption fluorescence spectroscopy performed on a half monolayer and a whole monolayer in combination with XRD on 30 nm thick films [141]. A further interesting result on the system sexiphenyl on Au(111) is that the formation of different nanoaggregates is dependent on the substrate temperature during the thin film growth. Flat film surfaces with weak crystallographic order appear at low substrate temperatures while elongated islands and needles with epitaxial character are formed by increasing the temperature [142].

It is known that the adsorption of molecules starts at step edges of surfaces [134]. These specifically adsorbed molecules can act as nucleation centers for the formation of crystallites; the role of step edges to the formation of in-plane oriented nanoaggregates is discussed in terms of "ledge-directed epitaxy" [143]. There are some reported examples in the literature for ledge-directed epitaxy [23, 28, 144], however, the observation of an overgrowth of step edges by nanoneedles reveals that the role of step edges is far to be understood.

6.6 Conclusion

Application-relevant properties like charge transport mobility and optical emission of organic nanoaggregates are strongly dependent on the crystallographic order on surfaces. The determination of crystallographic orientations (out-of-plane order) and the alignments of the crystallites (in-plane order) reveal the orientation of the molecules within the nanoaggregates. A further application-relevant parameter is the crystalline domain size in lateral and vertical direction relative to the substrate.

The combination of X-ray diffraction methods with transmission electron microscopy and diffraction works excellently for the correlation between morphology, crystal structure and crystal size of the nanoaggregates. The combination of X-ray diffraction with classical surface science tools gives further insight of the structural properties of the molecules at surfaces which are of crucial importance for the formation of the nanoaggregates.

The formation of nanoaggregates with specific morphology depends on following parameters: the type of molecule and its crystal structure type, the type of substrate, the purity of the substrate surface and its symmetry and finally on the film growth procedure itself which is the substrate temperature and the rate of molecular deposition. It is observed that the different growth procedures hot wall epitaxy and molecular beam deposition lead to similar results.

Acknowledgements

Financial support was thankfully received from the Austrian Science Foundation, the Austrian Scientific Exchange and by the European Union.

The authors thank the groups which provided carefully prepared samples for crystallographic investigations; by name A. Andreev, H. Sitter, Institute of Solid State and Semiconductor Physics, University Linz; M.G. Ramsey, G. Koller, J. Ivanco, Institute of Physics, University Graz; and A. Winkler, S. Müllegger, Institute of Solid State Physics, Graz University of Technology. Transmission electron microscopy studies were performed in close collaboration with B. Lotz and A. Thierry, Institute Charles Sadron, Strasbourg. Experimental support for the performance of X-ray diffraction studies using synchrotron radiation is greatly acknowledged to D. Smilgies, Cornell High Energy Synchrotron Source, Cornell University, Ithaca, New York; O. Konovalov and T. Schülli, European Synchrotron Research Facility, Grenoble. Finally diploma and PhD students of the Institute of Solid State Physics, Graz University of Technology are acknowledged for their individual contributions so that this book chapter could be realized; by name K. Erlacher, H.J. Brandt, H. Plank, M. Oehzelt, I. Salzmann, O. Lengyel, K. Matoy and O. Werzer.

References

1. F. Gutmann, L. Lyons, *Organic Semiconductors* (Wiley New York, 1967)
2. M. Pope, C. Swenberg, *Electronic Processes in Organic Crystals and Polymers*, 2nd. edn. (Oxford University Press, New York, Oxford, 1999)
3. T. Mikami, H. Yanagi, Appl. Phys. Lett. **73**, 563 (1998)
4. F. Balzer, V. Bordo, A. Simonsen, H.G. Rubahn, Appl. Phys. Lett. **82**, 10 (2003)
5. N. Karl, *Charge-Carrier Mobility in Organic Crystals* (Springer, Berlin Heidelberg New York, 2003)
6. E. Silinsh, V. Capek, *Organic Molecular Crystals* (American Institute of Physics, New York, 1994)
7. J. Bredas, J. Calbert, D. de Silva, J. Cornil, Proc. Nat. Acad. Sci. **99**, 5804 (2002)
8. J. Laquindanum, H. Katz, A. Lovinger, A. Dodabalapur, Chem. Mater. **8**, 2542 (1996)
9. G. Horowitz, M. Hjlaoui, Adv. Mater. **12**, 1046 (2000)
10. K. Herwig, J. Newton, H. Taub, Phys. Rev. B **50**, 15287 (1994)
11. G. Desiraju, A. Gavezzotti, Acta Cryst. B **45**, 473 (1989)
12. B. Stevens, Spectrochim. Acta **18**, 439 (1962)
13. D. Williams, Y. Xiao, Acta Cryst. A **49**, 1 (1993)
14. V. Klymenko, V. Rozembaum, J. Chem. Phys. **110**, 5978 (1999)
15. A. Kitaigorodsky, *Molecular Crystals and Molecules* (Academic Press, New York, 1973)
16. K. Sato, J. Phys. **D26**, B77 (1993)

17. J. Bernstein, *Polymorphism in Molecular Crystals* (Calendron Press, Oxford, 2002)
18. P. Erk, *Towards Control of Polymorphism in Organic Pigments*, Paper presented at the 17th European Crystallographic Meeting Instituto Superior Tecnico, Lisboa, Portugal, 24–28 August 1997
19. G. Heimel, K. Hummer, C. Ambrosch-Draxl, W. Chunwachirasiri, M. Winokur, M. Hanfland, M. Oehzelt, A. Aichholzer, R. Resel, Phys. Rev. B **73**, 024109 (2006)
20. M. Moebus, N. Karl, T. Kobayashi, J. Cryst. Growth **116**, 495 (1992)
21. N. Uyeda, M. Ashida, E. Suito, J. Appl. Phys. **36**, 1453 (1965)
22. M. Ashida, N. Uyeda, E. Suito, Bull. Chem. Soc. Jpn. **39**, 2616 (1966)
23. J. Osso, F. Schreiber, V. Kruppa, H. Dosch, M. Garriga, M. Alonso, F. Cerdeira, Adv. Funct. Mater. **12**, 455 (2002)
24. F. Biscarini, R. Zamboni, P. Samori, P. Ostoja, C. Taliani, Phys. Rev. B **52**, 14868 (1995)
25. D. Gundlach, T. Jackson, D. Schlom, S. Nelson, Appl. Phys. Lett. **77**, 3302 (1999)
26. R. Ruiz, B. Nickel, N. Koch, L. Feldman, R. Haglund, A. Kahn, G. Scoles, Phys. Rev. B **67**, 125406 (2003)
27. O. Boehme, C. Ziegler, W. Goepel, Adv. Mater. **6**, 587 (1994)
28. S. Ikeda, M. Kiguchi, Y. Yoshida, K. Yase, T. Mitsunaga, K. Inaba, K. Saiki, J. Cryst. Groth **265**, 296 (2004)
29. N. Koch, R. Resel, F. Meghdadi, G. Leising, K. Reichmann, Synthetic Met. **84**, 649 (1997)
30. B. Müller, T. Kuhlmann, K. Lischka, H. Schwer, R. Resel, G. Leising, Surf. Sci. **418**, 256 (1998)
31. G. Hlawacek, C. Teichert, S. Müllegger, R. Resel, A. Winkler, Synthetic Met. **146**, 383 (2004)
32. A. Andreev, G. Matt, C. Brabec, H. Sitter, D. Badt, H. Seyringer, N. Sariciftci, Adv. Mater. **12**, 629 (2000)
33. H. Yanagi, Y. Araki, T. Ohara, S. Hotta, M. Ichikawa, Y. Taniguchi, Adv. Funct. Mater. **13**, 767 (2003)
34. F. Balzer, H. Rubahn, Surf. Sci. **548**, 170 (2004)
35. A. Andreev, C. Teichert, G. Hlawacek, H. Hoppe, R. Resel, D.-M. Smilgies, H. Sitter, N. Sariciftci, Org. Electron. **5**, 23 (2004)
36. C. Teichert, G. Hlawacek, A. Andreev, H. Sitter, N. Sariciftci, Appl. Phys. A **82**, 665 (2006)
37. J. Fryer, J. Phys. **D26**, B137 (1993)
38. K. Baker, A. Fratini, T. Resch, H. Knachel, W. Adams, E. Socci, B. Farmer, Polymer **34**, 1571 (1993)
39. S. Forrest, Y. Zhang, Phys. Rev. B **49**, 11297 (1994)
40. D. Hooks, T. Fritz, M. Ward, Adv. Mater. **13**, 227 (2001)
41. R. Laudise, C. Kloc, P. Simpkins, T. Siegrist, J. Cryst. Growth **187**, 449 (1998)
42. W. Helfrich, W. Schneider, Phys. Rev. Lett. **14**, 229 (1965)
43. R. Mason, Acta Crystallogr. **17**, 547 (1964)
44. A. Mathieson, J.M. Robertson, V. Sinclair, Acta Crystallogr. **3**, 245 (1950)
45. F. Ahmed, D. Cruickshank, Acta Crystallogr. **5**, 852 (1952)
46. D. Cruickshank, Acta Crystallogr. **9**, 915 (1956)
47. V. Ponomarev, G. Shilov, Kristallografiya (Crystallogr. Rep.) **28**, 674 (1983)

48. C.P. Brock, J.D. Dunitz, Acta Crystallogr. B **46**, 795 (1990)
49. B. Marciniak, V. Pavlyuk, Mol. Cryst. Liq. Cryst. Sci. Technol. A **373**, 237 (2002)
50. R. Campell, J. Robertson, Acta Cryst. **15**, 289 (1962)
51. D. Holmes, S. Kumaraswamy, A.J. Matzger, K. Vollhardt, Chem. Eur. J. **5**, 3399 (1999)
52. Hertel, Bergk, Z. Phys. Chem. (Leipzig) **B33**, 319 (1936)
53. T. Siegrist, C. Kloc, J. Schön, B. Batlogg, R. Haddon, S. Berg, G. Thomas, Angew. Chem. Int. Edit. **40**, 1732 (2001)
54. C. Mattheus, A. Dros, J. Baas, A. Meetsma, J. deBoer, T. Palstra, Acta Crystallogr. C **57**, 939 (2001)
55. M. Oehzelt, R. Resel, A. Nakayama, Phys. Rev. B **66**, 174104 (2002)
56. R. Resel, M. Oehzelt, K. Shimizu, A. Nakayama, K. Takemura, Solid Sate Comm. **129**, 103 (2004)
57. C. Mattheus, A. Dros, J. Baas, G. Oostergetel, A. Meetsma, J. deBoer, T. Palstra, Synthetic. Met. **138**, 475 (2002)
58. E. Venuti, R.D. Valle, A. Brillante, M. Masino, A. Girlando, J. Am. Chem. Soc. **124**, 2128 (2002)
59. L. Drummy, D. Martin, Adv. Mater. **17**, 903 (2005)
60. S. Fritz, S. Martin, C. Frisbie, M. Ward, M. Toney, J. Am. Chem. Soc. **126**, 4084 (2004)
61. G. Leising, S. Tasch, W. Graupner, in *Handbook of Conducting Polymers 2nd Ed.: Fundamentals of Electroluminescence in Paraphenylene-Type Conjugated Polymers and Oligomers* (Marcel Dekker, New York, 1998)
62. H. Rietveld, E. Maaslen, C. Clews, Acta Cryst. **B26**, 693 (1970)
63. J. Baudour, Acta Crystallogr. B **28**, 1649 (1972)
64. J. Baudour, Y. Delugeard, H. Cailleu, Acta Crystallogr. B **32**, 150 (1976)
65. J. Baudour, H. Cailleau, W. Yelon, Acta Crystallogr. B **33**, 1773 (1977)
66. Hertel, Romer, Z. Phys. Chem. (Leipzig) **B21**, 292 (1933)
67. J. Baudour, L. Toupet, Y. Delugeard, S. Ghemid, Acta Crystallogr. C **42**, 1211 (1986)
68. J. Dejace, Bull. Soc. Fr. Miner. Crist. **92**, 141 (1969)
69. H. Saitoh, K. Saito, Y. Yamamura, H. Matsuyama, K. Kikuchi, M. Iyoda, I. Ikemoto, Bull. Chem. Soc. Jpn. **66**, 2847 (1993)
70. Y. Delugeard, J. Desuche, J. Baudour, Acta Crystallogr. B **32**, 702 (1976)
71. J.L. Baudour, Y. Delugeard, P. Rivet, Acta Crystallogr. B **34**, 625 (1978)
72. C. Toussaint, G. Vos, J. Chem. Soc. **B**, 1002 (1966)
73. C. Toussaint, Acta Cryst. **21**, 1002 (1966)
74. R. Resel, N. Koch, F. Meghdadi, G. Leising, L. Athouel, G. Froyer, F. Hofer, Cryst. Res. Technol. **36**, 47 (2001)
75. D. Smilgies, Acta Cryst. **B61**, 357 (2005)
76. J. Baudour, Acta Cryst. **B47**, 935 (1991)
77. A. Lenstra, C. van Alsenoy, K. Verhulst, H. Geise, Acta Cryst. B **50**, 96 (1994)
78. L. Athouel, G. Froyer, M. Riou, Synthetic Met. **55–57**, 4734 (1993)
79. B. Toudic, P. Limelette, G. Froyer, F.L. Gac, A. Moreac, P. Rabiller, Phys. Rev. Lett. **95**, 215502 (2005)
80. H. Koezuka, A. Tsumura, T. Ando, Synthetic Met. **18**, 699 (1987)
81. G. Horowitz, D. Fichou, X. Peng, Z. Xu, F. Garnier, Sol. Stat. Commun. **72**, 381 (1989)

82. F. van Bolhuis, H. Wynberg, E. Havinga, E. Meijer, E. Starling, Synthetic Met. **30**, 381 (1989)
83. R. Azumi, M. Goto, K. Honda, M. Matsumoto, Bull. Chem. Soc. Jpn. **76**, 1561 (2003)
84. L. Antolini, G. Horowitz, F. Kouki, F. Garnier, Adv. Mater. **10**, 382 (1998)
85. T. Siegrist, C. Kloc, R. Laudise, H. Katz, R. Haddon, Adv. Mater. **10**, 379 (1998)
86. W. Porzio, S. Destri, M. Mascherpa, S. Brueckner, Acta Polymer **44**, 266 (1993)
87. T. Siegrist, R. Fleming, R. Haddon, R. Laudise, A. Lovinger, H. Katz, P. Bridenbaugh, D. Davis, J. Mater. Res. **10**, 2170 (1995)
88. G. Horowitz, B. Bachet, A. Yassar, P. Lang, F. Demanze, J.L. Fave, F. Garnier, Chem. Mater. **7**, 1337 (1995)
89. D. Fichou, B. Bachet, F. Demanze, I, Billy, G. Horowitz, F. Garnier, Adv. Mater. **8**, 500 (1996)
90. B. Servet, G. Horowitz, S. Ries, O. Lagorsse, P. Alnot, A. Yassar, F. Deloffre, P. Srivastava, R. Hajlaoui, P. Lang, F. Garnier, Chem. Mater. **6**, 1809 (1994)
91. J. Ivanco, J. Krenn, M. Ramsey, F. Netzer, T. Haber, R. Resel, A. Haase, B. Stadlober, G. Jakopic, J. Appl. Phys. **96**, 2716 (2004)
92. L. Schwartz, J. Cohen, *Diffraction from Materials* (Springer Berlin Heidelberg New York, 1987)
93. W. Kraus, G. Nolze, J. Appl. Crystallogr. **29**, 301 (1996)
94. I. Salzmann, R. Resel, J. Appl. Crystallogr. **37**, 1029 (2004)
95. I. Robinson, D. Tweet, Rep. Prog. Phys. **55**, 599 (1992)
96. G. Vineyard, Phys. Rev. B **26**, 4146 (1982)
97. M. Birkholz, *Thin Film Analysis by X-Ray Scattering* (Wiley-VCH, Weinheim, 2006)
98. A. Stokes, A. Wilson, Proc. Camb. Phil. Soc. **40**, 40 (1944)
99. G. Williamson, W. Hall, Acta Metallurgica **1**, 22 (1953)
100. M. Gebhardt, A. Neuhaus, in *Numerical data and Functional Relationships in Science and Technology, Vol. 8: Epitaxy Data of Inorganic and Organic crystals, ed. by Landoltd-Börnstein* (Springer, Berlin Heidelberg New York, 1972)
101. A. Koma, Thin Solid Films **216**, 72 (1992)
102. A. Koma, Prog. Cryst. Growth Charact. **30**, 129 (1995)
103. T. Haber, M. Oehzelt, A. Andreev, H. Sitter, A. Thierry, R. Resel, J. Cryst. Growth **284**, 209 (2005)
104. T. Haber, M. Oehzelt, A. Andreev, H. Sitter, A. Thierry, D.M. Smilgies, B. Schaffar, W. Grogger, R. Resel, J. Nanosci. Nanotechnol. **6**, 698 (2006)
105. S. Mannsfeld, K. Leo, T. Fritz, Phys. Rev. Lett. **94**, 056104 (2005)
106. G. Koller, S. Berkebile, J. Krenn, G. Tzvetkov, G. Hlavacek, O. Lengyel, F. Netzer, C. Teichert, R. Resel, M. Ramsey, Adv. Mater. **16**, 2159 (2004)
107. H. Plank, R. Resel, N.S. Sariciftci, A. Andreev, H. Sitter, G.Hlawacek, C. Teichert, A. Thierry, B. Lotz, Thin Solid Films **443**, 108 (2003)
108. M. Oehzelt, K. Koller, J. Ivanco, S. Berkebile, T. Haber, R. Resel, F. Netzer, M. Ramsey, Adv. Mat. **18**, 2466 (2006)
109. R. Resel, Thin Solid Films **433**, 1 (2003)
110. A. Kawaguchi, M. Tsuji, S. Moriguchi, A. Uemura, S. Isoda, M. Ohara, J. Petermann, K. Katayama, Bull. Inst. Chem . Res., Kyoto Univ. **64**, 54 (1986)
111. R. Resel, N. Koch, F. Meghdadi, G. Leising, W. Unzog, K. Reichmann, Thin Solid Films **305**, 232 (1997)

112. E. Zojer, N. Koch, P. Puschnig, F. Meghdadi, A. Niko, R. Resel, C. Ambrosch-Draxl, M. Knupfer, J. Fink, J. Bredas, G. Leising, Phys. Rev. B **61**, 16538 (2000)
113. J. Taborski, P. Vaeterlein, H. Dietz, U. Zimmermann, J. Electron Spectrosc. Rel. Phen. **101**, 627 (1999)
114. L. Athouel, R. Resel, N. Koch, F. Meghdadi, G. Froyer, G. Leising, Synthetic Met. **101**, 627 (1999)
115. C. Ludwig, B. Gompf, W. Glatz, J. Petersen, W. Eisenmenger, M. Moebus, U. Zimmermann, N. Karl, Z. Phys. B **86**, 392 (1992)
116. R. Resel, I. Salzmann, G. Hlawacek, C. Teichert, B. Koppelhuber, B. Winter, J.Krenn, J. Ivanco, M. Ramsey, Org. Electron. **5**, 45 (2004)
117. S. Müllegger, I. Salzmann, R. Resel, G. Hlawacek, C. Teichert, A. Winkler, J. Chem. Phys. **121**, 2272 (2004)
118. S. Müllegger, G. Hlawacek, T. Haber, P. Frank, C. Teichert, R. Resel, A. Winkler, Appl. Phys. A **87**, 103 (2007)
119. R. Resel, M. Oehzelt, T. Haber, G. Hlawacek, C. Teichert, S.Müllegger, A. Winkler, J. Cryst. Growth **283**, 397 (2005)
120. K.Yase, E.M. Han, K. Yamamoto, Y. Yoshida, N. Takada, N. Tanigaki, Jpn. J. Appl. Phys. **36**, 2843 (1997)
121. H. Plank, R. Resel, S. Purger, J. Keckes, A. Thierry, B. Lotz, A. Andreev, N. Sariciftci, H. Sitter, Phys. Rev. B **64**, 235423 (2001)
122. H.J. Brandt, R. Resel, J. Keckes, B. Koppelhuber-Bitschnau, N. Koch, G. Leising, Mat. Res. Soc. Symp. Proc. **561**, 161 (1999)
123. M. Oehzelt, S. Müllegger, A. Winkler, G. Hlawacek, C. Teichert, R. Resel, Mater. Struct. **11**, 155 (2004)
124. R. Ruiz, A. Mayer, G. Malliaras, B. Nickel, G. Scoles, A. Kazimirov, H. Kim, R. Headrick, Z. Islam, Appl. Phys. Lett. **85**, 4926 (2004)
125. K. Erlacher, R. Resel, J. Keckes, F. Meghdadi, G. Leising, J. Cryst. Growth **206**, 135 (1999)
126. W.S. Hu, Y.F. Lin, Y.T. Tao, Y.J. Hsu, D.H. Wei, Macromolecules **38**, 9617 (2005)
127. A. Andreev, H. Sitter, C. Brabec, P. Hinterdorfer, G. Springholz, N. Sariciftci, Synthetic Met. **121**, 1379 (2001)
128. F. Balzer, H.G. Rubahn, Appl. Phys. Lett. **79**, 3860 (2001)
129. C. Teichert, G. Hlawacek, A. Andreev, H. Sitter, P. Frank, A. Winkler, N. Sariciftci, Appl. Phys. A **82**, 665 (2006)
130. H. Yanagi, T. Morikawa, Appl. Phys. Lett. **75**, 187 (1999)
131. E. Kintzel, D.M. Smilgies, J. Skofronick, S. Safron, D. Van Winkle, J. Vac. Sci. Technol. A **22**, 107 (2004)
132. T. Haber, S. Müllegger, A. Winkler, R. Resel, Phys. Rev. B **74**, 045419 (2006)
133. R. Resel, M. Oehzelt, O. Lengyel, T. Haber, T. Schülli, A. Thierry, G. Hlawacek, C. Teichert, S. Berkebile, G. Koller, M. Ramsey, Surf. Sci. **600**, 4645 (2006)
134. F. Rosei, M. Schunack, Y. Naitoh, P. Jiang, A. Gourdon, E. Laegsgaard, I. Stensgaard, C. Joachim, F. Besenbacher, Prog. Surf. Sci. **71**, 95 (2003)
135. M. Schwartz, M. Ellison, S. Coulter, J. Hovis, R. Hamers, J. Am. Chem. Soc. **122**, 8529 (2000)
136. B. Winter, J. Ivanco, F. Netzer, M. Ramsey, Thin Solid Films **433**, 269 (2003)
137. B. Winter, J. Ivanco, I. Salzmann, R. Resel, F. Netzer, M. Ramsey, Langmuir **20**, 7512 (2004)

138. S. Müllegger, I. Salzmann, R. Resel, A. Winkler, Appl. Phys. Lett. **83**, 4536 (2003)
139. S. Müllegger, A. Winkler, Surf. Sci. **600**, 1290 (2006)
140. C. France, B. Parkinson, Appl. Phys. Lett. **82**, 1194 (2003)
141. S. Müllegger, K. Hänel, T. Strunskus, C. Wöll, A. Winkler, Chem. Phys. Chem **7**, 2552 (2006)
142. S. Müllegger, G. Hlawacek, T. Haber, P. Frank, C. Teichert, R. Resel, A. Winkler, Appl. Phys. A **87**, 103 (2007)
143. S. Bonafede, M. Ward, J. Am. Chem. Soc. **117**, 7853 (1995)
144. S. Lukas, S. Vollmer, G. Witte, C. Wöll, J. Chem. Phys. **114**, 10123 (2001)

7

Growth and Electronic Structure of Homo- and Hetero-epitaxial Organic Nanostructures

G. Koller and M.G. Ramsey

7.1 Introduction

It is now generally recognized that the interfaces in organic devices are crucial to their performance; however, the basic questions of how and why, and to what extent they can be controlled are still open. The fundamental issues that are important are both electronic and geometric: the relative position of the electronic levels (band alignment) is a prime determinate of the charge injection ability; molecular orientation and crystallinity control the charge transport and light emission/adsorption, while film morphology is generally important to device construction and function, as illustrated in Fig. 7.1. Moreover, if devices based on organic nanostructures are ever to be realized a basic understanding of the organic self-assembly process is needed. To address these issues requires controlled/reproducible investigations and in the last few years there has been increasing activity in this area involving model molecules on single crystal substrates, i.e., the surface science approach.

Here we will principally confine the discussion to our work on the device relevant model chain-like molecules sexiphenyl and sexithiophene. Over the last few years we have applied a wide range of in-situ ultra high vacuum (UHV) based surface science techniques to explore the electronic and geometric structure of these molecules on a variety of single crystal substrates, including clean and modified metals, semiconductors, and oxides. In doing so we have gone from submonolayer coverages up to organic structures of device relevant dimensions. To bridge the gap from pure surface science to the more device leaning research, these in-situ prepared and characterized systems have also been investigated ex-situ in air for their morphology and crystal structure (see Chap. 6). A reasonably broad database has thus been assembled that allows the understanding of some of the basic principles and parameters controlling the self-assembly of organic nanostructures.

In Sect. 7.2 the organic-on-inorganic interface formation and the influence it has on subsequent growth of organic films and nanostructures will be discussed. We start in Sect. 7.2.1 with molecular monolayer formation and the

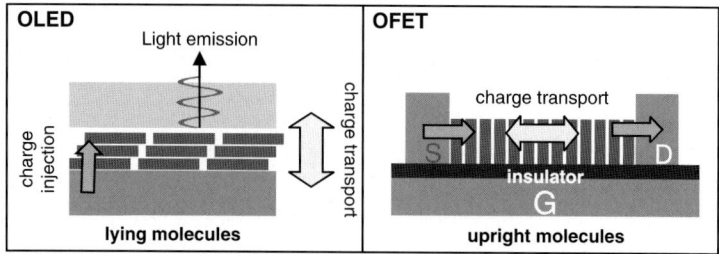

Fig. 7.1. Schematic of the desired molecular geometry in an OLED and an OFET

role of the substrate as a template for subsequent growth. It will be argued that the atomic details of the substrates have a profound influence on almost all aspects of the organic nanostructures that grow on them.

Unlike inorganic materials the basic building blocks of organic nanostructures, the molecules, are highly anisotropic. This together with the substrates ability to orient the molecules and give them preferred diffusion directions leads to the growth of different structures under different conditions. This and the role of diffusion vs. sticking anisotropy on growth is discussed in Sect. 7.2.2. In Sect. 7.2.3 the electronic level alignment at the organic–inorganic interface, important for charge injection in devices, is reviewed in general. The band alignment on nanoscopically patterned substrates is then considered in more detail in Sect. 7.2.4.

The growth of organic nanostructures on organic substrates with unique crystalline orientations is the topic of Sect. 7.3. Here organic substrates with two different terminations, the closed –CH and the more open and higher surface energy π-system termination, are contrasted in Sects. 7.3.1 and 7.3.2, respectively. With studies of molecular orientation, structure, and morphology, the ability of oriented organic films to act as substrate templates for the growth of a second organic layer is discussed. The organic substrates used are the highly crystalline films of sexiphenyl and sexithiophene of Sect. 7.2. The results show that not only does the orientation of the molecules in the first organic film determine the orientation in the heterolayer but this second film is also crystalline with a well-defined epitaxial relationship. Thus highly crystalline organic substrates can act as excellent growth templates similar to inorganic substrates, but seem to require less stringent ultra high vacuum growth conditions.

Finally, in Sect. 7.4 a short outlook into the future of organic nanotechnology is attempted.

7.2 Organic Films on Inorganic Substrates

The early work on films of device relevant molecules often displayed an extreme variety of crystallite orientations, film morphologies, and polymorphism. In hindsight this can be attributed to the then focus on film growth on technologically relevant substrates under technological conditions. Such substrates, ITO, glass, SiO_2, or evaporated metal films are often rough and inhomogeneous on the molecular length scale and cannot be prepared reproducibly, particularly under the poor vacuum conditions (High Vacuum 10^{-6} mbar) used for growth. Highly defined organic films and nanostructures necessary for the next generation devices will require well defined substrate templates for their growth. These are also prerequisites for controlled and reproducible growth studies required for the understanding of the basic principles and parameters controlling the organic self assembly process. What is illustrated in this section is that the substrate is not merely a passive support for growth but in fact can steer the growth kinetics that determines the molecular orientation and film morphology. Irrespective of the bonding mechanism, whether covalent or van der Waals, isolated molecules will lie parallel to a surface, maximizing their interaction to the substrate. It should be remembered that even alkylthiols with their anchoring head group lie parallel to a gold surface at low coverages. Whether the molecules reorient at higher coverages will depend on the delicate balance between molecule–substrate and molecule–molecule interactions.

7.2.1 The Substrate as a Template

The STM image of Fig. 7.2a shows isolated sexiphenyl molecules on a (2×1) oxygen reconstructed Ni(110) surface. The formation of this reconstruction

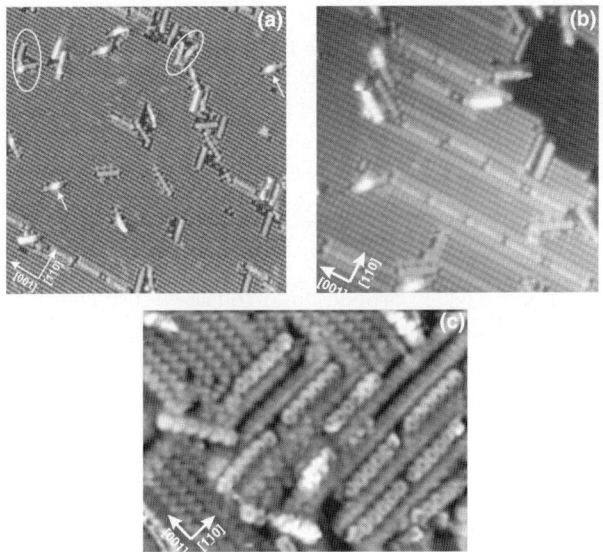

Fig. 7.2. STM images of submonolayer coverages sexiphenyl on Ni(110)-(2×1)O. (**a**) 400×320 Å^2 curved molecules are indicated. (**b**) 200×180 Å^2, showing strings of sexiphenyl molecules along the [001] azimuth. (**c**) 110×80 Å^2, sexiphenyl molecules oriented along the [1$\bar{1}$0] azimuth, due to substrate reduction. Reprinted with permission from [12]

effectively chemically polishes the Nickel making it a near perfectly flat surface, i.e., terraces of hundreds of nanometers in width separated by monoatomic steps. But as Pauli famously pointed out "the surface was created by the devil" and entropy does not allow a perfect surface and defects are always present. At very low exposures, on this relatively passive surface the 6P molecules have diffused to and adsorbed at these defects (Fig. 7.2a). The molecules appear as 27 Å long rod-like features with six maxima corresponding to the six phenyl rings of the molecule. Three principal orientations are observed: either parallel or perpendicular to the Ni–O rows of the surface or diagonal to the surface unit cell, i.e., aligned along the [001], [1$\bar{1}$0], or [112] surface azimuths. It is interesting to note that these rod-like molecules are not as rigid as often assumed and can sometimes be seen to bend around, or flow over atomic defects or steps. The exceptional resolution allows the molecular registry to be unambiguously assigned. The adsorption sites are those that maximize the number of fourfold hollow sites of the underlying nickel; for the diagonally aligned molecules all six phenyl rings are in such sites, while for the molecules oriented along the principal azimuths the first and fifth phenyl rings can, and do, occupy the fourfold hollow sites. Interestingly, such sites are the adsorption sites of the monomer, benzene, on Ni(110). For Ag(111) and Al(111) [4–6] the preferred benzene adsorption site is the threefold site and sexiphenyl aligns along the [1$\bar{1}$0] direction, where all six rings can occupy this site [7,8]. It would appear that the monomer adsorption sites might be a determinant for the longer oligomer adsorption and thus force the orientation of the oligomer in the first monolayer.

Going to higher coverages (Fig. 7.2b), beyond defect decoration, the molecules preferentially adopt the [001] orientation, but interestingly are arranged head to tail in molecular strings. This unexpected behavior results from the displacement of Ni–O rows of the substrate by the initial adsorption of sexiphenyl molecules, which opens up channels of bare Nickel, which are subsequently filled on further 6P exposure – a clear example that the substrate cannot be assumed to be a mere passive support. In addition to this molecule-induced modification of the substrate, the unavoidable hydrogen background during organic evaporation ($P_{H2} = 5 \times 10^{-10}$ mbar) reduces the Ni–O surface, resulting in strings of Nickel perpendicular to the Ni–O rows. These also act as nucleation sites and lead to molecules also oriented along [1$\bar{1}$0], as illustrated in Fig. 7.2c.

Finally, going beyond monolayer coverages small crystallites appear, whose orientations are determined by the three orientations of the molecules seen at the initial stages of monolayer formation. The Ni-(2×1)O example has illustrated the ability of the substrate, at the molecular level, to orient molecules and act as a template for further growth. However, the substrate was too sensitive to allow the growth of homogeneous molecular structures.

To obtain uniaxial structures we have gone to similar substrates, with rectangular unit cells of twofold symmetry that are more stable: Cu(110),

 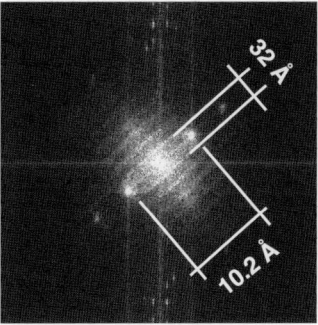

Fig. 7.3. (a) STM image (400 × 400 Å^2) of a sexiphenyl monolayer on Cu(110). A monoatomic step is seen in the top right hand corner. (b) Fourier transform of the surface showing the quasi periodicity

Cu(110)-p(2×1)O, or TiO$_2$(110)-(1×1). On these substrates, films and structures of chain-like molecules (sexiphenyl, sexithiophene, pentacene) could be grown with a high degree of crystallinity, where the molecules are uniaxially oriented parallel to the substrates and a principle substrate azimuthal direction from submonolayer coverage to device relevant thicknesses.

In Fig. 7.3 an STM image of the sexiphenyl monolayer (≈3 Å) on Cu(110) and the Fourier transform of the same surface are displayed. In the monolayer 90% of the molecules are aligned along the [001] substrate azimuth in a quasi-crystalline array reminiscent of a smectic liquid crystal. Although there is periodicity in the two principal directions, there is no long range two-dimensional order; the molecules along the [001] direction are in strings with a periodicity of 32 Å, while the spacing between the strings is 10.2 Å. The periodicity along the strings is not commensurate to the substrate and is determined by the van der Waals length of the molecule. In contrast the 10.2 Å between the molecules of neighboring strings is commensurate, being four times the copper spacing. This is much larger than the spacing between molecules in the bulk sexiphenyl structure. It is, however, not determined by commensurability considerations alone as a 3× structure would be commensurate and have a spacing similar to bulk sexiphenyl. Clearly the adsorbed sexiphenyl/surface moieties must be repulsive, due to the nature of the 6P/Cu bond (see later).

The behavior of sexiphenyl on Cu(110) as observed by STM highlights a number of other aspects of template formation. There is a belief that steps play an important role in orienting molecules, and indeed a number of groups are working with vicinal surfaces to control the growth – this is certainly not true for sexiphenyl on Cu(110) or TiO$_2$(110) [2]. As can be seen in Fig. 7.3a the monoatomic steps do not align the molecules and indeed numerous molecules can be seen flowing over the steps, again indicating their flexibility.

Interestingly, the sexiphenyl molecules are perpendicular to the atomic corrugation of the Cu(110) surface, unlike sexithiophene, where the molecules predominantly align parallel to the [1$\bar{1}$0] azimuth [10]. We and others have also been able to grow sexiphenyl parallel to [1$\bar{1}$0] [11]. The alignment of sexiphenyl along [001] observed here, we believe, may be driven by very small amounts of oxygen contamination. This leads to thin stripes of Cu–O rows parallel to [001] (apparent as dark stripes in Fig. 7.3), which appear to act as directors for the orientation of the molecules. On Au(110) and Cu(110) STM has shown pentacene orienting along the close-packed metal rows of the surface (along [1$\bar{1}$0]) [12–14] and similarly for sexiphenyl on Ag(111) [7]. In contrast on Au(111) sexiphenyl has been observed oriented between the close-packed rows ([11$\bar{2}$]-direction), but other alignments have also been found [15].

With the twofold symmetric TiO_2(110)-(1×1) and Cu(110)-p(2×1)O reconstructed substrate surfaces, the atomic corrugation is increased relative to the clean metals. The metal oxygen rows' (in the [001] direction) spacing is also increased to 6.5 and 5.1 Å, respectively. These are 15% larger and 8% smaller than the spacing of neighboring sexiphenyl molecules, for instance, in the (20$\bar{3}$) planes of bulk sexiphenyl. On these surfaces wetting monolayers of lying sexiphenyl, sexithiophene and pentacene form with the molecules being uniaxial aligned parallel to the substrate rows as evidenced by NEXAFS, ARUPS, and reflectance difference spectroscopy (RDS). Unfortunately, it is very difficult to image the monolayers on these substrates with STM at room temperature and LEED is too destructive. For instance, with room temperature-STM on TiO_2 only the substrate structure is imaged, and for sexiphenyl on Cu-(2×1)O the substrate is observed with a stress pattern radiating from occasional strings of molecules that are seen incorporated into the surface and parallel to [001], from submonolayer exposures up to the equivalent of tens of layers. Recently, with cryogenic STM (7 K) of this system we have observed a very well ordered sexiphenyl monolayer exhibiting a two-dimensional unit cell close to that of the (20$\bar{3}$) plane [10].

The atomically well ordered substrates and the wetting monolayers upon them all act as templates for the room temperature growth of highly oriented crystalline organic films or nanostructures with a unique crystallite orientation: 6P(21$\bar{3}$) on Al(111), Cu(110)-p(2×1)O, and TiO_2 yield 6P(20$\bar{3}$) and 6T (010) [2, 8, 16–19]. On all well ordered substrates the molecules are parallel to the surface irrespective of the surface bond – whether van der Waals (Al, TiO_2, Cu–O) or π bonding with significant charge transfer (e.g., 6P on Cu(110) [10]). The surface also orients the molecules azimuthally, leading to all molecules being uniaxial on the twofold symmetric substrates, while on the threefold symmetric surfaces the three equivalent molecular orientations in the monolayer lead to a multiplicity of symmetry equivalent crystallite epitaxial relationships. For growth at higher temperatures (≈400 K) the wetting monolayers are similar to at room temperature growth even though in the case of clean Al(111) and TiO_2(110) the second and higher layer molecules adopt a near vertical orientation – these epitaxial oriented films of 6P(001) grow on the monolayer of lying molecules.

7.2.2 Structure and Morphology Determinants: Sticking vs. Diffusion Anisotropy

The growth of sexiphenyl on $TiO_2(110)$ has been investigated by a large variety of techniques, probing both electronic and geometric structure, optical properties and morphology. It thus provides a reasonable comprehensive database for organic growth studies and the general conclusions that can be drawn from it are applicable to other molecule/substrate systems.

For growth at room temperature (RT) and below, crystalline nano-needles consisting of parallel molecules are formed, while at elevated growth temperatures large elongated islands of upright molecules result. The formation of the crystalline needles at room temperature cannot be understood in terms of thermodynamic growth, which would lead to near upright 6P(001) oriented crystallites (presenting the low surface free energy cleavage plane) and must be understood in terms of growth kinetics. The twofold symmetric surfaces not only orient the molecules but also give them a preferential diffusion direction. For $TiO_2(110)$ this is the [001] direction, i.e., perpendicular to the needle alignment, and thus diffusion anisotropy alone cannot explain this morphology. It can only be understood in terms of sticking to island anisotropy.

At low coverages grown at RT, the surface is decorated with needles such as shown in the STM image of Fig. 7.4a. These needles all have the same orientation along the [1$\bar{1}$0] substrate azimuth, i.e., perpendicular to the oxygen rows. As can be seen in the insert, the TiO_2 surface is relatively rough with numerous small terraces separated by monoatomic steps, which clearly have no influence on the growth direction of the needles. As we increase the coverage, the needles simply increase in number and size until the surface is completely tiled with 6P needles as illustrated in Fig. 7.4b. It is also interesting to note

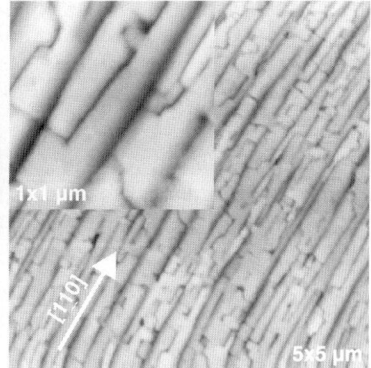

Fig. 7.4. (STM (**a**) and AFM ((**b**) z-scale 40 nm) images of p-6P films grown on the $TiO_2(110)$ substrate. The p-6P coverages are 3 Å (**a**) and 300 Å (**b**), respectively. The corresponding substrate azimuthal directions are indicated in the figures. The insets of (**a**) and (**b**) show higher magnification images

that they all have fairly uniform heights. Their regular shapes suggest that they are crystalline in nature, which has been confirmed by XRD and was found to be 6P($20\bar{3}$) (see Chap. 6). Assuming the bulk sexiphenyl structure, this implies that all the molecules are uniaxial, parallel to the substrate, and parallel to the oxygen rows of the substrate surface. XRD pole figures were also measured and demonstrate that these ($20\bar{3}$) oriented crystallites have a high degree of epitaxy with two symmetry equivalent domains [2].

The molecular orientation from the initial stages of formation up to device relevant thicknesses has been followed with C_K-edge NEXAFS for exposures from 3 Å to more than 300 Å. When the polarization of the incident X-rays is parallel to the surface and to the oxygen rows the absence of the $C_K \rightarrow \pi^*$ resonance indicates that the axes of the molecules are all parallel to the oxygen rows. Upon a detailed analysis of the angular behavior of the NEXAFS, an average angle of the aromatic planes with respect to the surface of $\pm 35°$ is found, which is in perfect agreement with the XRD and reflects the herringbone angle of the aromatic plane in the bulk structure [2].

The formation of these sexiphenyl nanostructures can only be understood in terms of sticking anisotropy. Here the key is the alignment of the molecules by the oxygen rows and their preferential diffusion in this direction. If we start off with a nucleus of lying molecules as shown in the schematic of Fig. 7.5, with preferential diffusion in the direction of the oxygen rows, i.e. [001], the weak interaction between the C–H groups at the end of the molecules will result in a low sticking probability when they approach at the side of a needle. While if the molecule comes in at the end of the needle, the stronger π–π interaction causes the molecules to stick and thus the needles grow perpendicular to the high diffusion direction. Therefore, sticking to island anisotropy drives the growth of these highly anisotropic needles.

For growth at 400 K (HT), in contrast to the needles obtained at RT and below, elongated islands are obtained, which run perpendicular to the needle

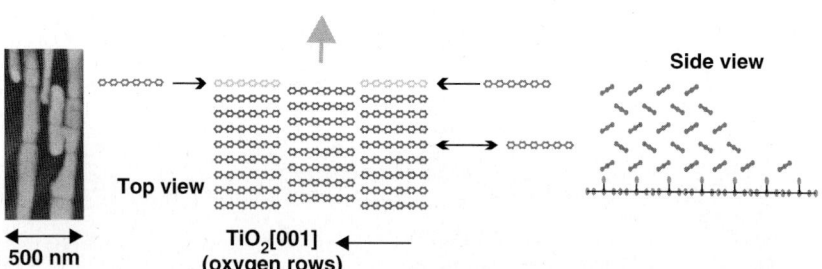

Fig. 7.5. The AFM image on the left is of 35 Å 6P grown at RT (z-scale: 30 nm). The schematic shows the 6P($20\bar{3}$) crystal structure as seen from the top and side of the RT grown needles. The orientation and diffusion direction of the molecules along the oxygen rows of the substrate are indicated. The resulting needle growth direction is shown by the large arrow above the top view. Reprinted with permission from [17]

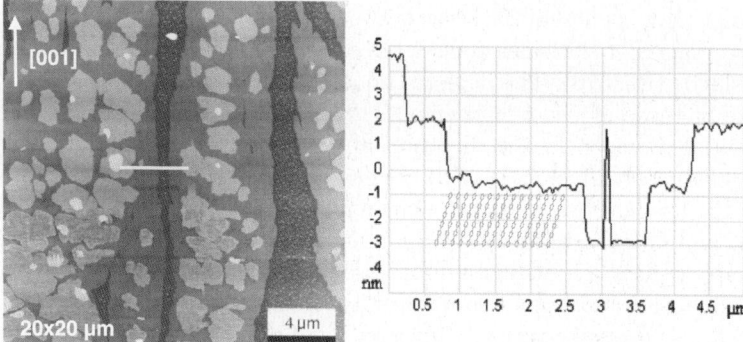

Fig. 7.6. 35 Å film of 6P grown on a $TiO_2(110)$ substrate held at an elevated temperature of 400 K, as observed by AFM (z-scale: 20 nm). The substrate [001] azimuthal direction is indicated by the large arrow. The linescan (**b**) is taken from the line in (**a**) and crosses a single crystallite between the large islands. Reprinted with permission from [17]

direction, i.e., parallel to the oxygen rows, as can be seen in the AFM of a 3.5 nm coverage in Fig. 7.6. In fact, while the widths of these islands do not exceed 10 µm, their lengths are much larger than 20 µm, the largest scan size of the instrument. In the line scan of Fig. 7.6, we can see that the steps are 2.7 nm high, approximately the height of an upright molecule. Second and third layers are also evident, but significantly, these layers no longer display the anisotropy that we see in the first upright layer and are more or less isotropic. Results from NEXAFS and XRD (see Chap. 6) show that films grown at HT indeed consist of upright molecules (6P(001) where the molecular axes is 13° from the surface normal). Additionally, the area between the larger islands is peppered with smaller crystallites, also consisting of upright molecules which we shall discuss in more detail later.

Like the needles that form at room temperature, there is also an azimuthal relationship of the sexiphenyl crystallites to the substrate, as revealed by monolayer sensitive XRD (see Chap. 6). Grazing incidence XRD (GID) shows that at a coverage of 35 Å, comparable to the AFM in Fig. 7.6, the crystallites are oriented with the (001) plane parallel to the surface, where the molecular axis is tilted 13° from the surface normal [20]. Four major equivalent domains are observed, where the molecules are not tilted along either of the major azimuths of the substrate, but rather between them. These four orientations allow half the bottom phenyl rings of the 6P, those in contact with the underlying surface, to line up with their aromatic planes parallel to the oxygen rows beneath, and with a spacing near commensurate to that between the oxygen rows. This orientation of the molecular plane to and the near commensurability with the substrate rows presumably leads to the domination of these particular domains.

Closer examination of the areas between the larger first layer islands grown at HT reveals that there is a zone depleted of the small crystallites around the large islands (Fig. 7.6). The depleted zones on the sides of the islands are half the size of those at the island ends. This is a clear indication that we have preferential diffusion in the direction of the oxygen rows (i.e., along [001]). This preferential diffusion leads to the elongation of these islands in the direction of the oxygen rows and thus we can understand the anisotropic islands of upright molecules in terms of diffusion anisotropy; naturally for near-upright molecules sticking anisotropy is suppressed. This is also supported by the fact that the second and higher layers are less anisotropic, as the high diffusion anisotropy of the substrate's lying monolayer is no longer present on top of the first layer of upright molecules. Diffusion anisotropy driven growth has been well documented in the study of inorganic on inorganic growth [21, 22], and a number of simulations to relate the width of the depletion zones to the diffusion anisotropy have been undertaken on systems such as homoepitaxial Si/Si(001)-(2 × 1) [21, 22]. These studies relate the width of the depletion zones W and the diffusion constant D with a scaling exponent γ as $W_f/W_s = (D_f/D_s)^\gamma$, where the subscripts f and s refer to the fast and slow diffusion directions, respectively. There is disagreement as to the value of γ. Mo et al. [21] find values of γ from 1/6 in the isotropic diffusion limit to 1/4 for completely anisotropic diffusion, while a newer theoretical treatment by Ebner et al. [22] suggests a value of 1/2, essentially independent of various conditions, including the magnitude of the diffusion anisotropy. Using the value of $W_f/W_s = 2$ extracted from the denuded zones around the islands of upright molecules [17], we can conclude that the 6P diffuses somewhere between 4 and 64 times faster in the direction of the [001] substrate azimuth depending on the choice of γ. While significant, it is much less than for inorganics such as Si/Si(001)-(2 × 1) and Ge/Si(001)-(2 × 1) where values in the order of 1000:1 are found [23].

The principal driving forces that determine the assembly of organic nanostructures and their morphology is the balance between diffusion and sticking anisotropy. In the organics, as opposed to inorganic systems, the basic building blocks, the molecules themselves, are highly anisotropic objects. This leads to an inherently strong sticking anisotropy and ultimately the formation of large aspect ratio needles that have been reported on a number of different surfaces [2, 24–26]. The explanation here for sexiphenyl needle formation on $TiO_2(110)$, where the needles are perpendicular to the high diffusion direction, can also explain the sexiphenyl needles on the alkali halide (100) surfaces [26, 27]. These are, however, fourfold symmetric substrates and thus the needles grow along the two orthogonal principle substrate azimuth directions.

A second factor that influences the island growth direction is the nature of the two-dimensional unit cells. In the case of sexiphenyl, a rectangular two-dimensional unit cell exists, that of the 6P($20\bar{3}$) plane, and growth purely orthogonal to the diffusion direction can occur. For sexithiophene (6T) this is not the case. On $TiO_2(110)$ [18] and Cu(110)-(2 × 1)O [19], 6T (010) oriented crystallites grow; 6T(010) has a similar packing to 6P(20$\bar{3}$) in that molecules in the planes parallel to substrate alternate in their herring bone angle. The 6T(010) planes two-dimensional unit cell, however, is rhomboidal with an angle of 23°. Consequently, the crystallites that grow have their needle directions ±23° from the orthogonal to the diffusion direction [18].

It is interesting to speculate on what growth would occur if the monolayer template fits another particular two-dimensional unit cell. For instance on Al(111) the monolayer of sexiphenyl has a rhomboidal unit cell with an angle close to that of the 6P(21$\bar{3}$) plane's two-dimensional unit cell. On this monolayer at room temperature only 6P(21$\bar{3}$) oriented crystallites do indeed grow, despite the fact that it has a very open surface and can be expected to have a higher surface energy than 6P(20$\bar{3}$) [8, 28].

Clearly, a number of factors will determine the dimensions of the organic three-dimensional structures: the particular substrate, the substrate temperature, the particular molecule, and the evaporation rate. This huge parameter space is as yet barely explored; however, a number of illustrative results can be given. For instance on the TiO_2(110) and Cu(110)-(2×1)O substrates 6P(20$\bar{3}$) structures form in a similar manner, but, for the same temperature and evaporation rate they have different crystallite sizes. For exposures that completely cover the surfaces, the 6P(20$\bar{3}$) crystallites on TiO_2 measure $\approx 2 \times 0.2\,\mu m^2$, while on Cu(110)-(2×1)O they are much smaller $\approx 0.5 \times 0.1\,\mu m^2$. A number of reasons for this can be speculated upon. It cannot be explained by substrate defect density, in terms of nucleation sites, as Cu(110)-(2×1)O is the less defective and a much smoother substrate. It may be a result of strain effects in the first monolayer given that, as mentioned before, the metal–oxygen rows spacings of the two substrates are different. It could also result from the much higher step density on the TiO_2, which perhaps allows the molecules to "change tracks" (the particular metal–oxygen row along which they diffuse) at step edges and reach the sites, where they can bond at the needle ends.

Till now only substrates with long-range atomic order have been considered, where structures made up of molecules lying parallel to the substrate are observed. Our results on disordered surfaces, such as oxidized Al(111) [1] and Si(111) [29] and sputtered TiO_2(110) [18], suggest that disordered substrates lead to films of upright molecules. The importance of substrate order, or lack of it, is vividly illustrated in Fig. 7.7. The results of Fig. 7.7 are from a high exposure of sexithiophene (350 Å) at RT on a $10 \times 10\,mm^2$ TiO_2(110) crystal. During surface preparation a crack developed in the crystal such that the annealing temperature was not uniform, causing the (1×1) substrate LEED pattern of one half of the crystal to be more diffuse than that of the other. The resulting molecular film morphologies and electronic structures are profoundly different. On the better ordered substrate half, large needle-like 6T(010) oriented crystallites of up to several microns in length and 100 nm in height are observed, which are oriented at approximately ±23° to the [1$\bar{1}$0] substrate azimuth direction (Fig. 7.7a) as discussed earlier. The morphology and electronic structure that forms on the less well ordered half of the substrate

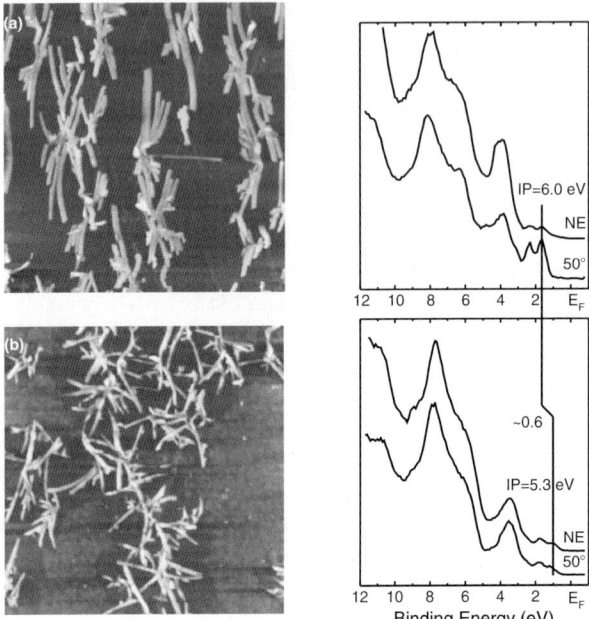

Fig. 7.7. AFM images ($10 \times 10\,\mu m^2$) from two regions of a $TiO_2(110)$ crystal after exposure to 350 Å of 6T at room temperature. The angle resolved UPS spectra in normal emission and 50° off-normal emission from the regions (**a**) and (**b**) expressing the differences in ARUPS behavior, HOMO alignment, and ionization potential. Reprinted with permission from [30]

(Fig. 7.7b) is very different. This surface also has large 6T crystallites of around 100 nm in height, but these have no preferred orientation. The majority of the surface is, however, covered by large terraces of between 5 and 15 nm in height, which can be assigned to 6T(100), i.e., terraces of near up-right 6T of 2–6 up-right monolayers in thickness. As surface sensitive techniques, such as photoemission and X-ray absorption spectroscopies, are dominated by the regions with the largest surface area, they will reflect the terraces of upright molecules and not the majority of molecules, which are contained in the needles. Significantly, growth on the disordered $TiO_2(110)$ substrate, despite being more "metallic" and reactive than the ordered, results in films of upright molecules. The dominant factor for molecule/crystallite orientation is the substrate order and not the chemical interaction/metallicity, as often thought [31] is also born out in organic–organic heterostructure growth (see Sect. 7.3).

The substrate temperature will change the molecular mobility and the degree of diffusion anisotropy. Interestingly, low temperatures do not change the substrates ability to orient the molecules [1, 32]. At 90 and 300 K, 6P(20$\bar{3}$) grows on $TiO_2(110)$. The crystallites grown at 90 K are very small

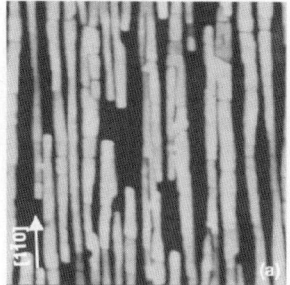 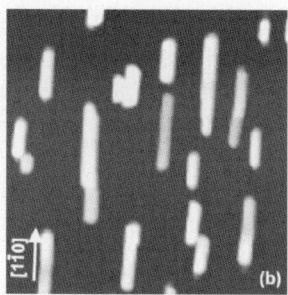

Fig. 7.8. $2.5 \times 2.5\,\mu m^2$ AFM images of 30 Å 6P on $TiO_2(110)$ deposited at (**a**) room temperature and by (**b**) sequential deposition of 10 Å at RT followed by 20 Å at 450 K. The Z-height greyscale range is 40 nm for (**a**) and 120 nm for (**b**). The $[1\bar{1}0]$ azimuthal direction of the substrate is indicated

at $0.1 \times 0.05\,\mu m^2$, compared to the $\approx 2 \times 0.2\,\mu m^2$ observed for room temperature growth. This can be easily understood in terms of difference in diffusion length/mobility with temperature. Growth at elevated temperatures can, but does not necessarily result in different crystallite orientations. As mentioned above, growth of 6P on TiO_2 at 400 K results in near upright molecules (6P(001)), whereas this does not occur in Cu(110)-(2 × 1)O or for 6T on TiO_2 (even near sublimation temperature).

A particularly interesting result is illustrated in Fig. 7.8, which shows AFM images of $6P(20\bar{3})$ needles formed by the same amount of 6P on $TiO_2(110)$ applied under different substrate temperature conditions (both nominally 30 Å). Figure 7.8a shows the long needles with 10 nm height that grow at room temperature, while those of Fig. 7.8b were created by first evaporating 10 Å at room temperature followed by 20 Å at 450 K. The lateral size of the crystallites of Fig. 7.8b is close to that expected for a 10 Å exposure at room temperature [2], but their heights, of 60–80 nm, suggest that the molecules applied at elevated temperature have grown on top of the initial needles and have adopted their structure. Thus, although the growth conditions for the thermodynamically favored 6P(001) orientation on TiO_2 were used, it did not arise. The initial $6P(20\bar{3})$ crystallites have acted as seeds for the further growth. Such seeding could allow variation of substrate temperature during growth to be used to control the three-dimensional shape of organic nanostructures.

Both 6P and 6T are chain-like molecules of comparable lengths that adopt the same orientation on $TiO_2(110)$. The size of their crystallites for the same substrate temperature and evaporation rate are different. While the lateral dimensions of the crystallites of the two molecules are comparable, their heights are vastly different. While the 6P has completely tiled the substrate with crystallites of 30 nm height, the 6T structures only cover a fraction of the substrate and are much higher at 150 nm (compare Fig. 7.4b and Fig. 7.7a).

It is yet too early to speculate on the reasons for these differences, or on growth in the third-dimension in general, more work needs to be done.

7.2.3 The Electronic Structure

UV-photoemission not only yields the valence band electronic structure and the important electronic level alignment, it also can elucidate the growth mode of the organic films/structures. Figure 7.9 shows the development of the valence band structure for increasing sexiphenyl exposure at room temperature on $TiO_2(110)$. As can be seen in Fig. 7.9a, despite the substrate's O_{2p} derived features being very strong, at only 3.5 Å exposure they are almost completely suppressed and, moreover, the change in work function is completed. Beyond this exposure the intensity of the sexiphenyl emissions grow very gradually and shift slightly (with respect to the Fermi level) until ≈ 180 Å, where the needles completely tile the substrate and beyond which there are no further changes. The saturation of the work function change and the suppression of the substrate features by 3.5 Å clearly indicate that a wetting monolayer of lying molecules is formed. It should be noted that neither room temperature STM nor LEED could detect this monolayer; weakly interacting large band gap materials are difficult to image in STM, while in LEED they are unstable due to electron beam damage. The UV-photoemission spectra were obtained with a standard angle integrating electron spectrometer (solid collection angle 16°) and an unpolarized light source; despite this, extreme differences are observed in spectra in different experimental geometries. This

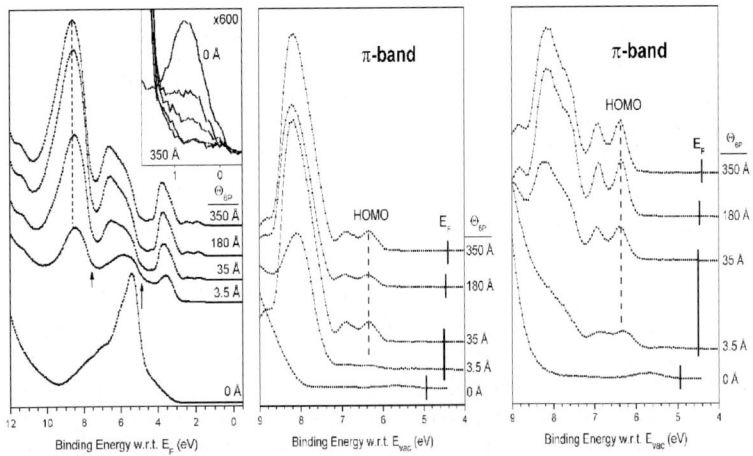

Fig. 7.9. Valence band structure for increasing sexiphenyl coverage on $TiO_2(110)$ grown at RT followed with He I UPS along the [001] substrate azimuth. The π-band has been enlarged and the spectra shifted by their work function in (**b**) and (**c**). The spectra of (**a**) and (**b**) were taken in normal emission, (**c**) in 45° off-normal emission geometry

is illustrated by the two panels that show the π-band for normal and 45° off-normal electron emission. From the monolayer in normal emission the HOMO and HOMO-1 orbitals are all but invisible, while the nonbonding orbitals at 8 eV below the vacuum level are very strong. In going to off-normal emission the relative intensities are reversed, with the nonbonding orbitals suppressed and HOMO, HOMO-1, and HOMO-2 being strong. These spectra were recorded in the [001] azimuth, i.e., the direction parallel to the molecules in the monolayer. The same experiment with the experimental plane perpendicular to the molecules (i.e., [1$\bar{1}$0]) shows different relative intensities, but no significant variation with take-off angle. Similar behavior is observed for other twofold symmetric substrates and other chain-like molecules (sexiphenyl, sexithiophene, pentacene). Clearly experimental geometry is an important parameter to note when recording spectra (often neglected in the literature) and even standard laboratory UV-photoemission can give valuable information on the molecular geometry.

It is noted in Fig. 7.9 that there are no significant shifts in the energy of the molecular emission features up to a coverage of 350 Å. In following the same amount of material on the Al(111) substrate, shifts to higher binding energy are observed. These arise from differences in growth morphology: on TiO$_2$ the surface is completely tiled by sexiphenyl, while on Al(111) small, but tall sexiphenyl islands grow and charging thus occurs at a lower exposure. Consequently, the morphology cannot be completely ignored when investigating the issue of electronic level alignment.

The band alignment question has been prominent in the literature of organic semiconductors in the last 10 years because of its importance to charge injection at the contacts of organic devices. The ability of photoemission to give the position of the HOMO with respect to the Fermi level and the work function have made it an important tool for studying charge injection barriers. Figure 7.10 schematically illustrates the three views of band alignment that can be found supported in the literature along with the relationships between the energy of the HOMO with respect to E_F (band alignment ϵ) to the substrate work function they would imply. The first is vacuum level alignment. A plot of the highest occupied molecular orbital (HOMO) binding energies with respect to E_F, measured by ultraviolet photoemission (UPS) from multilayer organic films, vs. ϕ_{sub} would yield a 45° line (the Schottky–Mott limit). There is no support for this simple picture although those involved in the measurement of macroscopic device properties more or less unwittingly still invoke the vacuum level (E_{vac}) alignment by virtue of their insistence on claiming to choose contact materials due to their work function. The second is Fermi level alignment with band bending, which would result in no variation in the band alignment irrespective of the substrate material. There is no evidence for this from controlled studies with pure molecules and clean substrates. This is not surprising as the band bending would require a high charge carrier concentration via doping with contaminants.

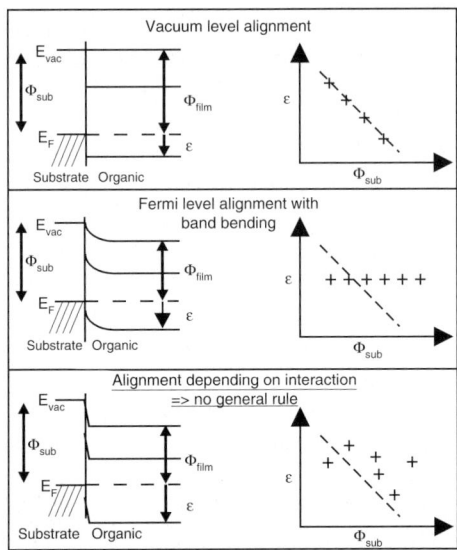

Fig. 7.10. Schematic of the three possible band alignment views together with the corresponding band alignment vs. substrate work function dependence. The dashed line indicates the Schottky–Mott limit

In contrast, controlled investigations prove that there is no general simple relationship and a scatter is found in plots of ϵ vs. the substrate work function. In our work with well controlled model systems (in-situ prepared oligomer films on single crystal substrates) we have shown that the ability to inject charge is not an intrinsic function of the contact material but rather a function of the specific interaction of the particular organic with the particular inorganic contact surface. Specifically, because of the interface dipole (Δ) arising from the interaction of the first molecular monolayer with the substrate. Naturally this can be heavily influenced by the nature and degree of contamination of the metal contact surface or, seen positively, the band alignment can be tailored by controlled modification of the surface of the inorganic [18, 33].

As a substrate work function is a result of the surface dipole arising from the spill out of charge from the surface, it is one of the most surface sensitive observables. A strong chemical bond with charge transfer is not necessary for a substantial change in the dipole. For instance, it has been shown theoretically and experimentally that the electrostatic bond of benzene to Al(111) leads to almost 0.5 eV lowering of the work function on completion of the monolayer. This results from a reduction of the electron density in the surface below the molecules due to Pauli repulsion [4], sometimes referred to as push-back in the organic electronics community.

Work functions are measured with area averaging techniques, and when growing molecular films it is seen to gradually change until the wetting monolayer is completed. For growth beyond the monolayer no further changes are observed. This monolayer work function determines the band alignment, rather than that of the clean substrates, and plots of ϵ against the monolayer work function do not yield a scatter but rather the 45° line similar to that of the Schottky–Mott limit of Fig. 7.10a. This is simply a reflection of the basic relationship between the HOMO binding energy with respect to the vacuum level, that is the ionization potential IP, and the Fermi level (ϵ) via the work function ϕ,

$$\mathrm{IP} = \epsilon + \phi.$$

For the chain-like molecules 6P and 6T such plots for various substrates and growth conditions do not yield a single 45° line, but rather two lines as displayed in Fig. 7.11, suggesting two distinct ionization potentials are possible for these molecules in the solid state. Invariably, when the molecules are oriented parallel to the substrate in either monolayers or three-dimensional crystalline structures (6P($20\bar{3}$), 6P($21\bar{3}$) or 6T(010)), they have a considerably higher ionization potential than when they are oriented near vertical to the substrate (6P(100) and 6T(100)). We believe this to be a result of differences in the conformation of the molecules on the different molecular crystal faces, i.e., a true ionization potential difference rather than a final state photoemission effect, due to differences in extra-molecular screening of the photoholes. These ionization potential differences can be seen in the UPS spectra for the two different morphologies 6T of Fig. 7.9 ($\Delta \mathrm{IP} = 0.7\,\mathrm{eV}$) or in the π band spectra of 6P in Fig. 7.11. The latter shows a comparison of the π bands of gas phase 6P with that of films of lying 6P and upright 6P on Al(111). The gas phase spectra have been shifted by 1.2 eV to account for the extra-molecular

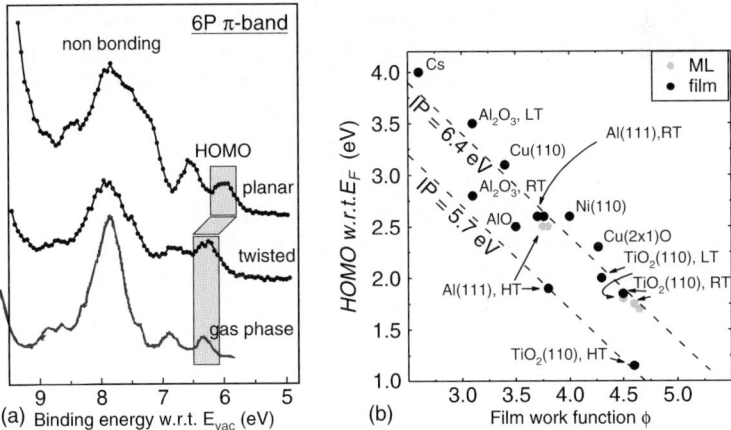

Fig. 7.11. (a) Sexiphenyl π-band for films of planar and twisted sexiphenyl compared to the gas phase spectrum. The UPS spectra are referenced to E_{vac}, and the two ionization potentials observed are indicated. (b) Band alignment vs. sexiphenyl film work function for condensed sexiphenyl films on different substrates. The two ionization potential are indicated by the lines

relaxation in the solid. The solid state spectra are referenced to the vacuum level by shifting them by the work function, which was the same for both, as both films are on the same wetting monolayer of lying 6P. The films of lying 6P have the same spectral shape, π band width, and ionization potential as the gas phase molecules, while the spectra from films of upright 6P have a greater π band energy width and 0.7 eV lower IP. Gas phase 6P is known to have a significant torsional angle between the phenyl rings, while in the solid state it is on average planar [34]. Electronically, in going from a twisted to a planar conformation increases the interring orbital overlap, the energy spread of the π band orbitals, and consequently results in a decrease in the first ionization potential. The change in IP has been calculated to be 0.5 eV, close to the differences measured [8,9]. The UPS from the organic crystal faces, where the molecules are parallel to the substrate, is that of the twisted molecules suggested to be a result of a novel form of surface reconstruction on these high surface energy faces. Unlike the surfaces of near upright molecules (6P(001) and 6T(001)), the molecules on these surface do not have the same packing as in the bulk and adopt a twisted conformation to lower the surface free energy. UPS is a very surface sensitive technique (5–10 Å or one or two lying molecular layers) and it is as yet an open question as to how deep into the bulk this change in conformation, and concomitant change in electronic structure, extends – this question is far from irrelevant when considering organic structures with nanometric dimensions.

7.2.4 The Electronic Band Alignment on Nanostructured Interfaces

Historically, the field of active molecular systems has been evolving from "plastic electronics" through "organic electronics" to "molecular electronics," i.e. polymers, better defined oligomeric films, to the holy grail of single molecule devices. Clearly, an understanding of the electronic level alignment on the molecular scale is vital for molecular electronics or devices based on organic nanostructures, but it can also be important in present day devices as their contact interfaces are unlikely to be uniform on this scale, an intricacy, which may lead to charge trapping at the interface and reduced device performance. The area of such defects can also be on the nanoscale and thus in the order of the here investigated organic nanostructures.

To investigate the electronic band alignment on nanoscopically patterned or inhomogeneous substrates requires highly controlled substrate templates such as the so-called Cu–CuO stripe phase, which consists of alternating stripes of clean Cu(110) and (2 × 1) oxygen reconstructed Cu(110), with respective stripe widths in the range of several nanometers depending on the amount of oxygen exposure; in the following respective stripe widths of 35 Å, as shown in the STM image of Fig. 7.12, were used.

In Fig. 7.13a, the UPS valence band spectrum of a monolayer coverage of sexiphenyl on clean Cu(110) and on the fully reconstructed Cu-(2×1)O surface

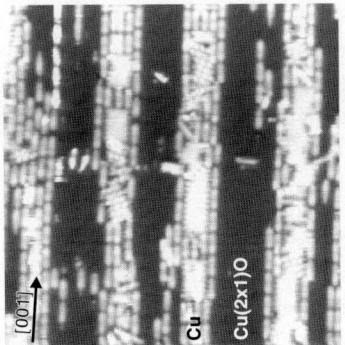

Fig. 7.12. STM image (25×250 Å^2) of half a monolayer sexiphenyl on a Cu–CuO stripe phase (1:1 ratio). The molecules have decorated the bare copper stripes and are preferentially aligned along the stripe direction

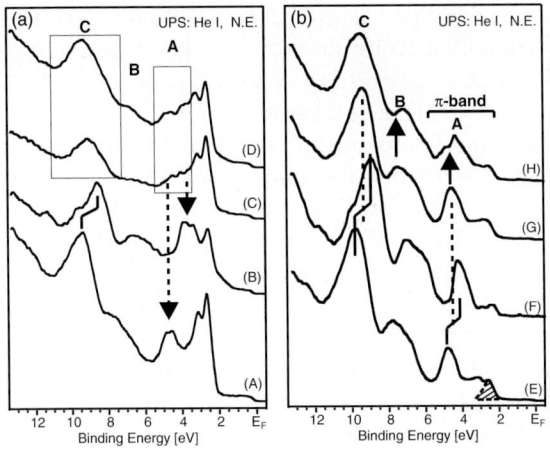

Fig. 7.13. Normal emission He I UPS spectra of sexiphenyl monolayers (**a**) and multilayers (**b**): (**a**) sexiphenyl monolayers on (A) clean Cu(110) ($\phi = 3.4$ eV), (B) on the fully reconstructed Cu(110)-p(2×1)O ($\phi = 4.3$ eV) surface, and (C) on the Cu(110) (2×1)O stripe phase ($\phi = 3.8$ eV). (D) weighted (1:1) sum of (A) and (B). Similarities between (C) and (D) of bands A and C are highlighted by boxes, the arrows indicate the position of band A on the homogeneous surfaces. (**b**) sexiphenyl multilayers on (E) clean Cu(110) ($\phi = 3.4$ eV), (F) on the fully reconstructed Cu(110)-p(2×1)O surface ($\phi = 4.3$ eV), and (G) on the Cu(110) (2×1)O stripe phase ($\phi = 3.8$ eV). (H) weighted (1:1) sum of (E) and (F). The major differences between (G) and (H) are marked by arrows. Reprinted with permission from [35]

is compared to a spectrum of a monolayer on the heterogeneous striped surface. As indicated in the figure all molecular orbital emissions of the homogenous surfaces are offset by 0.9 eV from each other (equal to the difference in the monolayer work function ϕ) due to the differences in the interface dipole.

It should be noted that the clean substrates work functions differ by only 0.4 eV. Adding these two spectra results in a composite spectrum, which is remarkably similar to the spectrum recorded for the monolayer on the heterogeneous Cu-(2×1)O stripe phase. This suggests that the molecular band emissions from two clearly distinct band alignments are observed in the photoemission spectrum and that the sexiphenyl molecules adsorbed on the bare copper stripes feel the local surface potential of the clean copper, while the sexiphenyl molecules on the oxygen reconstructed stripes experience the local surface potential of the Cu(2×1)O surface. In many respects the molecular monolayer emissions behave as a local work function probe, similar to the photoemissions of adsorbed Xenon (PAX) technique. Interestingly, the secondary electron cut-offs, used to determine the work function, of all surfaces were sharp and did not show a double cut-off that would reflect the two different local surface potentials, and these are thus no indicator for surface homogeneity.

In contrast the binding energy of the molecular orbitals of the multilayers on the striped substrate are situated between those obtained on the homogenous substrates, suggesting that we have a single band alignment with an average offset. This is illustrated in Fig. 7.13b, where the bottom three spectra show UV-photoemission spectra of a 50 Å thick sexiphenyl film, corresponding to ten molecular monolayers on clean Cu(110), fully (2×1) oxygen reconstructed Cu(110), and the Cu–CuO striped substrate (equal stripe width), respectively. Again on the homogenous surfaces there is a 1 eV difference in band alignment, which is due to the differences in the interaction of the molecules in their respective monolayers. Comparing the multilayer film spectrum on the striped Cu–CuO substrate to the weighted sum of spectra for multilayers on the homogeneous substrates shows that they are markedly different and, unlike for the monolayer, the spectra from the multilayer on the striped substrate are not due to a superposition of emissions offset by their respective interface dipoles. Therefore, the sexiphenyl molecules of the higher layers do not feel the local surface potential of the Cu and CuO stripes any longer, but rather are aligned to the average interface potential. With regard to when the paradigm shift from band alignment to the local interface potential as opposed to band alignment to the average interface potential occurs, the results suggest beyond the second layer, i.e., at a height less then the stripe widths that might be expected from a first approximation. This is illustrated in the schematic of Fig. 7.14.

The consequences of these results are that care must be taken in interpreting orbital shifts with increasing exposure on ill-defined/inhomogeneous substrates, such as contaminated substrates or even clean polycrystalline surfaces (consider different crystalline facets of metals can have 0.5 eV difference in work function). One should also be aware that substrate inhomogeneities, if significant enough, can only be detected in the first monolayer. Further implications include that future nanopatterned devices, relying on the local variation of the interface dipole, will ultimately require thin active organic layers only 1–2 molecules thick [35].

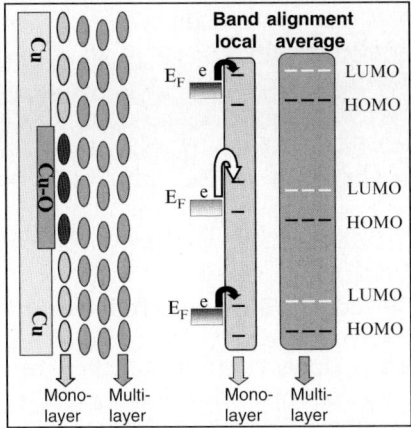

Fig. 7.14. Schematic illustrating the observed change from local to average band alignment on the nanopatterned substrate. The shaded discs represent sexiphenyl molecules, with the grey scale indicating the respective band alignment. Reprinted with permission from [35]

7.3 Organic–Organic Heteroepitaxy

Highly crystalline organic heteroepitaxial layers with controlled molecular orientations and morphologies are one of the keys for optimum organic device performance. However, up to now controlled studies have mainly focused on the growth of organic nanostructures on inorganic substrates, with few studies of organic nanostructure formation on organic substrates, so called organic heteroepitaxy being available. A possible reason for this deficiency in controlled studies of molecular heterostructures is the difficulty in producing uniform crystalline organic substrates with different well defined molecular orientations, which are a necessary prerequisite for such investigations.

As the very inhomogeneous morphologies with different molecular orientations suggests, studies of the interfaces and growth on organic films are fraught with difficulties. Clearly simplistic interpretations based on analysis with area averaging techniques can be misleading and should be treated with care. Even when films of a single molecular orientation are produced, these are not necessarily homogeneous. With care, and high enough exposures, nanocrystalline pyramidal and needle-like structures consisting of standing and lying molecules, respectively, can be grown such that they completely cover the substrates. However, the surfaces are laced with deep crystallite boundaries. In the case of homogeneous surfaces of upright molecules the resulting terraced surface has steps the height of the molecular length (≈ 25 Å). Not only are the step heights much larger than those on single crystal inorganic surfaces, but also the step faces and terrace surfaces are very different; the former present the reactive π system while on the latter the unreactive –CH ends of the molecules are exposed (Fig. 7.15) [16, 19].

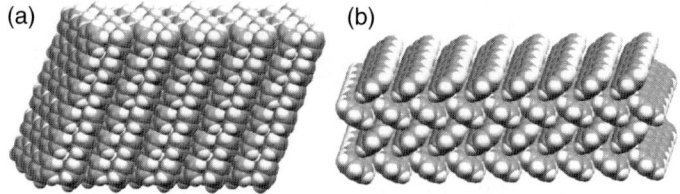

Fig. 7.15. Schematic models of the (**a**) 6P(001) and (**b**) 6P(20$\bar{3}$) crystallite surfaces

7.3.1 Growth on Closed –CH-terminated Organic Surfaces

Upright (001) oriented sexiphenyl substrate films, as a model for unreactive –CH terminated organic substrates (see Fig. 7.15a), can for instance be achieved by depositing 6P on an oxidized Al(111) surface at room temperature. Figure 7.16 summarizes the C_K-NEXAFS and AFM results of the growth of sexithiophene on 6P(001), which characterize the molecular orientation and morphology of this heterostructures, respectively. Figure 7.16a shows the spectra for 80 Å 6P grown at room temperature on the disordered surface of oxidized Al(111). The strongest π intensity occurs for normal incidence and the weakest for glancing incidence, suggesting a film of upright 6P. Analysis of the angular behavior yields an average angle of the aromatic planes with respect to the surface of $77° \pm 5°$, in agreement with a 6P(001) orientation, which on this substrate has also been confirmed by XRD. Clearly the agreement of the NEXAFS rules out any significant areas of 6P with another crystallite orientation in this film. 6T was then grown on this film in a stepwise fashion while monitoring both photoemission and NEXAFS. The signal of 6T only starts to dominate in the spectroscopies for exposures above 15 Å, indicating that wetting layer of lying 6T does not cover the upright 6P (note that this is contrary to what is often observed on ordered inorganic substrates). Analysis of the NEXAFS spectra after 165 Å 6T exposure, shown in Fig. 7.16b, indicates that the 6T molecules are near up-right with an angle of $75° \pm 5°$ to the substrate surface. This is very similar to the underlying 6P angle but significantly larger than that of 6T(001) of 66°. This result cannot be due to a mixture of other 6T orientations as these would lead to a lowering not an increase of the average angle. It is suggested that the 6T has adopted the orientation of the underlying 6P in this heterostructure.

The morphology of this heterostructures is shown in the AFM image of Fig. 7.16c. The surface primarily consists of angular-shaped stepped pyramidal islands with a base width of typically several hundred nanometers and an average height of 20 nm. These pyramidal structures have step heights of typically around 2.8 ± 0.3 nm and 2.2 ± 0.3 nm. These values are in good agreement with expected step heights for (001) oriented p-6P (2.7 nm) and 6T (2.2 nm), respectively. It should be noted that these interlayer spacings are slightly smaller than the molecular lengths due to the tilt angle of the molecules in their crystalline structure (Fig. 7.16d). The statistical analysis

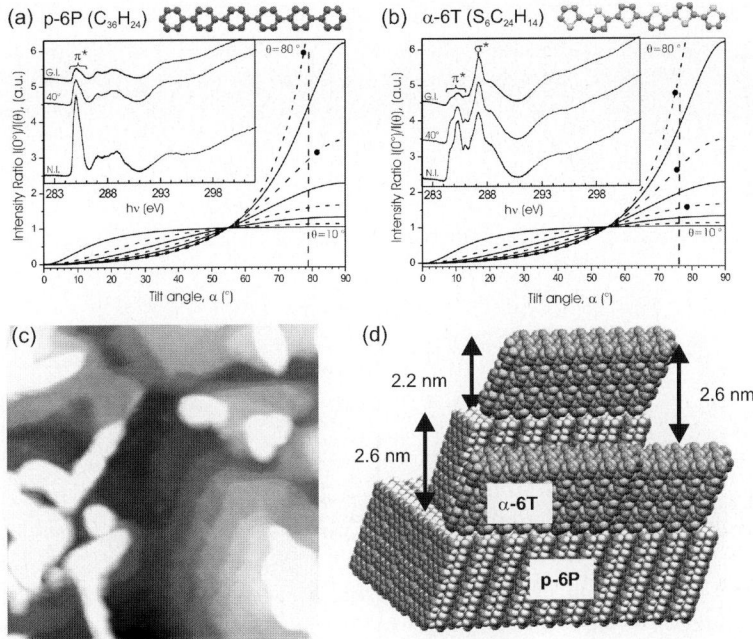

Fig. 7.16. Calculated intensity ratio plots for an (**a**) 80 Å thick p-6P film on AlO$_x$ and (**b**) after deposition of 165 Å 6T. The dashed lines indicate the concluded tilt angle of the respective molecular planes. The corresponding NEXAFS spectra are shown as inserts. (**c**) AFM image of the 165 Å 6T/80 Å 6P heterostructure. (**d**) The schematic model shows the near upright molecular orientations in the heterostructure. Reprinted with permission from [16]

performed on the angles observed between the terrace edges of the pyramids yields values that are expected between close-packed facets of 6P(001) and 6T(001) oriented crystallites and is thus a clear indication of the crystallinity of these pyramids [16].

7.3.2 Growth on Open π-terminated Organic Surfaces

The more challenging organic substrates are the open π-terminated surfaces, as they cannot be grown in thermal equilibrium or obtained by cleaving. Films of purely lying molecules are difficult to grow on inorganic substrates as defects or contaminated regions lead to the thermodynamically most stable phase with upright molecules, unless very stringent UHV substrate preparation is made (see Sect. 7.2.2).

Figure 7.17a shows the NEXAFS spectra from a 400 Å thick film of 6P grown on TiO$_2$ at RT for normal and glancing incidence in the two principle azimuths of the substrate. XRD and AFM of this film showed the substrate to be completely tiled with 6P(20$\bar{3}$) crystallites. For polarization parallel to

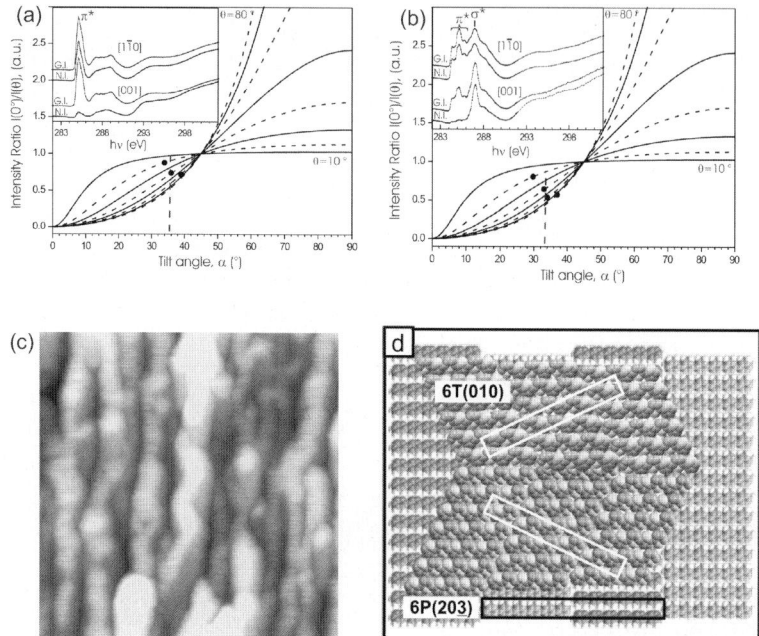

Fig. 7.17. Calculated intensity ratio plots for an (**a**) 40 nm thick p-6P film on TiO_2 and (**b**) after deposition of 40 nm 6T. The dashed lines indicate the concluded tilt angle of the respective molecular planes. The corresponding NEXAFS spectra for normal and gracing incidence and in the [001] and [1$\bar{1}$0] azimuths of the inorganic TiO_2 support are shown as inserts. (**c**) AFM image ($1 \times 1\,\mu m^2$) of the 6T/6P heterostructure. (**d**) Model of the uniaxially orientated heterostructure in top view, showing the two symmetry equivalent domains of 6T(010) on 6P(20$\bar{3}$). The respective surface unit cells are indicated. Reprinted with permission from [16]

the [001] azimuth in normal incidence the π intensity is very low, indicating that the molecules are aligned along this direction. Analysis of the intensity variation in the [1$\bar{1}$0] azimuth yields a tilt angle of the aromatic plane to the substrate surface of $36° \pm 5°$, in agreement with the herring-bone angle of bulk 6P. Stepwise growth of 6T on this 6P film was monitored up to an exposure of 400 Å. The NEXAFS of this heterostructure, displayed in Fig. 7.17b, shows the 6T molecules have the same orientation as the underlying 6P, i.e., both molecular species are uniaxially aligned throughout the whole heterostructure film. The molecular axes are parallel to the [001] azimuth of the underlying TiO_2 substrate and the angle of the aromatic plane to the substrate is 33°, slightly smaller than the underlying 6P but in good agreement with the herring bone angle in 6T [16].

The AFM image of Fig. 7.17c shows the morphology of this 40 nm thick 6T film on 40 nm thick 6P(20$\bar{3}$). The surface appears highly anisotropic with worm-like structures following the general direction of the underlying sexiphenyl crystallites. Typical dimensions of these 6T structures are up to a few micrometers in length and 80 ± 10 nm in width. As well as being in general thinner, the 6T structures are less regular than the underlying 6P crystals. On close inspection the elongated structures are seen to be segmented and often have a zigzag morphology in the direction of the underlying 6P needles, i.e., parallel to the substrates [1$\bar{1}$0] azimuth.

The epitaxial relationship between sexithiophene, sexiphenyl, and the inorganic support TiO_2 has been determined by an XRD pole figure measurement around the TiO_2(110) surface. This again confirms that the 6P(20$\bar{3}$) and the 6T(010) crystal planes are parallel with respect to each other and moreover it implies that the sexithiophene and sexiphenyl molecules are uniaxial throughout the whole heterostructure. As shown in the model of Fig. 7.17d this crystal orientation leads to two symmetry equivalent domains of 6T(010) on 6P(20$\bar{3}$), which explains the formation of the zigzag shaped crystallites observed in the AFM images. The intermolecular spacing of the 6P and the 6T molecules is with 5.57 and 5.52 Å very similar (mismatch $\approx 1\%$). However, the lattice mismatch between the 6P(20$\bar{3}$) and the 6T(010) unit cells (Fig. 7.17d) is around 6% for the molecules being all parallel to each other. High resolution synchrotron X-ray diffraction experiments determined the angle between the 6T and p-6P unit cells to be 22.5°, implying that the axes of the sexithiophene molecules are rotated by 1° relative to sexiphenyl. This is interesting because this very small 6T to 6P misalignment also leads to a better lattice match (mismatch $\approx 2\%$) in the 6P(010) direction than if the 6T molecules were perfectly parallel to the 6P molecules, and suggests that lattice match can be an issue even for weakly interacting organic heterostructures [16].

For the reverse system, the growth of sexiphenyl on lying sexithiophene (6T(010)), another interesting aspect of organic heteroepitaxy, is demonstrated. Grazing incidence diffraction (GID) measurements reveal that the 6P nanostructures grow with two different crystallite orientations, the 6P(20$\bar{3}$) and the 6P(21$\bar{3}$) orientation, on the uniaxially oriented 6T substrate. In both orientations the molecules lie parallel with respect to the substrate, but they are no longer exclusively uniaxial. The 6P(20$\bar{3}$) orientation on 6T(010) is simply the reverse of the system discussed before and here all molecules are uniaxially aligned. However, the 6P(21$\bar{3}$) orientation is different. The characteristics of this contact plane is that it is more open than the 6P(20$\bar{3}$) and that the aromatic planes of the molecules can lie flat on the substrate. In addition, two different epitaxial domains of crystallites are found, where the long molecular axes of the 6P are either parallel to or rotated $\pm 37°$ with respect to the 6T long molecular axes. Interestingly, for this second epitaxial orientation the projection of the cleavage plane of 6T(001) to the substrate is parallel to the projection of the cleavage plane of 6P(001) as illustrated in Fig. 7.18, which suggests that the 6P molecules grow at step edges of the 6T crystallites [19].

A central aspect of these organic templates is that they can be exposed to air without loosing their templating ability, unlike the inorganic templates,

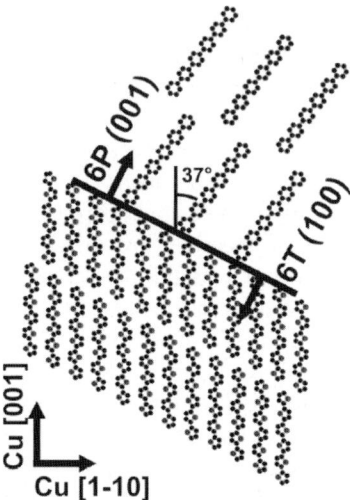

Fig. 7.18. Schematic showing the orientation of the 6P(21$\bar{3}$) crystallites relative to the 6T(010) substrate, where the planes of 6P(001) and 6T(001) are parallel to each other. Reprinted with permission from [19]

which are most sensitive to ambient gases. This is important for the technological applications of functional organics, in that less stringent conditions imply lower production costs. Another point is the ability to grow crystalline organic heterostructures with controlled molecular orientations. This will be very significant for the future of organic electronic devices as the molecular orientation within the active organic film is one of the key issues for optimum device performance.

The examples given illustrate that oriented organic films can act as excellent, stable templates for organic heteroepitaxial growth [16, 19]. For the growth on the thermodynamical lowest energy facet (e.g. the 6P(001) orientation), the heterolayer also adopts this orientation. Here, the 6P molecules on top of the 6T layer either line up with the substrate molecules or crystallize at step edges. However, the present findings clearly show that the epitaxial growth of organics on organics can not be simply understood in terms of lattice match. The organic growth appears to be mainly driven by the alignment of the overlayer molecules to the long molecular axes of the substrate molecules. It can be speculated that this behavior is also the driving force for the "preferred orientational growth" on so called rubbed organic surfaces [36]. A second directing feature appears to be the propensity to grow from step edges of organic single crystal substrates. As a consequence, not only the crystallite orientations of the second organic molecules but also the resulting morphologies of the heterostructures are determined by the structural features of the substrate molecules [19].

7.4 Outlook

A detailed understanding of the basic self organization processes that govern the growth of organics is a prerequisite for the advent of organic nanotechnology. Even though the literature of organics-on-inorganic systems is quite substantial, there are still large blank areas on this map, as the experimental space is vast: Naturally, many molecule substrate combinations will have to be considered to establish general trends for various classes of molecules. A second important parameter that needs to be investigated is the temperature of growth. In particular, the possibility of controlling the three-dimensional shape of the resulting crystallite structures by variation of the temperature during growth, as pointed out in Sect. 7.2.2, needs detailed exploration. Finally, the growth kinetics of the organics is complex, particularly due to their inherent anisotropy (see Sect. 7.2.2), and factors such as growth rate have barely been touched upon in the literature.

Once well defined structures, necessary for controlled surface science experiments, can be reproducibly obtained the growth of both inorganic and organics on them needs to be investigated. As pointed out in Sect. 7.3 the driving forces steering the growth on organic substrates can be significantly different to that on inorganic substrates. Once the formation of layered organic-on-organic heterostructures involving van der Waals interacting systems is understood the more demanding and possibly rewarding area of heterostructures with functional groups and directed chemical bonds between the species can be tackled. The authors believe such directed intermolecular bonds will be the key to create laterally patterned organic heterosystems via self-organization processes.

Acknowledgement

The authors thank the Austria Science Foundation (FWF) for financial support. The ESRF Grenoble (BM 32) and BESSY II Berlin (RG-BL, MUSTANG experimental station) synchrotron radiation facilities infrastructures and in particular the support at the respective experimental stations are gratefully acknowledged. We thank Falko P. Netzer for valuable and stimulating discussions. Finally, we thank S. Berkebile, J. Ivanco and M. Oehzelt for their individual contributions.

References

1. B. Winter, S. Berkebile, J. Ivanco, G. Koller, F.P. Netzer, M.G. Ramsey, Appl. Phys. Lett. **88**, 253111 (2006)
2. G. Koller, S. Berkebile, J. Krenn, G. Tzvetkov, C. Teichert, R. Resel, F.P. Netzer, M.G. Ramsey, Adv. Mat. **16**, 2159 (2004)
3. G. Koller, F.P. Netzer, M.G. Ramsey, Surf. Sci. Lett. **559**, L187 (2004)

4. R. Duschek, F. Mittendorfer, R.I.R. Blyth, F.P. Netzer, J. Hafner, M.G. Ramsey, Chem. Phys. Lett. **318**, 43 (2000)
5. I. Stensgaard, L. Ruan, E. Laegsgaard, F. Besenbacher, Surf. Sci. **337**, 190 (1997)
6. K.F. Braun, S.W. Hla, Nano Lett. **5**, 73 (2005)
7. S.W. Hla, K.F. Braun, B. Wassermann, K.H. Rieder, Phys. Rev Lett. **93**, 208302 (2004)
8. B. Winter, J. Ivanco, I. Salzman, R. Resel, F.P. Netzer, M.G. Ramsey, Langmuir **20**, 7512 (2004)
9. J. Ivanco, B. Winter, F.P. Netzer, M.G. Ramsey, Adv. Mat. **15**, 1812 (2003)
10. M. Oehzelt, L. Grill, S. Berkebile, G. Koller, F.P. Netzer, M.G. Ramsey ChemPhysChem **8**, 1707 (2007)
11. Y. Hu, K. Maschek, L.D. Sun, M. Hohage, P. Zeppenfeld, Surf. Sci. **600**, 762 (2006)
12. Q. Chen, A.J. McDowall, N.V. Richardson, Langmuir **19**, 10164 (2003)
13. V. Corradini, C. Menozzi, M. Cavallini, F. Biscarini, M.G. Betti, C. Mariani, Surf. Sci. **532**, 249 (2003)
14. S. Lukas, G. Witte, Ch.Wöll, Phys. Rev. Lett. **88**, 283011 (2002)
15. C.B. France, B.A. Parkinson, Appl. Phys. Lett. **82**, 1194 (2003)
16. G. Koller, S. Berkebile, J.R. Krenn, F.P. Netzer, M. Oehzelt, T. Haber, R. Resel, M.G. Ramsey, Nano Lett. **6**, 1207 (2006)
17. S. Berkebile, G. Koller, G. Hwalacek, F.P. Netzer, M.G. Ramsey, Surf. Sci. **600**, L313 (2006)
18. J. Ivanco, T. Haber, J.R. Krenn, F.P. Netzer, R. Resel, M.G. Ramsey, Surf. Sci. **601**, 178 (2007)
19. M. Oehzelt, G. Koller, J. Ivanco, T. Haber, J.R. Krenn, F.P. Netzer, R. Resel, M.G. Ramsey, Adv. Mat. **18**, 2466 (2006)
20. R. Resel, M. Oehzelt, O. Lengyel, T. Haber, T.U. Schülli, A. Thierry, G. Hlawacek, C. Teichert, S. Berkebile, G. Koller, M.G. Ramsey, Surf. Sci. **600**, 275 (2006)
21. Y.-W. Mo, J. Kleiner, M.B. Webb, M.G. Lagally, Surf. Sci. **268**, 4645 (1992)
22. C. Ebner, K.-B. Park, J.-F. Nielsen, J.P. Pelz, Phys. Rev. B **68**, 245404 (2003)
23. Y.-W. Mo, M.G. Lagally, Surf. Sci. **248**, 313 (1991)
24. F. Balzer, H.G. Rubahn, Surf. Sci. **548**, 170 (2004)
25. A. Andreev, G. Matt, C.J. Brabec, H. Sitter, D. Badt, H. Seyringer, N.S. Sariftci, Adv. Mat. **12**, 629 (2000)
26. H. Yanagi, T. Morikawa, Appl. Phys. Lett. **75**, 187 (1999)
27. E.J. Kintzel Jr., D.-M. Smilgies, J.G. Skofronick, S.A. Safron, D.H. Van Winkle, J. Cryst. Growth **289**, 345 (2006)
28. R. Resel, I. Salzmann, G. Hlawacek, C. Teichert, B. Koppelhuber, B. Winter, J. Krenn, J. Ivanco, M.G. Ramsey, Org. Electron. **5**, 45 (2004)
29. J. Ivanco, M.G. Ramsey, F.P. Netzer, T. Haber, R. Resel, J. Krenn, A. Haase, B. Stadlober, G. Jakobic, J. Appl. Phys. **96**, 2716 (2004)
30. G. Koller, S. Berkebile, J. Ivanco, F.P. Netzer, M.G. Ramsey, Surf. Sci. (2007), doi:10.1016/j.susc.2007.06.070
31. G.E. Thayer, J.T. Sadowski, F. Meyer zu Heringdorf, T. Sakurai, R.M. Tromp, Phys. Rev. Lett. **95**, 256106 (2005)
32. L.D. Sun, M. Hohage, P. Zeppenfeld, S. Berkebile, G. Koller, F.P. Netzer, M.G. Ramsey, Appl. Phys. Lett. **88**, 121913 (2006)

33. G. Koller, R.I.R. Blyth, A. Sardar, F.P. Netzer, M.G. Ramsey, Appl. Phys. Lett. **76**, 927 (2000)
34. K.N. Baker, A.V. Fratini, T. Resch, H.C. Knachel, W.W. Adams, E.P. Socci, B.L. Farmer, Polymer **34**, 1571 (1993)
35. G. Koller, B. Winter, M. Oehzelt, J. Ivanco, F.P. Netzer, M.G. Ramsey, Org. Electron. **8**, 63 (2007)
36. W.-S. Hu, Y.-F. Lin, Y.-T. Tao, Y.-J. Hsu, D.-H. Wei, Macromolecules **38**, 9617 (2005)

8

Mechanisms Governing the Growth of Organic Oligophenylene "Needles" on Au Substrates

K. Hänel and C. Wöll

8.1 Introduction

Today, 50 years after the first introduction of the transistor inorganic semiconductors, most importantly Si and GaAs are still the materials of choice for producing high performance, fast semiconductor devices. In the past decade, however, organic materials are attracting an increasing amount of interests with regard to use them as active compounds in semiconductor devices. These applications generally do not aim at high-end applications such as very fast switches, instead the current interest aims at developing to establish low cost or plastic electronics. The most important application presently are radio frequency identification devices (RFID), which will be used to identify not only individual objects like luggage on an airport but will also be used to identify products, e.g., to the cashier in the supermarket. Soft, organic materials made of either polymers or large molecules are under intense investigation by an increasing number of groups in many countries. Since organic materials have already made their way to electronic applications, for example, in connection with organic light-emitting diodes, it is rather likely that the first commercial products employing organic materials as active component in an electronic device, like an organic field effect transistor (OFET) will be introduced fairly soon.

Despite the pronounced interest in using organic or molecular materials for electronic applications several fundamental issues related to the injection of charge into molecular materials and, more importantly, the transport of electronic charge in soft matter remain to be unraveled. Since in particular the transport properties are unclear, the design principles that have to be used to synthesize molecules with high charge carrier mobility are still unknown. Although for later applications it will be essentially that polymeric materials are used in order to produce the RFID tags by a printing process molecular materials made out of smaller units are much better suited for fundamental investigation.

For this reason the growth of high quality organic adlayers on appropriate substrates, either on metal as a model for the electrodes where charge injection is carried out or on oxide substrates modeling the gate oxide of an organic field effect transistor are important areas of investigation. Compared to similar studies for conventional semiconductors, e.g., the growth of inorganic semiconductors on other substrates and the deposition of metal on semiconductors, the field of organic molecular beam epitaxy and of the deposition of molecules on substrates using other methods is still infancy.

In the first studies related to the growth of molecules on metals and inorganic substrates it has become clear that regular epitaxial growth of organic materials on any substrate is clearly the exception in most cases that the position of molecules on either metals, semiconductors, or insulators will lead to rather heterogeneous adlayers and the observation of dewetting in combination with the formation of three-dimensional islands has generally to be expected [1]. Of particular interest with regard to this class of materials are three-dimensional structures, which are formed when oligophenyls consisting of a sequence of phenyl units linked together in auto positions are deposited on solid substrates. For this class of materials organic molecular beam deposition (OMBD) on a number of substrates leads to the formation of organic "needles" with well defined properties, which have already found very interesting applications in a number of fields [2].

Currently, single-crystalline needle-like shaped crystallites formed when oligophenylenes (e.g. *para*-Quaterphenylene (P4P) and *para*-Hexaphenylene (P6P)) are deposited on solid substrates using OMBD are attracting a substantial amount of interest with regards to application in optoelectronics [2–6]. While originally the growth of these organic "needles" was discovered on mica substrates, more recent work has demonstrated that also on metal and metal oxide substrates appropriate growth conditions result in the formation of needle-like structures [6–8]. In the present paper we will review the results of a series of investigations where the growth of P4P needles has been studied on single crystalline and polycrystalline gold substrates. The aim of the study is to relate the orientation and the growth mechanism of single molecules to the orientation and structure of the mesoscopic needles. We will start by first discussing the deposition of P4P molecules on different types of gold substrates where we will mainly use Near X-ray Adsorption Fine Structure (NEXAFS) spectroscopy to determine the orientation of the molecules. These results will be related to earlier data obtained by low energy electron diffraction (LEED) and thermal desorption spectroscopy (TDS). Then, in the next chapter we will show how the orientation of the molecules and the precise structure of the gold substrate will influence the occurrence and orientation of the organic needles and than in the third part we will show how more detailed insides into the interface between the organic needles and the gold substrate can be obtained by manipulating the organic needles with a tip of an STM operated under SEM control.

8.2 Experimental

The deposition of P4P and P6P on different types of Au-substrates (differently oriented single crystals, recrystallized Au foil) was carried out in several different apparatuses under UHV conditions using organic molecular beam deposition (OMBD) [9]. Prior to the deposition of the oligophenyl-molecules, the substrates were cleaned by sputtering with Argon ions ($1\,\text{kV}$, $5 \times 10^{-3}\,\text{Pa}$) and subsequent annealing (at $1{,}000\,\text{K}$) cycles in UHV until no contamination of the surface could be seen in X-ray photoelectron spectroscopy (XPS). In case of the polycrystalline Au-substrates the sputtering/annealing procedure (repeated cycles) resulted in a recrystallized gold surface comprising single crystal grains with a diameter of about $100\,\mu\text{m}$ [9, 10].

Thin layers of P4P with an average thickness of $20\,\text{nm}$ were deposited onto the clean substrate surfaces by evaporation from a home-made quartz glass Knudsen cell. In most cases the amount of deposited P4P as well as the evaporation rate (typically $3\,\text{Å}\,\text{min}^{-1}$) were measured in-situ using a quartz crystal microbalance.

The SEM and STM measurements were performed at room temperature with an SEM/STM combination instrument (JEOL JSPM-4500S) [11]. The SEM column is equipped with a Schottky-type W(1 0 0)/Zr emitter and the measurements were performed with an accelerating voltage of $12\,\text{kV}$. The images were recorded using a standard secondary electron detector. The SEM is operated under grazing incidence of the electron beam on the surface, so that STM and SEM can be performed simultaneously.

The NEXAFS experiments were conducted at the HE-SGM beamline (resolution $\Delta E = 0.4\,\text{eV}$ at $300\,\text{eV}$) of the BESSY II synchrotron facility (Berlin, Germany). In these experiments, the analysis chamber was equipped with a sputter gun for ion bombardment, a gas dosing line for argon and oxygen, a low energy electron diffraction (LEED) optics (Vacuum Generators), a quadrupole mass spectrometer (Balzers), a twin X-ray source (VG), and a CLAM2 energy analyzer (VG). A high-purity Au(1 1 1) single crystal ($10\,\text{mm}$ in diameter) served as the substrate. The sample could be rotated around two axes making NEXAFS measurements with different polar and azimuthal angles possible. All NEXAFS spectra were recorded in the partial electron yield mode (retarding voltage $150\,\text{V}$) using a homemade electron detector based on a double channel plate (Galileo). For the energy calibration of the NEXAFS spectra, the photocurrent of a carbon-contaminated gold grid with a characteristic peak at $284.9\,\text{eV}$ calibrated against the π^*-resonance of HOPG at $285.38\,\text{eV}$ was recorded simultaneously with each spectrum. This grid was also used as a radiation flux monitor. The C1s NEXAFS raw data were normalized by a procedure consisting of several steps. First the spectra were normalized by division through spectra recorded for a clean gold substrate and finally the spectra were normalized to yield an edge-jump (difference in intensity between 270 and $330\,\text{eV}$) of one. For a more detailed description of the data analysis of the NEXAFS spectra see the paper by Reiß et al. [12].

8.3 The Importance of Molecular Conformations in P4P

When addressing the question on the orientation of single P4P or P6P molecules on a solid substrate, one has to consider the molecular conformation of the individual oligophenylene molecules. As will be shown in the next section, this question is critical with regard to determining the molecular orientation of P4P and P6P-molecules adsorbed on a metal substrate, an issue of pronounced importance for the organic molecular beam epitaxy of oligophenylenes on solid substrates. In fact the question on the presence of a dihedral or twist angle between the phenyl-ring planes in oligophenylenes in the bulk phase is an interesting one. Molecular geometries obtained from a crystal structure data basis like CSD (Cambridge structural data basis) typically propose an all-planar arrangement or one with rather small twist-angles between the phenyl-rings, and in fact most of the papers published presently refer to this molecular geometry. A more detailed study, however, of the original papers describing the X-ray structure analysis reveals [13–17] that the planar structure rather corresponds to an average geometry. In none of the previous papers on the bulk structure of P4P and P6P a precise value of the twist angle between the phenyl rings has been provided, instead typically large thermal displacements and/or low energy torsional modes are quoted. Note, that for many substituted P4P and biphenyl (P2P) derivatives bulk structures have been reported with significant (30°–40°) twist angles between the phenyl rings. Although the question on the precise bulk structure of P4P and P6P has not been settled yet, it is probably safest to start with the structure of the free molecules.

A geometry optimization of the free, i.e., gas phase P4P molecule using a commercial software package, where electron correlation effects were treated on the Moeller–Plesset perturbation theory (MP2), yielded an average inter-ring twisting angle of 48°, see Fig. 8.2a,b. These calculations also reveal that the twisted P4P, visualized in Fig. 8.2a, is about 0.73 eV more stable than a (hypothetical) planar P4P [18].

8.4 Molecular Orientation and Conformation within Ultrathin P4P Films Grown on Gold Substrates: Studies using Soft X-ray Absorption Spectroscopy

In the case of P4P two different kinds of monolayer species were found on the Au(1 1 1)-surface as evidenced by the occurrence of two different saturated peaks in the TDS data [7]. Further, the more strongly bound species will be referred to as α, the more weakly bound monolayer species as β. In the next sections an ultrathin P4P layer consisting of a completed α and β phase will be referred to as a monolayer (1 ML) following the definition by Müllegger et al. [7]. This monolayer exhibits a nominal thickness of 0.3 nm [7].

8.4.1 Ultrathin Layer Containing only the α-Species

To determine the orientation of the P4P-molecules on the substrates, P4P adlayers of different thickness were deposited on the substrates and then investigated using NEXAFS-spectroscopy. NEXAFS spectra recorded for different P4P coverages as a function of photon angle of incidence of the linearly polarized X-ray photons are shown in Fig. 8.1. The NEXAFS spectra for a 1/2 ML film (see Fig. 8.1a) are dominated by a sharp π^*-resonance at 284.9 eV with an FWHM of 1.0 eV. This resonance can be attributed to the excitations of localized C 1s electrons into delocalized unoccupied π^* molecular orbitals in the phenyl-subunits [19]. The π^*-resonance does not show a major broadening compared to the data recorded for the P4P multilayers (see Fig. 8.1c), where the FWHM amounts to 0.8 eV. This observation implies the absence of any significant chemical interaction between the nP and the Au(1 1 1) surface. This observation indicates that the interaction of the individual phenyl subunits with the gold substrate is different from for benzene with the more reactive Rh(1 1 1) and Pt(1 1 1) surfaces, where a significant broadening is seen [20,21]. The sharp π^*-resonance at 284.9 eV is followed by Rydberg resonances, the edge-jump, and then the rather broad σ^* resonances at higher energies. A detailed assignment of the different peaks in the NEXAFS spectra is provided

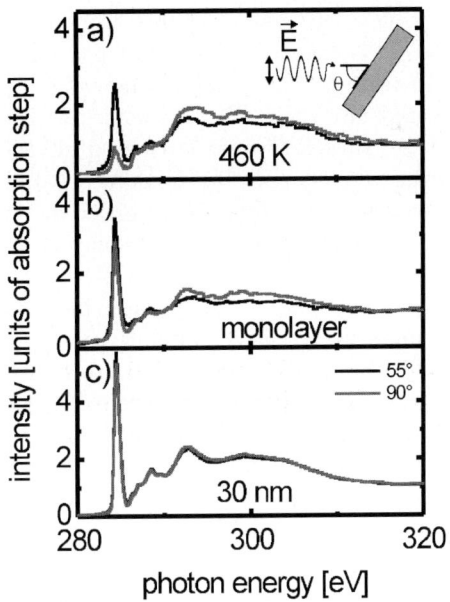

Fig. 8.1. (a) NEXAFS spectra recorded for a half monolayer of P4P on Au(1 1 1). Prior to recording the measurements the substrate was heated to 460 K, yielding a half monolayer of P4P containing only the α phase. The inset shows a schematical sketch of the experimental geometry. (b) Monolayer NEXAFS spectra. (c) Series of C1s NEXAFS spectra recorded for a 30 nm thick P4P film (bulk) grown on Au(1 1 1)

Table 8.1. Assignment of the different NEXAFS resonances following [10] for the different features observed in the data recorded for P4P and P6P adlayers on gold substrates

Photon energy (eV)	Assignment
284.9	π^*
287.5	Rydberg
288.9	π^*/π^* shake up
290.1	σ^* (C–H)
293.0	σ^*
299	Shape resonance
303	Shape resonance

in Table 8.1. The intensity of the π^*-resonance at 284.9 eV is used for the determination of the angle between the transition dipole moment (TDM), which is orientated perpendicular to the phenyl ring plane, and the surface normal.

A quantitative analysis of the variation of resonance intensity with photon angle of incidence can be used to determine the molecular orientation within the organic thin layers [12,19]. The spectra were recorded for different incident angles, the results for $\theta = 55°$ and $\theta = 90°$ are shown in Fig. 8.1. In this case the incident angle θ is taken as the angle between the direction of light incidence and the substrate surface plane as shown in the inset of Fig. 8.1a.

Only flat-on oriented molecules (i.e., molecules with their molecular axis parallel to the substrate surface) would have zero intensity at normal X-ray photon incidence of the π^*-resonance like NEXAFS data recorded for strictly planar aromatic molecules, e.g., pentacene adsorbed on Au(1 1 1), have shown [22]. The remaining intensity of the half monolayer of P4P under normal incidence, however, is clearly nonzero. A simple explanation for the finite intensity of the π^*-resonance at normal photon incidence would be that a small fraction of the molecules already populate the β phase with an edge-on orientation (see below). Due to the preparation procedure of heating to 460 K since the thin films for which the data shown in Fig. 8.1c were taken, were heated to 460 K before deposition of the NEXAFS data, the presence of the β phase, however, can be safely ruled out since at this temperature the β species has already desorbed from the surface [7]. A very plausible explanation for the nonzero π^* intensity at normal X-ray photon incidence is based on the geometry of the free P4P molecules that exhibit at a dihedral angle, i.e., a twist angle between the phenyl planes, of around 40° [13]. This angle is also reproduced by precise ab-initio calculations using commercial program packages for quantum chemistry (see below).

Upon adsorption on the Au(1 1 1) surface this tilt angle may be slightly reduced, but it is rather unlikely that this angle approaches zero since the energy needed to fully planarize a P4P molecule amounts to 0.73 eV, according to the quantum chemical calculations (see Sect. 8.3). This energy corresponds to a substantial fraction of the total binding energy of these molecules to the gold surface, which has been determined by temperature programmed desorption [23] (2.6 eV for P4P, 3.6 eV for P6P).

An adsorption geometry where the phenyl rings of the P4P molecules are twisted with respect to each other by about 40° in an alternate fashion very nicely explains the fact that at normal incidence a finite intensity for the π^*-resonance is observed, even for molecules in the most strongly bound, low coverage α phase.

If we assume a dihedral angle present in the adsorbed molecules the nonzero π^* intensity can be explained in a straight forward fashion. In previous STM work an evidence has been found for the presence of a twisted P4P conformation for P6P monolayers adsorbed on a Ag(1 1 1) surface [24]. Also in the case of rubrene adsorbed on different metal (Au) and oxide (SiO_2)-surfaces, the presence of aplanar distortions in molecules adsorbed in the monolayer regime has been reported [25]. In this case it has been found that in the first monolayer the molecules adopt the gas-phase conformation, which is characterized by a significant aplanar distortion of the tetracene backbone, whereas in the bulk exhibit a tetracene backbone, which is almost planar. A comparison of a simulation and measured data yielded in the best agreement of an angle of $\alpha = 15$–$20°$. This implies that adjacent phenyl rings in a single molecule are twisted by about 30° to 40°, roughly the same value as calculated for the free molecule (see Sect. 8.3). Compared to the P4P phase molecules with a reported interring-twisting between 30° and 40°, no significant reduction of the interring-twisting angle (dihedral angle δ) upon adsorption of a single molecule on the surface can be inferred (as long as the coverage does not exceed that of a monolayer).

In the case of a single P6P-molecule on Ag(1 1 1) an interring twist angle of 11.4° has been derived from STM data at 6 K [24]. The quantitative values for the interring twist angle derived from STM data are, however, can only be regarded as a rough guide. It would be highly desirable to obtain a more accurate value, e.g., from theoretical calculations for this rather interesting system where intramolecular forces compete with molecule/substrate interactions. Unfortunately, the presently most frequently used approach to carry out ab-initio electronic structure calculations for molecules weakly bound to solid substrates suffers from fundamental problems (or DFT) [26,27]. Another theoretical approach, where electronic correlation effects are treated using perturbation theory applied to accurate wave-functions, has recently been shown to provide reliable results for such systems where dispersion forces (or van der Waals interactions) contribute significantly [28, 29]. Unfortunately, however, these calculations are computationally extremely expensive and can at present not be carried out for molecules significantly exceeding the size of benzene.

8.4.2 Full Monolayer Containing α- and β-Species

The NEXAFS spectra of P4P and saturated monolayer films on Au(1 1 1) recorded under normal and under grazing incidence are shown in Fig. 8.1b. The nominal film thickness amounts to about 0.3 nm, the deposition was carried out at a substrate temperature of 390 K. In contrast to the spectra recorded for the thin P4P monolayer, discussed in the previous section, now the intensity of the π^*–resonance depends only weakly on the X-ray photon angle of incidence.

The comparison to the corresponding data for thick P4P films (which adopt the bulk structure according to previous results [30]) reveal a close similarity and suggest that the molecular packing within the full monolayer (containing the α and the β phase) is similar to the bulk structure. However, a quantitative comparison of the relative π^* intensities between monolayer and bulk films shows that the resonance intensity at normal incidence is slightly decreased for the monolayer films. This would imply average side tilt angles of somewhat smaller than 66° for the edge-on oriented molecules in the monolayer phase.

8.4.3 Multilayers

Typical NEXAFS spectra recorded for a P4P multilayer with a nominal thickness of 30 nm also deposited on an Au(1 1 1) single crystal substrate are also shown in Fig. 8.1. In contrast to the NEXAFS spectra recorded for the thin layers discussed in the previous section, the variation of the π^* intensity with the angle of photon incidence is rather small.

Previous structure investigations of Müllegger et al. have revealed that the P4P bulk films grown at room temperature on Au(1 1 1) are highly crystalline, exhibiting a predominant P4P(2 1 1) orientation of the crystallites, i.e., the bulk (2 1 1) planes are oriented parallel to the substrate surface [30]. Figure 8.2c schematically illustrates the herringbone-type packing structure of the P4P molecules in the monoclinic bulk structure at room temperature [17, 31]. In the structure model proposed by Müllegger et al. [30] the P4P(2 1 1) crystal plane is oriented parallel to the substrate surface, yielding a structure where at the surface 50% of the P4P molecules are oriented with their phenyl planes more parallel to the surface whereas the other half is oriented in a more edge on position exhibiting an angle of 66° between the (2 1 1) surface plane and the molecular backbone.

In the case of two differently oriented but otherwise identical molecular P4P species (edge-on and flat-on) the interpretation of the corresponding NEXAFS spectra will be complicated by the effect that the spectra of the individual molecules will be superimposed. As a result the tilt angle of the transition dipole moment (which is oriented perpendicular to the individual phenyl ring planes of the two P4P molecules in the unit cell of the structure model) as determined by NEXAFS is an average value [19]. Note, that because of this averaging process NEXAFS spectroscopy cannot distinguish between

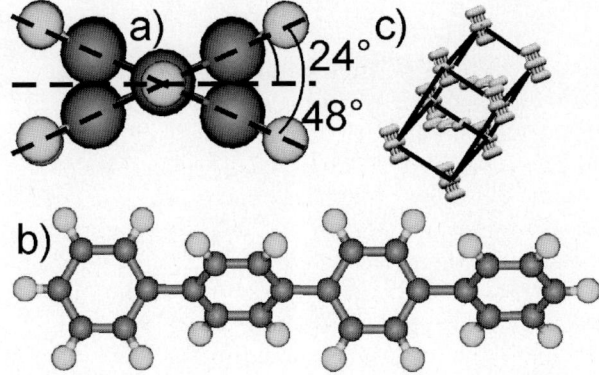

Fig. 8.2. The calculated structure of a P4P molecule based on ab-initio perturbation theory calculations (MP2 level) (**a**) along the long molecular axes to visualize the interring twisting. (**b**) Side view of the P4P. (**c**) Schematical sketch of the P4P bulk structure [17]

a structure where the phenyl ring planes are rotated in a random fashion in a situation where 50% of the phenyl ring planes are strictly parallel to the surface and the other 50% are strictly oriented particular to the surface [19]. Because of this averaging it is not possible to determine the tilt-angle of the two individual molecules separately. By Müllegger et al. the consistency of a structural model is checked by a comparison with the experimental data successfully [20].

8.5 The Orientation of Organic Oligophenylene "Needles" on Gold Substrates

After the structure of ultra thin P4P layers deposited on a gold substrate has been described in the previous section, we now turn our attention to the organic needles that are formed when larger amounts of P4P molecules are deposited on a gold surface. Very often these organic needles show a preferred orientation with regard to the substrate. To unravel the microscopic mechanism for the formation of the needles and the mechanisms giving rise to their orientation the deposition of P4P has been studied on polycrystalline gold substrates [9]. Employing such a polycrystalline substrate offers the advantage that many different possible surface orientations of the gold can be studied at the same time and under exactly the same conditions. Corresponding studies using a number of differently orientated gold single crystals require a much larger experimental effort. The deposition of P4P on a polycrystalline gold foil yields a rather complex morphology [9]; the image shown in Fig. 8.3 reveals the presence of a number of different oriented P4P crystallites. The SEM image demonstrates that the P4P grew in a variety of different morphologies

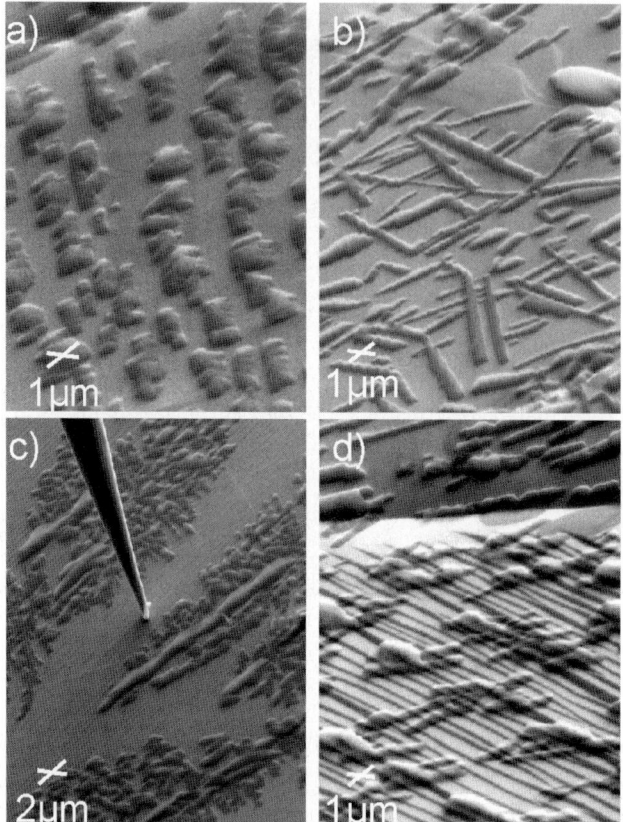

Fig. 8.3. Different morphologies of P4P grown on a polycrystalline Au foil imaged by SEM. The images show (**a**) chains of isotropic grown P4P islands forming needles, (**b**) needle shaped P4P islands with random distribution of orientation, (**c**) dendritic formed P4P islands, (**d**) isotropic islands on a facetted Au surface

on the polycrystalline Au foil, with the precise shape and size of the crystallites being different for the different crystallites. The different morphologies include needle-shaped islands, as presented in Fig. 8.3b, and isotropic islands, see Fig. 8.3d. An interesting combination of isotropic islands forming chains is presented in Fig. 8.3a. Also P4P islands with a dendritic shape could be observed, see Fig. 8.3c.

On a polycrystalline surface the different crystallites exhibit different surface orientations and step densities, which obviously control the growth mode of the deposited P4P. Using EBSD and SEM the substrate orientation of a single grain of the polycrystalline substrate could be determined; at the same time the P4P islands on the particular grain could be imaged.

Although, from the EBSD experiments, the orientation of the single grains is known, the step density on the surfaces of the grains could not be determined in a straight forward fashion using this diffraction technique. Since we expect the density of steps and in particular their orientation to have a major influence on the density and orientation of the P4P islands, we have determined the step density and orientation of steps on several parts of the polycrystalline surface prior to the deposition of the P4P using STM. With a precision of better than 50 nm the same spot on the surface could be investigated before and after P4P deposition even though the substrate had to be moved to a different apparatus for the P4P depositions.

The STM data recorded on the plain Au substrate prior to the film deposition clearly reveal the stepped surfaces of the Au grain [9]. This STM micrograph shows a 30 nm×30 nm STM image of the plain Au surface, which has been recorded at the same position as shown in Fig. 8.4a, i.e., on grain A located left of the grain boundary. Almost parallel rows of bunched steps are

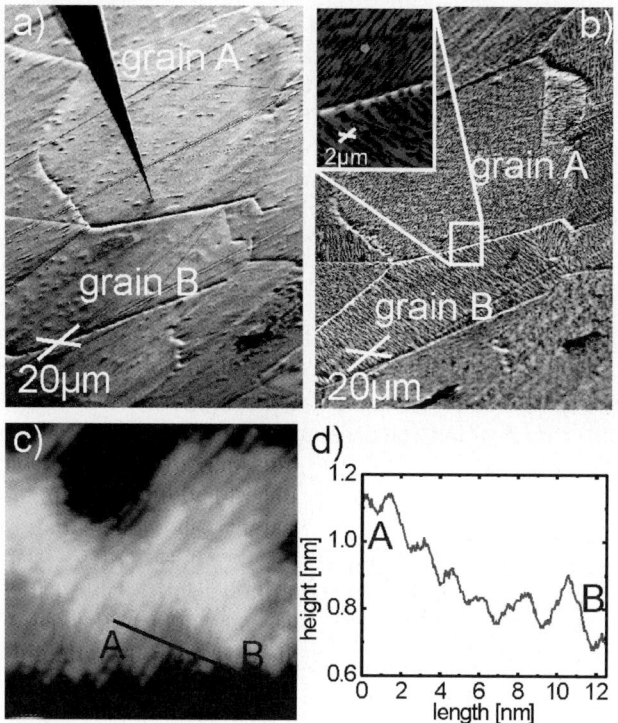

Fig. 8.4. (a) SEM image showing the position of the STM tip relative to the Au surface for the combined STM/SEM measurements. (b) SEM image of the ex-situ prepared P4P covered gold foil. The film thickness corresponds to nominal 20 nm. The inset presents at higher resolution of the anisotropically grown P4P islands (a). (c) $30 \times 30 \, \text{nm}^2$ STM image ($U_{Sample} = 2.0 \, \text{V}$, $I = 2.0 \, \text{nA}$) taken on grain A at the tip position shown in (a). The height profile corresponding to the line AB (c). Reprinted with permission from [9]

visible, with an inter-row distance of 1–1.5 nm, as demonstrated in the height profile in Fig. 8.4d, and this height profile has been taken along the line AB indicated in Fig. 8.4c. After the ex-situ deposition of the P4P islands, SEM micrographs with high resolution have been recorded at the same sample position as shown in Fig. 8.4a, the procedure used to reposition the STM on the same spot has been described above; basically it uses the same characteristic landmarks and a stepwise increase of the resolution of the SEM. This is presented in Fig. 8.4b, where P4P grown in needle-like shape can be observed. On the grain labeled A the P4P needles are grown perpendicular to the direction of the corresponding steps of the Au surface underneath. From a comparison of the SEM data recorded after the deposition with the STM data recorded before the deposition now the orientation of the needles can be related to the orientation and the density of steps on the surface prior to the deposition. For example, this comparison reveals that the P4P needle visible in Fig. 8.4B has grown perpendicular to the direction of the steps that were present on the Au substrate before deposition. This of course is an interesting and somewhat counter-intuitive observation since one may have expected that the orientation of the needles is in fact parallel to the rose. By a systematic comparison with data recorded on other parts of the surface, it could in fact be demonstrated that the mechanism governing whether the needles are oriented perpendicular or parallel to the steps depends on the width of the terraces before deposition. As shown in a previous publication [9], the critical length of the terraces is <2 nm. If the terraces are smaller, the needles tend to be arranged perpendicular to the rose, if the terrace width is larger than 2 nm, the needles are oriented in a parallel fashion. This critical size of the step terraces is related to the length of the P4P molecule, obviously for small terrace lengths the orientation of the single molecules is forced to be parallel to the steps, which then in turn – given the molecular arrangement of the P4P molecules within the needles [9] – implies that the needles steps are then oriented perpendicular to the steps.

8.6 Manipulation of Organic Needles Using an STM Operated under SEM Control

As described in the previous section, previous work has demonstrated that on polycrystalline Au substrates the P4P needle growth is favored on terraces with a width less than 2 nm or on crystalline surfaces consisting of (1 1 0) terraces. The needles are oriented predominantly normal to the Au $\langle 1\bar{1}0\rangle$ direction above a critical terrace width and perpendicular to the steps below a critical terrace width [9]. In this previous study the question remained open whether the area between the needle-like structures is covered by P4P molecules with the structure described in Sect. 8.1, or whether a thicker organic film is present between the organic needles.

In the following we will describe experiments where we have used the tip of an STM operated under SEM control to selectively remove single nanometer-sized organic P4P needles and to subsequently investigate the former contact area using the STM.

In these experiments we have employed the fact that the simultaneous operation of SEM and STM allows to observe a bending of the STM tip as a result of forces between the STM tip and the organic P4P structured imaged with the STM. We will show that, if these forces exceed a certain value, it becomes possible to manipulate a single SEM needles, removed from the substrate, and reposition them on any given part of the substrate surface.

Before presenting these results, we would like to emphasize the importance of beam damage when imaging the P4P needles in the SEM. The presence of beam damage could be demonstrated by heating the substrate to a temperature above the desorption temperature of P4P multilayers after imaging in the SEM, whereas on parts of the surface that had not been imaged with the SEM, no residual molecules were found on the heated surface, and parts which were imaged with the SEM frequently showed the presence of P4P crystallites with the same morphology as before. From the fact that these structures have remained on the surface despite heating above the P4P desorption temperature, we conclude that the P4P molecules were polymerized as a result of the incident high energy electrons. This is an interesting observation, since it may provide the basis for a positive e-beam lithography and may allow to use the P4P material directly as it is. With regard to the experiments presented in the following we have made sure by several control experiments that none of the conclusions about the growth of the P4P adlayers presented below are affected by this electron beam induced polymerization of the adlayers.

8.6.1 STM Studies of P4P Needles

In the studies described in the previous section, we were only able to characterize the substrate with STM before deposition of the P4P islands. After the MBE deposition the P4P structures could only be imaged with SEM in a straightforward fashion. In the STM mode no stable tunneling conditions could be achieved, which is a result from the fact that the conductivity of P4P is too low for establishing a current STM tip, even if the tip mechanically touches the P4P crystallite. The touching of the STM tip could, however, be monitored nicely by watching the STM tip in the SEM. When the STM tip touches any solid material, mechanical forces start to act leading to a deformation or bending of the tip, which can be clearly be seen in the corresponding SEM images (see Figs. 8.6a and 8.7a). When interpreting this SEM image one has to note that due to the low secondary electron emission coefficient of carbon the secondary electrons seen by the SEM detector mainly originate from the metal substrate. Thick P4P layers lead to a larger attenuation than thin P4P layers and, as a result, appear darker in the SEM image. SEM images recorded after the attempts to image the sample by STM at an area covered

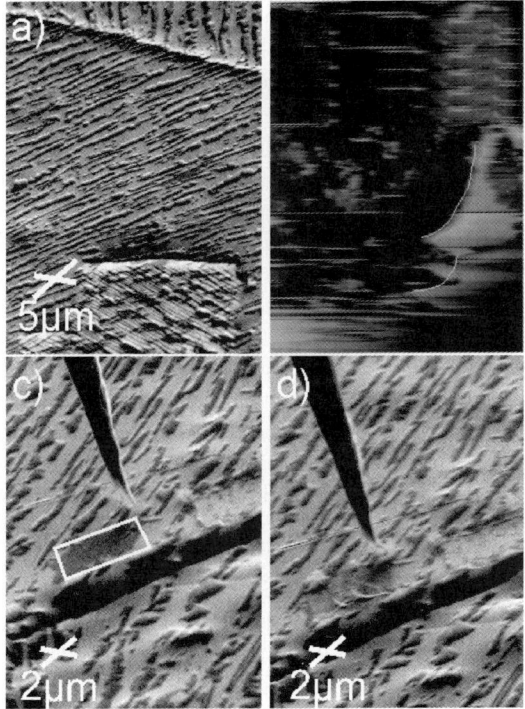

Fig. 8.5. (a) SEM image recorded for a P4P covered polycrystalline Au surface. The P4P forms elongated needles, (b) STM image performed on a homogeneous P4P covered area of the Au sample ($2{,}100\times 2{,}100\,\text{nm}^2$, $U_{\text{sample}} = -2.8\,\text{V}$, $I = 5\,\text{pA}$, $Z \approx 297\,\text{nm}$), (c) SEM image before the STM experiment (the rectangle marks the position of the STM experiment), and (d) after the STM experiment

with different island types revealed that the P4P needles were severely damaged by the STM tip (Fig. 8.5d), although very small tunneling currents of 5 pA together with high tunneling voltages (all voltages given are with respect to the sample) of up to -2.8 V were used. The STM image in Fig. 8.5b shows two protrusions, which were created when part of the P4P film was displaced by the tip of the STM during scanning. This displaced material is also visible in the SEM micrograph shown in Fig. 8.5d.

Attempts to image a P4P needle by gradually shifting the STM field of view towards such an object gave rise to interesting STM imaging artifacts, this is demonstrated in Fig. 8.6c, which shows an STM micrograph where the P4P needle is positioned at the lower left. Unexpectedly, the edge of the needle appears as a depression in the STM micrograph. By studying the scanning of the tip simultaneously with the SEM the reason for the apparent depression could be identified. When approaching the needle sideways, the STM tip is initially in a stable tunneling position. When the tip reaches the edge of the needles, it is mechanically hindered from moving further sideways and

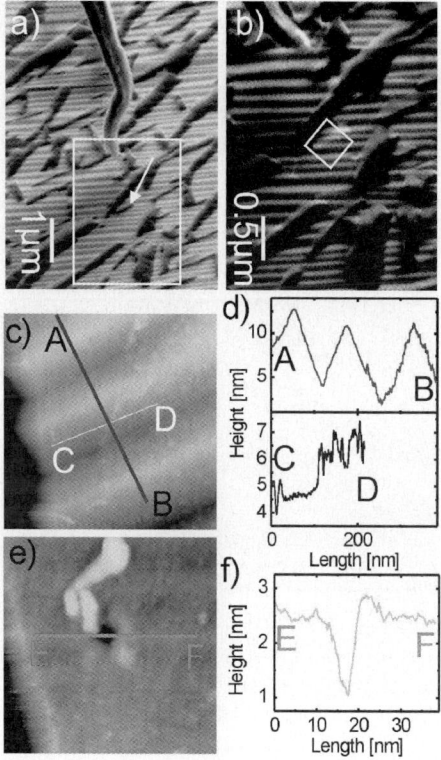

Fig. 8.6. (a) SEM image of a facetted polycrystalline gold surface covered with P4P islands. The arrow marks a small P4P needle. (b) Detailed SEM image zoomed in (a) after the marked needle was removed. (c) STM image of a P4P covered Au surface at the position marked by a white square in (b) ($400 \times 400 \, \text{nm}^2$, $U_{\text{sample}} = 2.0 \, \text{V}$, $I = 0.1 \, \text{nA}$). (d) Height profiles AB and CD of the STM image of (c). (e) STM image of the former contact area ($50 \times 50 \, \text{nm}^2$, $U_{\text{sample}} = 2.0 \, \text{V}$, $I = 0.1 \, \text{nA}$). (f) Height profile EF of the STM micrograph in (e)

the tips starts to bend (see the schematic drawing in Fig. 8.8). At the same time the feedback mechanism pushes the tip closer to the surface in order to compensate the loss in length of the tip caused by the bending. In the lower part of the schematic drawing the resulting profile line along the fast scan direction is shown. This proposed bending of the STM tip can be directly seen in the SEM images reproduced in Fig. 8.7. In Fig. 8.7a the tip is approached and in direct contact with the P4P needle. To investigate the reversibility of the tip deformation process the tip was retracted and again an SEM image was recorded (Fig. 8.7b). The shape of the STM tip is almost identical to that recorded before the tip approach, indicating that the mechanical deformation was of mainly elastic character.

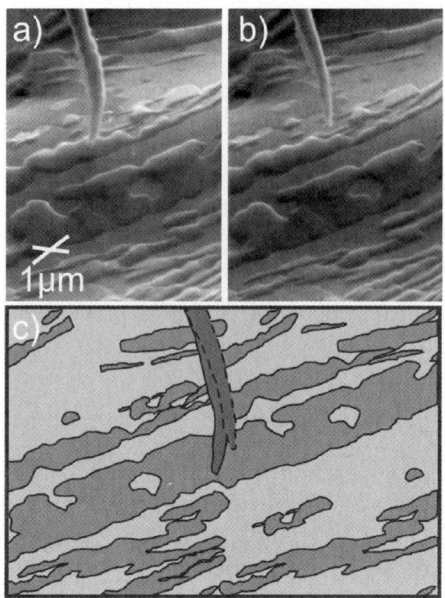

Fig. 8.7. (a) SEM image of an STM tip in contact with the P4P needle grown on a polycrystalline gold substrate. The tip has penetrated the P4P structure and is bent as a result of the lateral forces. (b) SEM image recorded after the tip shown in (a) has been attracted; the bending of the tip has now almost completely been removed. (c) Schematic drawing illustrating a difference in bending of the tip in (a) and (b) shown above

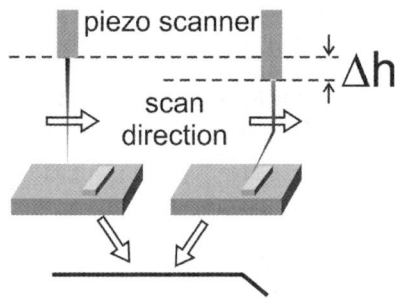

Fig. 8.8. A schematic illustration to demonstrate how the bending of a tip can lead to an apparent depression in the corresponding STM data

The SEM images reproduced in Fig. 8.7a,b directly demonstrate the (mainly elastic) deformation of the STM tip. As shown by the schematic drawing in Fig. 8.7 the tip is bent in Fig. 8.9a with respect to the situation in Fig. 8.7b.

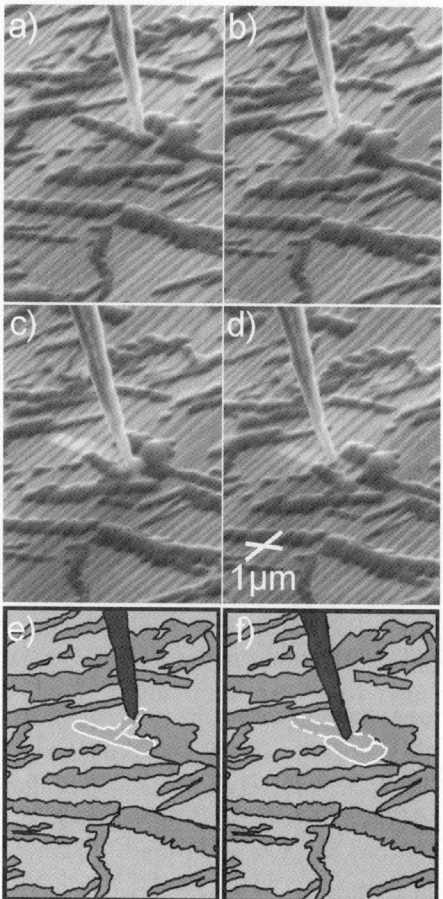

Fig. 8.9. SEM images of a manipulation of a P4P covered facetted Au surface. (**a**) STM tip is approached to a P4P needle. (**b**) The P4P needle is sticking to the STM tip and is half lifted. (**c**) The needle is dragged to another position, but the STM tip is still sticking to the needle. (**d**) The STM tip is now free. (**e**) Schematic drawing of the lifting of the needle. (**f**) Schematic drawing of the movement of the needle

8.6.2 Manipulation of STM Needles

The observation reported above that attempts to obtain stable tunneling conditions on a P4P-needle lead to the build-up of forces actually strong enough to detach the needles from the substrate directly, suggesting that the operation of the STM under SEM control can be used to manipulate the needles and redeposit them on different parts of the substrate or, possibly, even on another substrate. The method employed to manipulate and to transfer the needles is described in the following section.

The SEM image in Fig. 8.9a was recorded for a facetted gold surface covered with needle-shaped P4P islands. After approaching the P4P covered Au surface the STM tip was positioned on top of a P4P needle under SEM control. Figure 8.9b clearly reveals that, after retracting the tip, the right part of the P4P needle is not at the same position than before and must have been removed from the substrate. A more detailed investigation of the image reveals that the material missing from the substrate now adheres to the tip of the STM. The left part of the needle is still adhered to the gold substrate. Under the imaging conditions of Fig. 8.9b the removed part of the needle adhering to the tip of the STM is hard to see since it appears with the same brightness as the gold substrate. To facilitate the interpretation of the SEM images we provide the schematical drawings shown in Fig. 8.9e. In the schematic drawing the P4P islands are illustrated by grey areas that are surrounded by black solid lines. The manipulated island is surrounded by a white line. The original situation, i.e., the needles being attached to the surface, is represented by the solid white line. The dashed white line represents the situation of the half-lifted needle adhering to the tungsten tip (see Fig. 8.9b).

To further test the possibility to manipulate the P4P structures in the next step the STM tip was displaced laterally, thus pulling the P4P needle towards the lower right (Fig. 8.9c). Now it becomes possible to investigate the previous contact area, where the P4P structure adhered to the gold substrate, more thoroughly. This inspection reveals that the previous contact area appears brighter than the surrounding area, indicating that the remaining organic adlayer must be rather thin (at least thinner than the organic adlayer between the P4P needles before starting the experiment). The removed needle is folded, the left half is lying directly on the (P4P covered) gold surface whereas the right part is bent by about 90° and still adheres to the tip.

In the next step of the experiment the STM tip was moved back to the former contact area (Fig. 8.9d). Now the displaced needle can be seen as an "L"-shaped structure on the substrate. The lifting of the P4P needle is shown schematically in Fig. 8.9e, whereas the lateral displacement is represented in Fig. 8.9f.

The manipulated needle is again marked by a grey area surrounded by a white line. The dashed line shows the position of the P4P needle before the manipulation and also of the initial contact area, whereas the white solid line denotes the position and the shape of the folded needle after the movement of the needle.

Another series of SEM images showing the manipulation of a P4P needle are shown in Fig. 8.6a,b. First, the STM tip was approached to the P4P covered substrate in the area between the needles until stable tunneling conditions were achieved. Then the tip was moved laterally under SEM control along the surface until a P4P needle was reached. When the tip touched the P4P crystallite the SEM images clearly revealed a bending of the tip, demonstrating the build-up of a significant force between tip and P4P needle.

For a sufficiently conducting crystallite the STM feedback electronics would withdraw the tip in order to achieve stable tunneling conditions on top of the organic structure. Here, the conductivity of the organic multilayer is so poor that no stable tunneling conditions can be achieved on top of the P4P crystallite. Consequently, the feedback electronics keeps the tip at a position with sufficiently high conductivity, i.e., at the thin organic adlayer at the edge of the P4P needle. As a result, a significant mechanical force is exerted by the tip on the needle, which initially leads to an elastic deformation of the tip.

If the force exerted on the P4P crystallites by the tip exceeds a critical value, the needle is disconnected from the substrate. In some cases it was observed that the needle adhered to the STM tip as shown in Fig. 8.9. The SEM images in Fig. 8.6a,b shows the removal of a relatively small P4P needle, marked by an arrow in Fig. 8.6a. The white rectangular shows the position of the higher magnified SEM image in Fig. 8.6b recorded after the removal of the P4P needle. The schematic drawing in Fig. 8.10 illustrates the manipulation of the P4P needles on the gold surface. The STM tip approaches the P4P needle

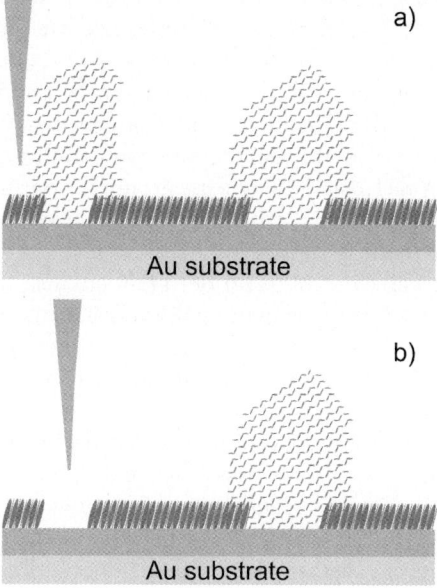

Fig. 8.10. Manipulation of the P4P needles adhering to a gold substrate. When pushing the STM tip against the needle the mechanical force can get so large that the needle is detached from the substrates. The schematic drawing also illustrates the P4P structures on the gold surface. On top of the interface layer with unknown molecular orientation the needles grow with P4P molecules with their axes orientated parallel to the substrate, whereas in the area between the needles we propose an upright orientation of the molecular axes

sideways. When the applied force exceeds a critical value the P4P needle is disconnected from the substrate, see Fig. 8.10b. After the removal of a P4P needle the former contact area could again be clearly identified, see the bright area in the SEM image presented Fig. 8.6b.

8.6.3 STM Investigations of the Former Contact Area

To obtain more detailed information about the anchoring of the P4P crystallites to the substrates, the former contact area and pristine areas between the P4P needles were investigated by STM (see Fig. 8.6c). The STM data in Fig. 8.6c shows the facetted gold surface, with widths of the facets of 70 nm and heights of 9 nm, as shown in the line profile AB in Fig. 8.6d. Note that in this region the organic adlayer is sufficiently thin to allow for a stable STM imaging (measured with $U_{sample} = 2.0$ V and $I = 100$ pA) of the organic adlayer. The height profile CD presented in Fig. 8.6d reveals a depth of the former contact area with respect to the surrounding pristine organic adlayer of 1.5 nm, demonstrating directly that the thickness of the organic layer in between the crystallites must be equal to or larger than this value. The bottom of the former contact area of the crystallites appears rather smooth and homogeneous in the STM data, see Fig. 8.6c and the corresponding height profiles Fig. 8.6d.

To resolve the question whether the bottom of the former contact area actually corresponds to the plain Au substrate or whether there is still an organic layer present we used special tunneling conditions, which are known to allow for imaging metals surfaces with atomic resolution, namely going to a small gap resistance. When using large STM currents of 2.0 nA together with small voltages (5 mV), it was found that the tip created defects in the contact area. A typical STM induced defect, a circular hole with a width of 6 nm and a depth of 1.5 nm is shown in the STM image in Fig. 8.6e together with a height profile (Fig. 8.6f). Since Au (and other metal) substrates are stable under these tunneling conditions, it can be concluded that the bottom of the contact area still consists of a layer with a thickness of 1.5 nm.

On the basis of the available data we can not decide whether this material was part of the P4P needle that was removed or whether it is a (possibly amorphous) layer present between the needles and the Au substrate. From the fact that a pronounced correlation exists between the P4P needles and the Au substrate [9], we tentatively propose that the P4P molecules present in the former contact area originally were part of the tip but adhered to the substrate when the P4P needle was pulled away from the substrate.

Investigations of different contact areas created by removing other P4P needles from the substrate did not indicate significant differences, indicating a similar nucleation process during the growth of the needles on different parts of the surface.

A schematic model of the structure of the P4P needles and the surrounding P4P organic adlayer based on previous data and the present results is shown in Fig. 8.10. On the gold surface we have first an interface layer (wetting layer) of molecules with their molecular axes oriented parallel to the Au substrate [7]. This adsorption geometry is also described for *para*-hexaphenyl on Ag(1 1 1). It is proposed by Braun et al. [24] that the phenyl rings of the *para*-hexaphenyl are alternatingly twisted, whereas the long axis of the molecule is oriented parallel to the substrate. Also based on spectroscopic data, i.e. UPS, a reduction of the twist angle caused by the adsorption is suggested by Koch et al. [14]. On top of this wetting layer there is a thin P4P film with a thickness of at least 1.5 nm [14]. The P4P crystallites (in which the molecules are oriented parallel to the substrate) grow on top of this interface layer. The area between the P4P needles is covered by a second layer of upright oriented molecules, as shown in Fig. 8.10. This model is consistent with the fact that in the XRD data for this system [7] in addition to the diffraction peaks related to the needles also other phases are seen, which are consistent with upright-oriented molecular axes. Note that the model shown in Fig. 8.10 is also consistent with the TPD data [20], since molecules desorbing from the organic layer between the needles will have basically the same binding energy as molecules incorporated into the P4P needles.

Acknowledgments

We would like to thank A. Winkler and S. Müllegger, Graz, Austria, for a very successful collaboration.

References

1. G. Witte, C. Wöll, J. Mater. Res. **19**, 1889 (2004)
2. F. Balzer, H.G. Rubahn, PhiuZ **36**(1) (2005)
3. F. Balzer, H.G. Rubahn, Adv. Funct. Mater. **15**, 17 (2005)
4. F. Balzer, J. Beermann, S. Bozhevolnyi, A. Simonsen, H.G. Rubahn, Nano Lett. **3**, 1311 (2003)
5. F. Balzer, H.G. Rubahn, Surf. Sci. **548**, 170 (2004)
6. F. Quochi, F. Cordella, R. Orru, J. Communal, P. Verzeroli, A. Mura, G. Bongiovanni, A. Andreev, H. Sitter, N. Sariciftci, Appl. Phys. Lett. **84**, 4454 (2004)
7. S. Müllegger, I. Salzmann, R. Resel, G. Hlawacek, C. Teichert, A. Winkler, J. Chem. Phys. **121**, 2272 (2004)
8. G. Hlawacek, C. Teichert, S. Müllegger, R. Resel, A. Winkler, Synthetic Met. **146**, 383 (2004)
9. S. Müllegger, S. Mitsche, P. Pölt, K. Hänel, A. Birkner, C. Wöll, A. Winkler, Thin Solid Films **484**, 408 (2005)
10. S. Müllegger, O. Stranik, E. Zojer, A. Winkler, Appl. Surf. Sci. **221**, 184 (2004)
11. T. Sato, S. Kitamura, M. Iwatsuki, J. Vac. Sci. Technol. A **18**, 960 (2000)
12. S. Reiß, H. Krumm, A. Niklewski, V. Staemmler, C. Wöll, J. Chem. Phys. **116**, 7704 (2002)

13. H. Yanagi, T. Morikawa, Appl. Phys. Lett. **75**, 187 (1999)
14. N. Koch, G. Heimel, J. Wu, E. Zojer, R. Johnson, J.L. Brédas, K. Müllen, J. Rabe, Chem. Phys. Lett. **413**, 390 (2005)
15. P. Puschnig, C. Ambrosch-Draxl, Phys. Rev. B **60**, 7891 (1999)
16. K. Baker, A. Fratini, T. Resch, H. Knachel, W. Adams, E. Socci, B. Farmer, Polymer **34**, 1571 (1993)
17. Y. Delugeard, J. Desuche, J. Baudour, Acta Cryst. B **32**, 702 (1976)
18. G. 03, (Gaussian, Inc., Pittsburgh PA, 2003)
19. J. Stöhr, *NEXAFS Spectroscopy*, Springer Series in Surface Science, vol. 25 (Springer, Berlin Heidelberg New York, 1992)
20. S. Müllegger et al., To be published
21. K. Weiss, S. Gebert, M. Wühn, H. Wadepohl, C. Wöll, J. Vac. Sci. Technol. **16**, 1017 (1998)
22. G. Beernink, T. Strunskus, G. Witte, C. Wöll, Appl. Phys. Lett. **85**, 398 (2004)
23. S. Müllegger, A. Winkler, Surf. Sci. **600**, 1290 (2006)
24. K.F. Braun, S.W. Hla, Nano Lett. **5**, 73 (2005)
25. D. Käfer, L. Ruppel, G. Witte, C. Wöll, Phys. Rev. Lett. **95**, 166602 (2005)
26. A. Hauschild, K. Karki, B. Cowie, M. Rohlfing, F. Tautz, M. Sokolowski, Phys. Rev. Lett. **95**, 209602 (2005)
27. R. Rurali, N. Lorente, P. Ordejon, Phys. Rev. Lett. **95**, 209601 (2005)
28. P. Bagus, K. Hermann, C. Wöll, J. Chem. Phys. **123**, 184109 (2005)
29. P. Bagus, V. Staemmler, C. Wöll, Phys. Rev. Lett. **89**, 096104 (2002)
30. S. Müllegger, I. Salzmann, R. Resel, A. Winkler, Appl. Phys. Lett. **83**, 4536 (2003)
31. C.C.D.C. Mercury Software, (2003)

Part III

Optics

9

Nanooptics Using Organic Nanofibers

K. Thilsing-Hansen, S.I. Bozhevolnyi, and H.-G. Rubahn

9.1 Morphology and Optical Response

9.1.1 Static Response

The possibility to separate nanofibers from each other by distances that are larger than the wavelength of the emitted light allows one to investigate in detail the influence of morphological changes in the nanometer-range on the optical properties. As an example we show in Fig. 9.1 spectra obtained from a single *para*-hexaphenylene (p6P) nanofiber (circles) and from an ensemble of p6P nanofibers (solid line). The spectrum from the single nanofiber has been obtained by illuminating the nanofiber inside a microscope with UV light and sampling the emitted light also inside the microscope with an optical fiber, connected to a miniature spectrometer. The well-separated spectral lines are due to a vibronic progression of the exciton emission (perpendicular lines on top of the graph). In the case of the single nanofiber spectrum the highest energy (0,0) band is not visible due to a cut-off filter in the microscope. Nevertheless, comparison with the spectra from the needle ensembles reveals that the light emission becomes more focussed to a narrow color range (namely 420 ± 5 nm) if an individual nanoaggregate is considered. More extended spectroscopic measurements along a nanofiber show that the spectral width of this residual line depends on the morphology of the aggregate and that it becomes narrower if the nanofiber width decreases, e.g., at the tip of the nanofiber [1].

It is interesting to note that the width of the spectral lines does depend only weakly on substrate temperature or functionalization of the molecular building blocks. Thus for nanofiber-based devices there is no need for extensive cooling. Functionalization for optimized optoelectronic properties shifts the spectral lines but in many cases does not affect the narrowness of the emitted spectra.

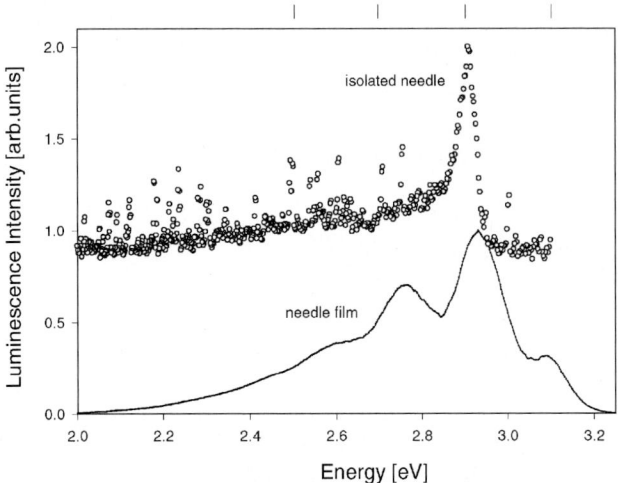

Fig. 9.1. Room temperature luminescence spectra obtained from an isolated nanofiber (*open circles*) and an ensemble of nanofibers (*solid line*) [2]. The equidistant lines on top of the graph represent the expected vibronic progression due to the C–C stretching vibrations of all carbon atoms of the individual molecules in the nanofiber. Due to reabsorption the highest energy (0–0) mode is relatively weak. It becomes stronger if one cools down the sample (Fig. 9.2). Reprinted with permission from [2]

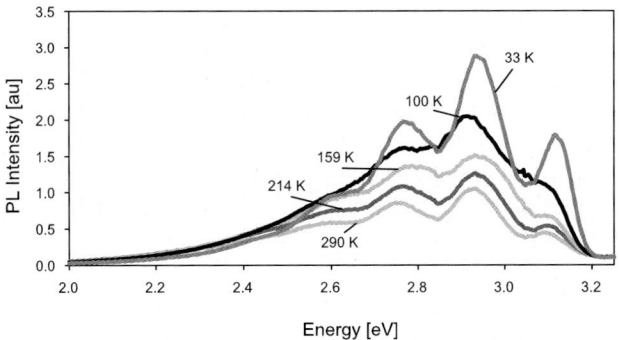

Fig. 9.2. Luminescence spectra of p6P nanofibers as a function of surface temperature. (F. Balzer, A. Pogantsch, H.-G. Rubahn, unpublished results)

9.1.2 Dynamic Response

In contrast to investigations of the static optical properties of the nanofiber films or of individual nanofibers the *temporal* emission characteristics of organic nanoaggregates is largely unexplored. The finite dimensions of the aggregates are expected to have an influence on the localization of excitonic photoexcitations. In quasisingle crystalline nanofibers the luminescence lifetime should be significantly longer as compared to the case of extended, nonoriented films, where one finds values below one nanosecond. The quality of the nanofibers should also affect the emission spectra seen as a function of decay time following initial excitation. In measurements on continuous p6P films a deterioration of the initially rather well-resolved exciton spectrum and the appearance of a red-shifted defect emission peak have been reported [3].

Time-resolved measurements on nanofiber-coated mica samples were performed after inducing luminescence in the samples by ultrashort 100 fs laser pulses at a wavelength of 266 nm. The luminescence from the samples was collected and focused at the entrance slit of a spectrometer, after which it was focused into the slit of the input optics of a Streak Camera with 2 ps resolution. The laser spot size at the sample was of the order of 50 µm in diameter.

The luminescence of molecules in the bulk of a solid is affected by exciton–exciton scattering, i.e., radiationless transitions. Hence the exciton density n, which is proportional to the total luminescence intensity, can be described by a first-order differential equation of the form [4]

$$\dot{n}(t) = -\frac{1}{\tau} n(t) - \gamma n(t)^2, \qquad (9.1)$$

where τ denotes the monomolecular decay constant and γ is interpreted as an annihilation decay rate caused by exciton–exciton scattering.

Figure 9.3a shows the measured decay for a p6P nanofiber film of 40 nm nominal mass thickness, excited with $86\,\mathrm{nJ\,cm^{-2}}$ per pulse. Note that, e.g., for a nominal 5-nm thick p6P film the height of individual nanofibers would still be of the order of 50 nm as deduced from simultaneously performed AFM measurements.

The dashed curves in Fig. 9.3 are the solutions of (9.1) fitted to the decay. The dotted curves are the monomolecular decay alone. From the linearity of $\dot{n}(t)/n(t)$ plotted vs. $n(t)$ it is concluded that τ and γ are time independent. Only a small splitting between the dashed and dotted line is observed in Fig. 9.3a, indicating a monomolecular decay with little annihilation from

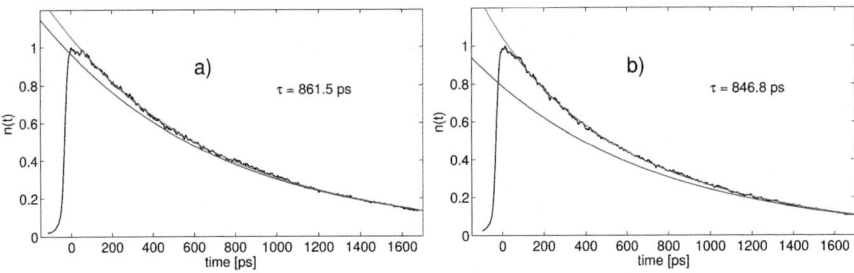

Fig. 9.3. (a) Fitted luminescence decay for a 40-nm thick p6P nanofiber film excited with $86\,\mathrm{nJ\,cm^{-2}}$ per pulse. The *dotted line* represents the monomolecular decay alone. (b) Same, but for $420\,\mathrm{nJ\,cm^{-2}}$ per pulse. Reprinted with permission from [5]

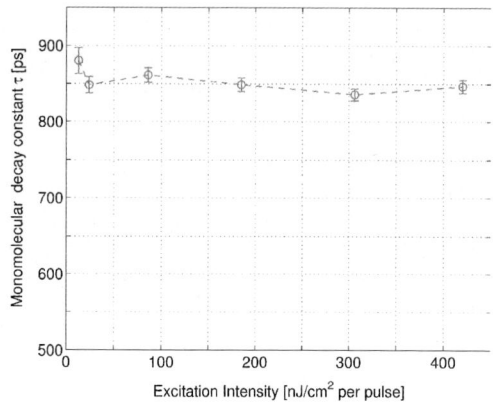

Fig. 9.4. The change in monomolecular decay constant τ as a function of excitation intensity for a 40-nm thick p6P nanofiber film. Reprinted with permission from [5]

exciton–exciton scattering. In Fig. 9.3b the excitation intensity is $420\,\text{nJ}\,\text{cm}^{-2}$ per pulse. Here, a clear splitting between the two curves is observed, indicating a significant exciton–exciton scattering rate for higher excitation intensities.

Figure 9.4 shows the monomolecular decay constant τ vs. the excitation intensity for the 40 nm nanofiber film. The value of τ for the 40 nm film lies around 850 ps and is independent on the excitation intensity since the monomolecular decay constant does not depend on the exciton density. However, γn_0 increases by a total value of $3.5 \times 10^8\,\text{s}^{-1}$ for increasing excitation intensities from 13 to $420\,\text{nJ}\,\text{cm}^{-2}$. At high excitation intensity γn_0 is found to be of order $5 \times 10^8\,\text{s}^{-1}$.

The monomolecular decay lifetime for thick films (above 20 nm) is rather independent of film thickness. As one decreases the nanofiber film density into the regime of individual nanofibers, this picture changes. Atomic force microscopy (AFM) images of three different p6P nanofiber films are shown in Fig. 9.5, where (a) is the 40 nm, (b) the 21.4 nm, and (c) the 7.5 nm thick p6P nanofiber film. As seen in Fig. 9.5a the 40 nm film consist of both large "continuous film like" nanofibers and smaller "needle like" nanofibers, with widths ranging between 0.5 and 2.5 µm and heights between 100 and 200 nm. The 21.4-nm thick nanofiber film (Fig. 9.5b) consists of very rough and densely packed "needle like" nanofibers (widths of 300–700 nm and heights of 60–80 nm). Finally, the 7.5-nm thick nanofiber film (Fig. 9.5c) is made of well separated, smooth, and similarly shaped nanofibers (widths of 200–400 nm and heights of 60–80 nm).

Note that the homogeneity of the needles depends very much on the nominal film thickness. In general, the needles in thicker films are very inhomogeneous and resemble more rough films.

The upper part of Fig. 9.6 shows the monomolecular decay constant τ for the three nanofiber films shown in Fig. 9.5 and additionally for a 5-nm thick nanofiber film with similar morphology as the 7.5 nm film (Fig. 9.5c).

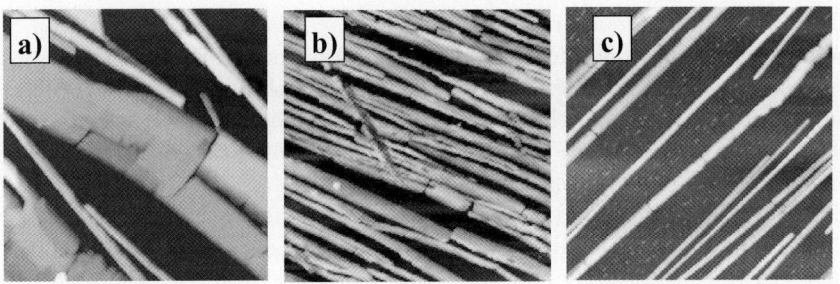

Fig. 9.5. Atomic force microscopy images ($10 \times 10\,\mu m^2$) of three different p6P nanofiber films. (**a**) 40 nm thickness, (**b**) 21.4 nm thickness, (**c**) 7.5 nm thickness. Reprinted with permission from [5]

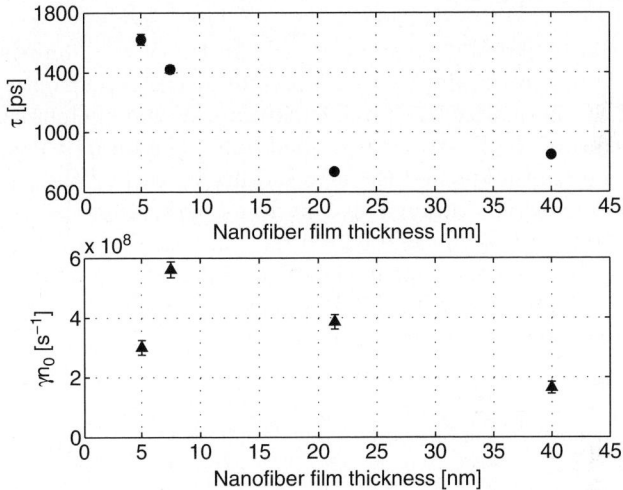

Fig. 9.6. Measured data for p6P nanofibers on mica excited with $185\,nJ\,cm^{-2}$ per pulse. *Top*: τ as a function of nanofiber thickness. *Bottom*: γn_0 as a function of nanofiber thickness. Reprinted with permission from [5]

In the individual nanofiber regime the monomolecular luminescence lifetime increases by more than a factor of two as compared to the thick film regime since the number of defects in the nanofibers decreases at which the excitons could relax. This is consistent with the AFM measurements, which reveal homogeneous nanofibers in the individual nanofiber regime (Fig. 9.5c). The lower part of Fig. 9.6 shows the annihilation decay rate vs. nanofiber thickness. With decreasing film thickness the annihilation decay rate increases for fixed excitation intensity since the relative excitation rate at fixed laser intensity increases. In the individual nanofiber regime the annihilation decay rate is very sensitive not only on the amount of material, but also on the homogeneity of the film.

In order to obtain information about the localization of the excitation, the polarization of the photoluminescence from a 5 nm p6P nanofiber film (with morphology similar to Fig. 9.5c) was measured with a polarizer being placed between the sample and the spectrograph. The nanofibers were excited with 185 nJ cm^{-2} per pulse and the E-field of the laser pulse was parallel to the long molecular axis, i.e., the excitation was p-polarized. Photoluminescence with E-field parallel and E-field perpendicular to the long molecular axis was measured, i.e., p-polarized and s-polarized emission, respectively. The monomolecular decay constant for p-polarized emission ($\tau = 1{,}622$ ps) was found to be larger than that for s-polarized emission ($\tau = 1{,}270$ ps), indicating that there is a higher degree of delocalization for P-polarized exciton emitters.

Time-Resolved Spectra

Figure 9.7 shows photoluminescence (PL)-spectra of the luminescence from a 40 nm p6P nanofiber film excited with 306 nJ cm^{-2} and obtained at five different decay times after the initial excitation. A sum of Gaussian functions is fitted to each of the PL-spectra (dashed line). The ratio of the peak values for the exciton transitions and the fitted width for each of the peaks remains approximately constant throughout the decay, indicating that the temporal decay is independent of vibronic transition.

There is no hint for the appearance of green emission at larger delay times. However, as seen also in Fig. 9.7, the PL-spectra shift slightly to the blue for increasing decay time. Within error the average blue shift of 6.7 ± 2.6 meV ns^{-1}

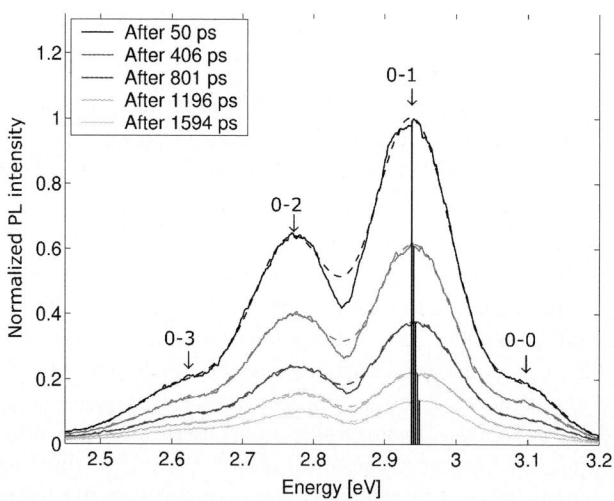

Fig. 9.7. Photoluminescence spectra taken at different time delays (uppermost curve: after 50 ps, lowest curve: after 1,594 ps) after excitation with 306 nJ cm^{-2} per pulse for a 40 nm p6P nanofiber film. The *dashed line* is a fit curve; see text. Reprinted with permission from [5]

does not depend on neither excitation intensity nor on sample morphology (nanofiber thickness). The reason for the lack of a red shift in the emission spectra apparently is the crystalline quality of the nanofibers which suppresses possible green emissions. Temperature-dependent photoluminescence studies for p6P (Fig. 9.2) recorded at 290 and 214 K give a 0–1 exciton transition at 2.9376 and 2.9540 eV, respectively, i.e., the PL-spectrum at 214 K is blue shifted 16.4 meV (a shift of -0.21 meV K^{-1}). This might explain the *blue* shift in the present spectrum. The sample is electronically heated by the laser pulse at the beginning of the decay. A rather fast (picosecond) electron–phonon coupling can be expected, leading to a lattice heating which subsequently cools down on a timescale of several hundred picoseconds, resulting in the blue shift of the PL-spectrum at later decay times.

9.2 Guiding of Electromagnetic Waves

Besides propagation and decay of excitons organic nanofibers there might also be a propagation of launched electromagnetic waves if the morphological conditions are fulfilled.

Due to a difference in indices of refraction of the underlying substrate ($\sqrt{\epsilon_s} = 1.58$) and the nanofiber ($\sqrt{\epsilon_{iso}} = 1.7$ [7]) the fiber can act as a waveguide if a critical width is overcome. In the absence of scattering from surface irregularities, if one assumes negligible damping of the evanescent field in the substrate and for wavelengths large enough that absorption can be neglected (for *para*-hexaphenylene absorption goes to zero at wavelengths larger than 500 nm), propagation losses in the fiber are very small. Here, we investigate the waveguiding behavior for a wavelength that is close to the lower limit for low-loss guiding, namely $\lambda = 425.5$ nm (Fig. 9.8). At this wavelength some losses are expected due to reabsorption of light within the nanofiber itself.

Inside a fluorescence microscope 365 nm UV light is focussed onto a selected fiber (focal radius 15 µm). The UV light transfers population into the S_1 electronically excited state, from which it relaxes predominantly into the first vibrationally excited level of the electronic ground state S_0. Following initial excitation, the 425.5 nm light propagates via transfer of excitation between neighboring p6P molecules. In order to study the propagation losses fibers with breaks have been investigated. The widths of the breaks are less than 150 nm which is much narrower than the wavelength of the propagating light. The breaks do not affect the straightness of the nanometric waveguide. Breaks are visible in the fluorescence microscope images as bright spots and are characterized morphologically by AFM (Fig. 9.8). Both the induced luminescence intensity at the point of excitation and the scattered luminescence intensity at the position of a break are measured quantitatively with the help of a CCD camera. The distance between excitation and scattering points is varied by a movable focusing lens and by moving the sample with micrometer precision.

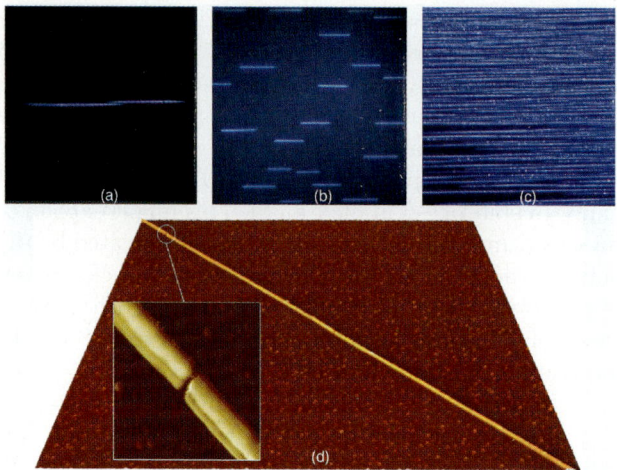

Fig. 9.8. Fluorescence images of two long, isolated nanofibers ((**a**), $322 \times 322\,\mu m^2$), a bunch of nearly monodisperse short fibers ((**b**), $115 \times 115\,\mu m^2$) and densely packed long fibers ((**c**), $0.5 \times 0.5\,mm^2$). (**d**) Force microscopy of the isolated fiber on the left-hand side of (**a**) ($21 \times 21\,\mu m^2$). The radiating region around a break is shown in high resolution ($1.6 \times 1.6\,\mu m^2$) as an inset. Reprinted with permission from [6]

A quantitative understanding and improved control of the waveguiding properties requires an analytical theory for optical waveguiding in nanometer-scaled aggregates.

In local approximation electromagnetic waves can propagate in a nanofiber waveguide only as TM–modes [8], where the number of possible modes, $m = 1, 2, 3 \ldots$, is restricted by the condition

$$m < \frac{2a}{\lambda} \sqrt{\frac{\epsilon_\perp}{\epsilon_\parallel}} \sqrt{\epsilon_\parallel - \epsilon_s}, \qquad (9.2)$$

with λ the wavelength, ϵ_s the dielectric constant of the substrate, and a the fiber width. We assume the nanofiber to have a rectangular cross section and to be an optically uniaxial medium with the dielectric tensor component $\epsilon_\parallel$ along the large molecular axis. The other two components are equal to $\epsilon_\perp$. The cutoff wavelength for the guided modes is

$$\lambda_c = \frac{2\sqrt{\epsilon_\perp}\,a}{m}. \qquad (9.3)$$

In the present case we have $\epsilon_\perp = 1.9$ and $\epsilon_\parallel = 4.8$. With $\lambda = 425.5\,nm$ the minimum fiber width for which at least one mode can propagate is $a_1 = 222\,nm$. The second mode appears for widths larger than $a_2 = 444\,nm$. This finding agrees quantitatively with experimental results that show waveguiding for fibres with 400 nm diameter, but not for fibres with diameters smaller than

222 nm [9]. The cutoff wavelengths for nanofibers of different widths range from $\lambda_c = 1{,}103$ nm for $a = 400$ nm to $\lambda_c = 689$ nm for $a = 250$ nm.

Damping of the intensity of the propagating light is dominated by the imaginary part of the waveguides dielectric tensor. The reabsorption of light by p6P molecules takes place if the electric field vector of the light is directed along the molecular axes. Quantitative values of the component $\epsilon''_\parallel$ can be determined from a comparison of measured and calculated losses of luminescence intensity as a function of distance along the fiber axis.

Figure 9.9 shows measured distance dependencies of scattered luminescence intensity for a fiber on mica (open symbols) and on NaCl (filled circles), respectively. The measured characteristic widths are 400 nm (fiber on mica) and 300 nm (fiber on NaCl). From a variety of measurements on selected individual fibers on a single growth substrate we obtain a value of $\epsilon''_\parallel = 0.012 \pm 0.002$ for the imaginary part of the dielectric function of nanoscaled fibers. Measurements on other substrates result in slightly different values, see below.

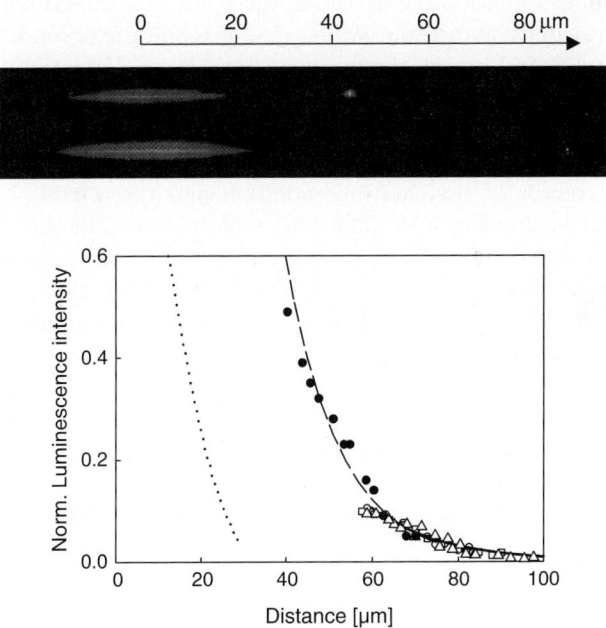

Fig. 9.9. Luminescence intensity as a function of distance from the excitation point for an individual, 400 nm wide fiber on mica (*open symbols*) and for a 300 nm wide fiber on NaCl (*filled circles*). The *dotted line* represents the spatial excitation profile. Theoretical predictions are given by *dashed* and continuous *solid lines*. In the upper part of the picture we show fluorescence images along the fiber on mica for two distances between excitation and radiating region. Reprinted with permission from [8]

9.3 Spatial Distribution of Molecular Emitters

The polarized spatial intensity distribution of photons generated in the nanofibers is obtained from two-photon microscopy [10]. Local orientations of the hexaphenylene molecules along the nanofibers can be deduced from a comparison of the intensity distributions at different combinations of polarization of exciting and detected light and employing the tensorial nature of the respective optical response.

The experimental setup for two-photon microscopy is explained in Chap. 2. The intensity of two-photon luminescence, I_{TPL}, is determined by $I_{\text{TPL}} \propto |D_{\text{if}} M_{\text{fi}}|^2 I^2$ [11] with

$$M_{\text{fi}} = \sum_s \frac{(\mathbf{d}_{\text{fs}} \cdot \mathbf{e}^\omega)(\mathbf{d}_{\text{si}} \cdot \mathbf{e}^\omega)}{\hbar(\omega - \omega_{\text{si}})} \tag{9.4}$$

the two-photon transition matrix element for the transition between the states $|i\rangle$ and $|f\rangle$ of an aggregate of molecules, $D_{\text{if}} = (\mathbf{d}_{\text{if}} \cdot \mathbf{e}^{\omega'})$ and I the intensity of the incoming radiation. Here, $\mathbf{d}_{\text{jk}}$ is the transition dipole matrix element for the intermediate transition $|k\rangle \to |j\rangle$, ω_{jk} is the corresponding transition frequency and ω and $\mathbf{e}^\omega$ are the frequency and the polarization vector of the laser radiation, whereas $\mathbf{e}^{\omega'}$ is the polarization vector of the luminescence light.

Let us choose the x-axis of the laboratory frame parallel to the p-polarization vector of the exciting radiation and the y-axis parallel to the s-polarization vector Fig. 9.10. The p6P molecules within the nanofiber are oriented with their long axes at an angle θ with the x-axis. We assume that for all transition dipole moments the value of the component along the long molecular axis considerably exceeds that of the other two components. In such a case the matrix element M_{fi} can be expressed in terms of the transition matrix element in the coordinate frame associated with the molecular axes, $M_{\text{fi}}^{\text{mol}}$, as $M_{\text{fi}} = M_{\text{fi}}^{\text{mol}} \cos^2 \theta$ for p-polarized exciting radiation and as $M_{\text{fi}} = M_{\text{fi}}^{\text{mol}} \sin^2 \theta$ for s-polarized radiation. Similar expressions hold for $D_{\text{if}} = D_{\text{if}}^{\text{mol}} \cos \theta$ for p-polarization and $D_{\text{if}} = D_{\text{if}}^{\text{mol}} \sin \theta$ for s-polarization.

The ratios of the total signal intensities in two different excitation polarizations are thus

$$\frac{I_{\text{sp}}}{I_{\text{pp}}} = \frac{I_{\text{ss}}}{I_{\text{ps}}} = \tan^4 \theta , \tag{9.5}$$

allowing one to easily deduce the angle θ of the molecular axis with respect to the laboratory x-axis. Note that besides the solution $+\theta$, (9.5) allows also the solution $-\theta$. This uncertainty can be removed by either rotating the analyzer and obtaining a continuous polarization dependence or by setting the sample at two different, fixed angles with respect to the incoming and outgoing electric field vectors.

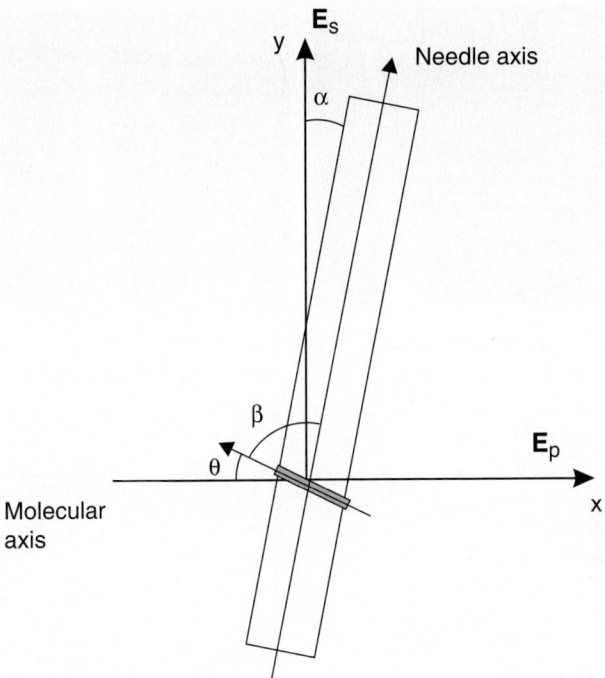

Fig. 9.10. Laboratory and molecular coordinate frames. The angle α is the misalignment of the needle axis with respect to the s-polarized light field. The angle β denotes the orientation of the molecular axis with respect to the long needle axis and θ is the angle of the molecular axis with respect to the x-axis. Reprinted with permission from [12]

Figure 9.11 shows $10\times10\,\mu m^2$ two-dimensional plots of two-photon intensity obtained with p-polarized detection. The exciting light was p-polarized (Fig. 9.11a) and s-polarized (Fig. 9.11b).

Following background subtraction the ratio $I_{\rm pp}/I_{\rm sp}$ is deduced from these plots for several hundred points along the long needle axes. From that the angle θ is determined with respect to the x-axis and is corrected for the measured misalignment of the individual needles with respect to the y-axis (angle α in Fig. 9.10). Figure 9.12 shows the deduced molecular orientation angle β (Fig. 9.10) as a function of position on the needle axis. The different symbols denote needles on various samples.

As seen, the deduced orientation angles are between 78° and 69° with a dominant orientation of $\beta = 77 \pm 1°$, in agreement with the angle of approximately $\beta = 75°$ found in previous femtosecond work [13] and also in agreement with polarized fluorescence measurements that show that the main linear transition dipole moment is oriented nearly perpendicular to the

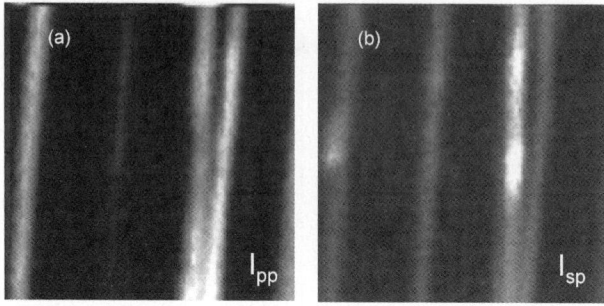

Fig. 9.11. TPI images ($10 \times 10\,\mu m^2$) of nanofibers. Detection p-polarized, excitation light p-polarized (**a**) and excitation light s-polarized (**b**). Reprinted with permission from [12]

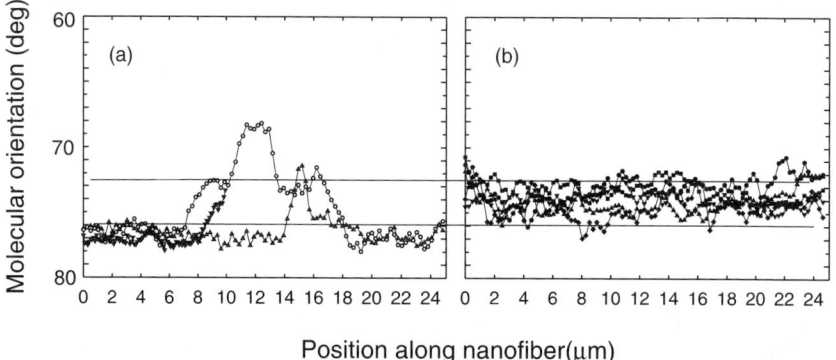

Fig. 9.12. Variation of molecular orientation angle β along needle axes as determined from the local ratios I_{pp}/I_{sp}. The different symbols denote different needles. The *parallel lines* are expected molecular orientations of 75.9° (*lower curve*) and 72.7° (*upper curve*). Reprinted with permission from [12]

needle axes. The next preferred discrete orientation is $\beta = 73 \pm 1.5°$, whereas one of the needles shows a $3\,\mu m$ wide contribution with $\beta = 69 \pm 1.5°$.

Para-hexaphenylene molecules grow with their bulk lattice constants either with the $(1\bar{1}\bar{1})$ or (010) faces parallel to the mica surface. The short axis is parallel to one of the unit cell vectors of the substrate, whereas the angle of the long axis of the molecules relative to the long needle axis is either 75.9° ($(1\bar{1}\bar{1})$-face) or 72.7° ((010)-face). These angles have been plotted as parallel lines in Fig. 9.12.

Diffraction measurements on ensembles of needles confirm the $(1\bar{1}\bar{1})$-face and thus $\beta = 75.9°$ to be the preferred orientation of the molecules. Due to the rather strict confinement of the p6P molecules to the mica surface dipole orientation in the initial growth process, it is most likely that $\beta = 72.7°$ results from inhomogeneities on the needle surface such as hillocks or valleys, which might also give rise to a certain degree of depolarization. At the same instance, $\beta = 69°$ might be due to a certain degree of polycrystallinity on thick needles.

9.4 The Optical Near Field of Nanofibers

Next the possibility of optical investigations of nanofibers using near-field microscopy techniques is exploited. Due to the rather low-detection sensitivity of near-field optical probes and due to the low-photobleaching threshold of nanoaggregates which consist only of a small number of organic molecules, this task is nontrivial.

By using focussed far-field excitation and near-field detection out of this focus, the waveguiding efficiency and thus the imaginary part of the dielectric function of nanofibers is directly determined. These results are a valid benchmark for the determined dielectric functions that have been discussed in the previous paragraphs on the basis of far field measurements.

9.4.1 Single Photon Tunneling Microscopy

The experimental set up consisted of a scanning near-field optical microscope (SNOM) operated in collection mode and an inverted epifluorescence microscope mounted below the SNOM on a (x, y, z)-movable table for focussed illumination of the sample. The sample emits photons, which are sampled in the near field by the scanning optical fiber tip.

It should be noted that the detection efficiency of even an uncoated fiber in an SNOM is typically [14] less than 10^{-3}, making it necessary to strongly focus the exciting UV light, which limits the achievable signal-to-noise ratio due to photobleaching effects.

In Figs. 9.13–9.15 we present topographic and optical images of p6P nanofibers on mica. The UV light was focussed a few micrometers outside the

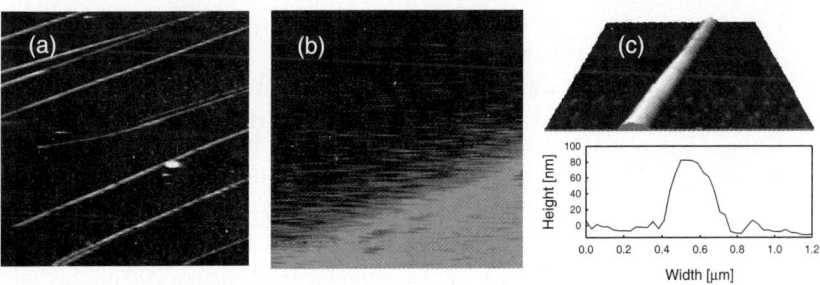

Fig. 9.13. (a) Topographic image ($30 \times 30\,\mu\mathrm{m}^2$) of p6P nanofibers on mica, obtained using SNOM. (b) Optical image. Illumination on the lower right perpendicular to the nanofiber direction. (c) AFM image of a typical nanofiber of the investigated samples. A linescan is shown below the image. Reprinted with permission from [15]

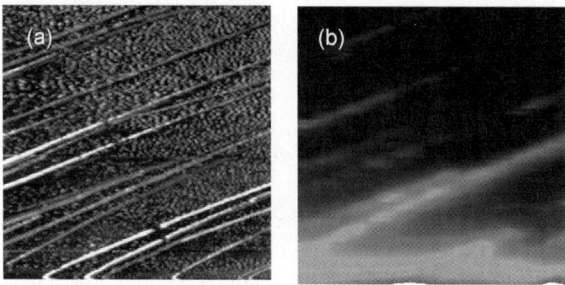

Fig. 9.14. (a) Topographic image (60 × 60 µm^2) of p6P nanofibers on mica. (b) Optical image. Illumination from the lower left along the nanofibers. Reprinted with permission from [15]

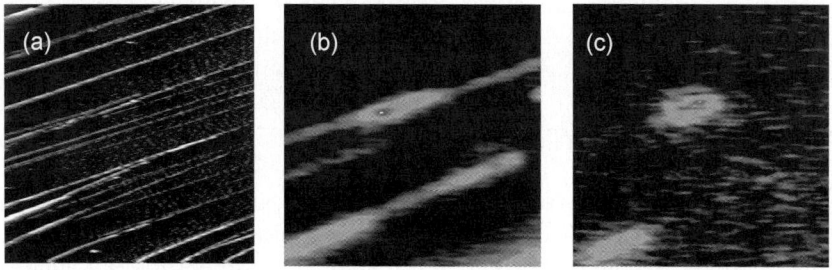

Fig. 9.15. (a) Topographic image (45 × 45 µm^2) of p6P nanofibers on mica. (b) Shear force mode optical image. Illumination from the left-hand side. (c) Optical image, taken at a distance of 200 nm above the sample, i.e., outside the near-field. Reprinted with permission from [15]

scanning range of the SNOM through the thin mica sample onto its frontside, which was covered with the nanofibers. In Fig. 9.13a, b data are shown with parallel illumination and in Figs. 9.14 and 9.15 with perpendicular illumination. The topographical images show in all cases well-separated nanofibers, which also have been detected via far-field fluorescence microscopy through the same objective that served for illumination.

The optical images show in the case of illumination parallel to the nanofiber long axis no waveguiding out of the directly illuminated area (Fig. 9.13b), whereas in the case of perpendicular illumination 425 nm light propagates along the nanofibers (Fig. 9.14b).

In order to separate that part of the light from the optical image that radiated into the far-field we have compared measurements in shear force mode (Fig. 9.15b) with measurements at an offset distance of 200 nm above the sample, i.e., outside the near-field (Fig. 9.15c). Apparently, the crossing points between nanofibers and their end points are those areas where scattering into the far-field occurs.

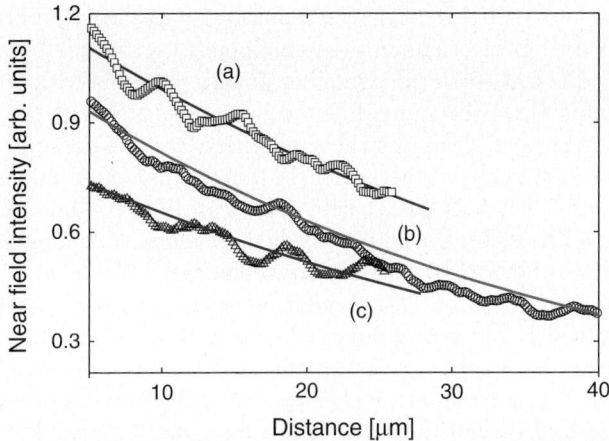

Fig. 9.16. Decrease in luminescence intensity, measured in the near-field for three selected nanofibers [16]. The lines are theoretical curves, resulting in dielectric functions of $\epsilon_\parallel = 0.0038 \pm 0.0001$ (a), $\epsilon_\parallel = 0.0045 \pm 0.0001$ (b) and $\epsilon_\parallel = 0.0040 \pm 0.0001$ (c), respectively. Reprinted with permission from [16]

Next, we have quantitatively determined the decrease of luminescence intensity, measured in the near-field as the distance z from the excitation point z_0 along the surface is increased (Fig. 9.16). We find an exponential decrease of near-field intensity,

$$I(z) = I(z_0) \exp\left[-\alpha(z - z_0)\right]. \tag{9.6}$$

Independent AFM investigations revealed typical dimensions of individual nanofibers of 300–400 nm width and 60–120 nm height. Hence the nanofibers represent single-mode waveguides allowing propagation of the mode $m = 1$ only, with

$$\alpha = \frac{2\pi \epsilon_\parallel''}{\lambda \sqrt{\epsilon_\parallel'}} \left[1 - \frac{1}{\epsilon_\perp} \left(\frac{\lambda}{2a} \right)^2 \right]^{1/2}. \tag{9.7}$$

As discussed in the previous sections we suppose that only the $\epsilon_\parallel$ component has an imaginary part ($\epsilon_\parallel = \epsilon_\parallel' - i\epsilon_\parallel''$) and $\epsilon_\parallel'' \ll \epsilon_\parallel'$.

From a fit of the decay curves plotted in Fig. 9.16 we obtain values of $\epsilon_\parallel''$ that are about three times smaller than those obtained from far-field measurements, although the needle widths are comparable in both cases. In order to test whether this difference is due to the near-field detection mode, we made far-field waveguiding measurements on the same sample as used in the present near-field experiments. These measurements [9] provided a value of $\epsilon_\parallel'' = 0.0034 \pm 0.0006$, in agreement with the fit to the data in Fig. 9.16. Thus the difference in the $\epsilon_\parallel''$ values is due to the different samples. In addition to the dielectric losses there is also an attenuation of the waveguiding mode due to scattering at imperfections in the nanofibers.

A remarkable feature of the nanofiber images obtained with SNOM is that individual nanofibers show quite different brightness although they are seen in the far-field images with almost the same brightness. A possible explanation is that we are observing in the near-field images just those nanofibers which show waveguiding whereas the far-field images are due to a direct UV illumination of all nanofibers. This would imply that the nonvisible nanofibers possess widths less than the critical width for waveguiding, which is about 220 nm. A systematic investigation of 29 individual nanofibers showed no clear correlation between width and waveguiding efficiency. However, we have used topographic information for that investigation that has been obtained directly by the SNOM in shear force mode, whose operation is still poorly understood [14] and not optimized for faithful topographic imaging. It is well known that the topographic images of nanostructures are always too wide due to the convolution with the tip end and that the height information is influenced by the nature of friction. In addition, charging effects between the dielectric tip of the SNOM and the dielectric functions of mica and of the nanofibers might falsify the topographic information, i.e., broaden the observed features. Hence it is well possible that the near-field detection provides a way of determining those nanofibers which are waveguiding.

9.4.2 Two-photon Near Field Microscopy

In order to obtain information about the orientations of the molecular emitters in the nanofibers in a way similar to the far field approach discussed before one needs to combine local polarized excitation with polarized detection. This is done in the present approach with the help of a two-photon SNOM.

For two-dimensional maps of the two-photon intensity the p-polarization vector is chosen to be parallel to the x-axis of the laboratory frame. From a comparison of the intensity distributions at different combinations of polarization of exciting and detected light we deduce local orientations of the hexaphenylene molecules along the nanofibers. The ratios of the total signal intensities in two different excitation polarizations are given by (9.5). These ratios allow one to deduce the angle θ of the molecular axis with respect to the x-axis (Fig. 9.10).

The measurements on nanofibers shown on Fig. 9.17 were carried out using the two-photon SNOM setup with a step size of 50 nm and an excitation wavelength of 790 nm. We define p- and s-polarization as being horizontal (along the x-axis) and vertical, respectively. Three individual parallel nanofibers can be easily identified. By analyzing cross sections perpendicular to the fibers through the two-photon images it is possible to determine a minimum value of the resolution capabilities of the two-photon SNOM. On Fig. 9.18 a cross section obtained from the center of the two-photon image for pp polarization

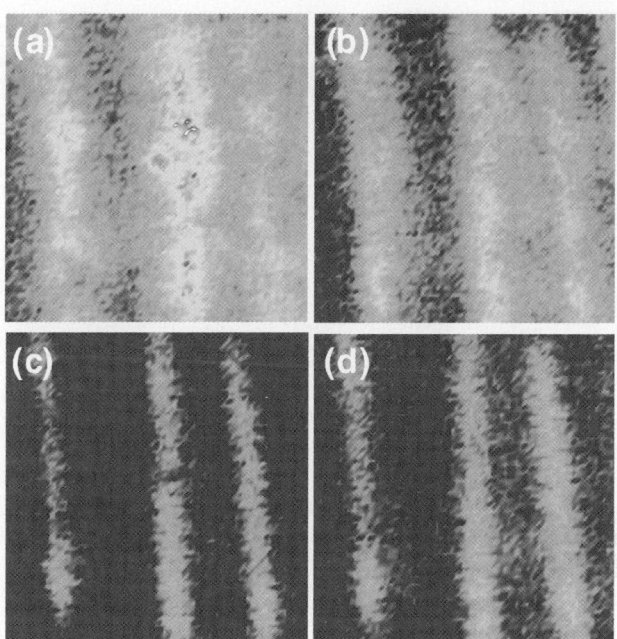

Fig. 9.17. Two-photon (790 nm) induced near-field images ($4 \times 4\,\mu m^2$ with 80×80 points) of nanofibers obtained for four perpendicular polarizations of illumination: (**a**) pp, (**b**) ps, (**c**) sp, and (**d**) ss. Reprinted with permission from [15]

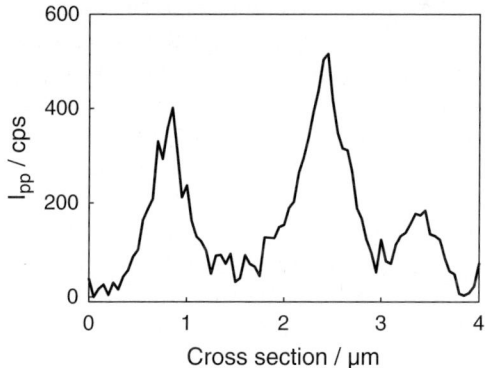

Fig. 9.18. Cross section from the TPI SNOM image after background correction, obtained for pp polarization combination and averaged over three adjacent lines. Based on the cross section the resolution in the TPI images can be verified as being better than $0.4\,\mu m$. Reprinted with permission from [15]

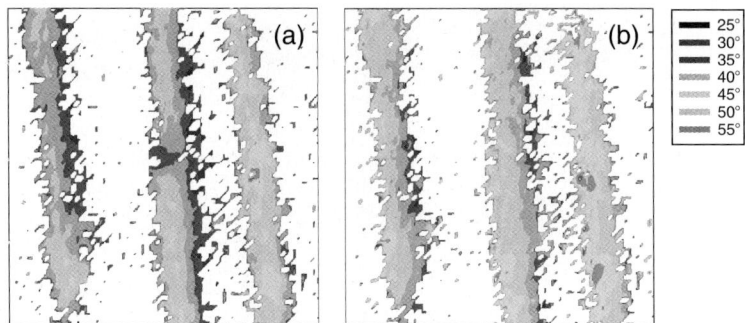

Fig. 9.19. Molecular orientation angles relative to the x-axis as obtained from the TPI-images ($4 \times 4\,\mu m^2$) shown in Fig. 9.17. For each image the signal is averaged over three adjacent lines and calculated from $I_{\rm sp}$ and $I_{\rm pp}$ for (**a**), $I_{\rm ss}$ and $I_{\rm ps}$ for (**b**). Reprinted with permission from [15]

can be seen. The resolution in the two-photon images is much better than $0.4\,\mu m$.

Figure 9.19a shows the resulting molecular orientation for the combination sp/pp and Fig. 9.19b for the combination ss/ps. Because of the birefringence of muscovite mica the absolute value of the molecular angle has not been determined. In addition, a slight depolarization might occur during the coupling of light to the fiber, implying that the light polarization is never perfectly oriented.

Despite these limitations we observe that the orientation angle is fairly constant along the fiber on the right-hand side whereas small-scale deviations of almost 10° from the mean angle are seen for the other two fibers. This observation has to be attributed to inhomogeneities of individual fibers as being confirmed by polarized linear fluorescence measurements. From a complete polarization measurement (unpolarized excitation with a high-pressure mercury lamp, $\lambda \approx 365$ nm, polarized detection) the angle of the transition dipole with respect to the small fiber axis has been determined via a fit to Malus law. Both, a constant angle at the expected value of 15° as well as a variation between the two maximum angles of ±15° are observed for different fibers. The latter might also be related to the observation that some of the fibers in fact resemble fiber bundles and not individual aggregates.

9.5 Conclusions

In this chapter we have visualized the influence of morphological changes on a nanoscale on the static and dynamic spectroscopic response of light emitting nanoaggregates. By choosing an appropriate width of the needle-like aggregates waveguiding of blue light has been observed, giving rise to a direct measurement of the imaginary part of the dielectric function of the nanoaggregates both in the far and in the near field. As expected, the dielectric function depends on the aggregates environment, too.

Due to the large oscillator strengths of the dye molecules the organic nanofibers are made of, nonlinear spectroscopies become easily accessible, especially if ultrashort laser pulses are used. These nonlinear optical measurements allow for spatially resolved investigations, e.g., of the orientation of transition dipole moments and molecular axes. Also, direct near field investigations could be performed mainly due to the brightness of the nanoscaled light emitting aggregates.

It is noted that a huge advantage of the light emitting organic nanofibers besides their morphological flexibility is the possibility to transfer them onto arbitrary substrates. Hence one is able to generate either isolated entities at well-localized spots on a surface or dense arrays of mutually parallel-oriented waveguides. The optical response of these two extrema will obviously be different. In addition one might transfer the nanofibers on dedicated substrates such as materials with low index of refraction (homogenizing the environment around them from an optical point of view) or micron-scaled structures, which might facilitate easier coupling of light or even enhance anisotropic light emission.

Acknowledgment

H.-G. Rubahn is indebted to his collaborators F. Balzer and V.G. Bordo. He thanks the Danish research foundations FNU and FTP for financial support.

References

1. A. Simonsen, H.G. Rubahn, Nano Lett. **2**, 1379 (2002)
2. F. Balzer, H.G. Rubahn, Adv. Funct. Mater. **15**, 17 (2005)
3. E. Faulques, J. Wery, S. Lefrant, V. Ivanov, G. Jonussauskas, Phys. Rev. B **65**, 212202 (2002)
4. T. Kobayashi, S. Nagakura, Mol. Phys. **24**, 695 (1972)
5. K. Thilsing-Hansen, M. Neves-Petersen, S. Petersen, R. Neuendorf, K. Al-Shamery, H.-G. Rubahn, Phys. Rev. B **72**, 115213 (2005)
6. F. Balzer, V.G. Bordo, A.C. Simonsen, H.-G. Rubahn, Appl. Phys. Lett. **82**, 10 (2003)
7. A. Niko, S. Tasch, F. Meghdadi, C. Brandstätter, G. Leising, J. Appl. Phys. **82**, 4177 (1997)
8. F. Balzer, V. Bordo, A. Simonsen, H.G. Rubahn, Phys. Rev. B **67**, 115408 (2003)
9. F. Balzer, V. Bordo, R. Neuendorf, K. Al-Shamery, A. Simonsen, H.G. Rubahn, IEEE Trans. Nanotechn. **3**, 67 (2004)
10. K. Pedersen, S. Bozhevolnyi, Phys. Stat. Sol. (A) **175**, 201 (1999)
11. Y. Shen, *The Principles of Nonlinear Optics* (Wiley, New York, 1984)
12. J. Beermann, S. Bozhevolnyi, V. Bordo, H.G. Rubahn, Opt. Comm. **237**, 423 (2004)

13. F. Balzer, K. Al Shamery, R. Neuendorf, H.G. Rubahn, Chem. Phys. Lett. **368**, 307 (2003)
14. C. Courjon, *Near Field Microscopy and Near Field Optics* (Imperial College Press, London, 2003)
15. J. Beermann, S. Bozhevolnyi, F. Balzer, H.G. Rubahn, Laser Phys. Lett. **2**, 480 (2005)
16. V. Volkov, S. Bozhevolnyi, V. Bordo, H.G. Rubahn, J. Microscopy **215**, 241 (2004)

10

Optical Gain and Random Lasing in Self-Assembled Organic Nanofibers

F. Quochi, F. Cordella, A. Mura, and G. Bongiovanni

10.1 Introduction

Organic solids have shown great potential for the realization of thin-film based light emitters, optical amplifiers, and lasing devices [1]. Polymer-based material systems have been demonstrated to yield low-threshold amplified spontaneous emission (ASE) and lasing under optical pumping. Various strategies have been adopted to achieve well-controlled lasing emission, among which is the principle of distributed feedback based on photonic gratings [2], where the lasing mode(s) can be tuned in wavelength by varying the grating pitch. Owing to the red shift produced in the emission/gain spectrum, with consequent reduction of the optical losses induced by self-absorption, dye-doped organics are in general advantageous over intrinsic polymers for achieving low-threshold laser action [3]. The major drawback of the technology based on dye-doped organics resides in the fact that, at high dye concentrations, the molecular interactions lead to the formation of dimers and higher-order aggregates that decrease the optical emission efficiency (concentration quenching). This puts an upper limit to the dye concentration that can be used – typically a few percent – and thus to the amount of optical gain that can be obtained in such material systems.

A rather different strategy for achieving low-threshold device operation is represented by the molecular organic film approach. Small oligomers have the potential to form well-ordered structures with superior optoelectronic properties. In fact, highly ordered conjugated oligomer materials feature high luminescence quantum yields [4] and high charge carrier mobilities [5]. Unlike single crystals, which can be grown in the centimeter length scale, molecular films tend to exhibit small crystalline domains. Microcrystallinity is detrimental to the optical performance of organic thin films: a high density of surface states affects the luminescence efficiency, while large propagation losses can arise from light scattering at the grain domain interfaces [6]. One way to circumvent this problem is to inhibit microcrystallization by growing amorphous molecular films. For instance, ASE and lasing with distributed feedback gratings have been achieved in thin amorphous films based on spirolinked molecules such as spirophenyls [7].

It has become clear that the morphological properties of organic films depend sensitively on the substrate and the deposition technique. Deposition of *para*-sexiphenyl (p-6P) and similar oligomers on substrates exhibiting strong surface dipoles, such as muscovite mica and potassium chloride, leads to the self-assembly of highly ordered aggregates through a dipole induced-dipole interaction mechanism. Needle-shaped aggregates are formed with lengths of up to 1 mm and cross-sectional dimensions (widths and heights) of the order of 100 nm [8–10]. X-ray diffraction studies show a high degree of epitaxial alignment; the p-6P long molecular axes are nearly parallel to the substrate and perpendicular to the needles' axes [11].

These linear nanoaggregates, usually referred to as *organic nanofibers*, display a number of important optical and photonic properties, i.e., strong anisotropy both in absorption and emission [9], light waveguiding [12,13], two-photon processes [14], Raman gain amplification [15], photoinduced spectral narrowing [16], ASE and random lasing [17–20]. Upon substrate contamination and subsequent modulation of the surface electric field, morphologically diverse nanoaggregates can be obtained, e.g., ring-shaped structures (*microrings*) [21], which further enlarge the horizon of the potential applications of self-assembled organic molecular nanoaggregates.

In this chapter, we review the presently available information on waveguided ASE and random laser action in organic (p-6P) nanofibers epitaxially grown on muscovite mica [17–20]. The chapter is structured as follows: the basics of random lasing are briefly reviewed in the following Sect. 10.2; Sect. 10.3 gives a description of the experimental techniques; in Sect. 10.4 we report the experimental results on random lasing and ASE in thin films of close-packed nanofibers; Sect. 10.5 is devoted to random lasing and ASE in isolated nanofibers; further considerations on potential applications of random lasing organic nanofibers in photonics and optoelectronics are reported in Sect. 10.6; the final Sect. 10.7 presents a brief and conclusive summary.

10.2 Overview on Random Lasing

By random lasing one refers to the process of lasing in a system where optical feedback is not provided by well-defined reflectors, but rather by disorder-induced light scattering [22]. Though generally detrimental to laser action, light scattering can serve as the basic mechanism to increase the gain path length in the active medium, thereby enhancing optical amplification by stimulated emission. In a disordered system, a light field undergoes a number of scattering events that change its amplitude, phase, and direction. If the light field does not return to its initial position before it escapes the system (open-loop path), phase information is lost: amplification turns out to be incoherent and nonresonant, resulting in narrowing of the emission spectrum towards the center of the material gain spectrum (ASE narrowing or *incoherent* random lasing). On the contrary, for a light field returning to its original position after one round trip, i.e., walking on a closed-loop path, spatial and frequency resonances of the electromagnetic field are realized, provided that the total phase acquired by the light field in the round trip is equal to a multiple of 2π. The lower the loss (and so the spectral width) of a resonant mode, the longer is the dwell time of light in such a mode. As the excitation level is raised, gain progressively compensates the loss of the resonance modes until the width of some mode drops to zero. This indicates the occurrence of laser instability (*coherent* random lasing) [23].

Both coherent and incoherent random lasing have so far been observed in many different material systems, including inorganic semiconductor powders [24], nanorods [25] and needles [26], epitaxially grown inorganic semiconductor layers [27], dye infiltrated synthetic opals [28] and biological tissues [29], and high-gain organic films based on polymers [30] and small molecules [31].

Micrometer-sized systems based on strongly scattering media have been demonstrated to yield coherent random lasing with strong photon localization, with the linear extension of the mode field intensity being of the order of the light wavelength in the medium [32]. Potential applications of random lasing in the strong localization regime have then been envisioned in nanophotonics and remote sensing [33].

It is the scope of this chapter to extend the phenomenology of coherent random lasing and ASE to the case of thin films of self-assembled molecular nanoaggregates based on strongly emissive, highly ordered p-6P crystals, as reported in recent papers [17–20].

10.3 Experimental Techniques

Para-sexiphenyl nanofiber films are grown on freshly cleaved, (001)-oriented muscovite mica by hot-wall epitaxy (HWE) [9] and organic molecular beam deposition (OMBD) [10].

In the HWE technique, p-6P is purified by threefold sublimation under dynamic vacuum. The base pressure during growth is about 6×10^{-6} mbar and the p-6P source temperature is fixed at 240°C. The substrate temperature is 130°C. The growth time is varied between 10 s and 120 min. Further details can be found in [9, 11, 34].

In the OMBD technique, sheets of muscovite mica are cleaved in air and are transferred immediately after cleavage into a high-vacuum apparatus (base pressure of 5×10^{-8} mbar). Before organic material is deposited, the samples are outgassed at a temperature of around 130°C such that low energy electron diffraction shows the well-known hexagonal surface structure of clean mica with electric surface dipoles present. *Para*-sexiphenyl is deposited from a home-built Knudsen cell by vacuum sublimation; during the deposition the pressure inside the vacuum system rises to 2×10^{-7} mbar. Long p-6P needles grow for deposition rates of $0.1 \, \text{Å s}^{-1}$ and at substrate temperatures around 150°C.

The p-6P films are optically pumped using ultrashort (~150 fs) laser pulses of a frequency-doubled (380–390 nm) Ti:sapphire regenerative amplifier running at a repetition frequency of 1 kHz. The UV beam is focused to circular spots ranging from 120 to 180 μm in diameter on the samples. The pump field polarization is set perpendicular to the axis of the nanofibers (and thus parallel to the long axis of the p-6P molecules) for maximum optical absorption. The optical emission spot is focused onto the input slit of a single imaging spectrometer equipped with a liquid N_2 (LN_2) cooled charge-coupled device (CCD) for high-sensitivity measurements. A low-pass ($\lambda > 400$ nm) filter is used to reject the scattered pump UV light.

For ensemble-averaged measurements in films of close-packed nanofibers (Sect. 10.4), the samples are photoexcited at normal incidence from the front surface (where the p-6P films are deposited). The samples are placed inside a recirculating-loop, cold-finger cryostat to perform measurements in the 30–300 K temperature range. The emission is collected at various angles with no optical magnification.

For spatially resolved measurements in films of sparse, isolated nanofibers (Sect. 10.5), the samples are excited at normal incidence from the back surface of the substrate. The emission is collected from the front surface using a 32× microscope objective, which focuses it onto the input slit of the spectrometer. Setting the spectrograph to zeroth-order diffraction and fully opening the input slit, one can image a ~65 (horizontal) × 205 (vertical) μm² area with a linear spatial resolution of about 2 μm. Tuning the spectrometer to first-order diffraction and closing the input slit to ~100 μm, the detection system becomes a microspectrometer that allows one to resolve the emission of a single nanofiber aligned vertically (thus parallel to the input slit) with a spectral resolution of about 0.2 nm, while the vertical spatial resolution is still ~2 μm.

Continuous-wave (cw) epifluorescence measurements are performed in an inverted microscope with a high-pressure Hg lamp as the excitation source. Complementary morphological characterization of the nanofibers is carried out by scanning-probe atomic force microscopy (AFM) using Si probes in tapping mode in air.

10.4 Random Lasing and Amplified Spontaneous Emission in Close-Packed Organic Nanofibers

Ensemble-averaged optical measurements are performed on p-6P films grown by HWE [17,19]. AFM topographic studies of the surface morphology of films prepared with increasing growth times demonstrate that nanofibers are assembled on muscovite mica by regrouping of individual crystallites that originate at the very early growth stages (for deposition times <10–25 s). For growth

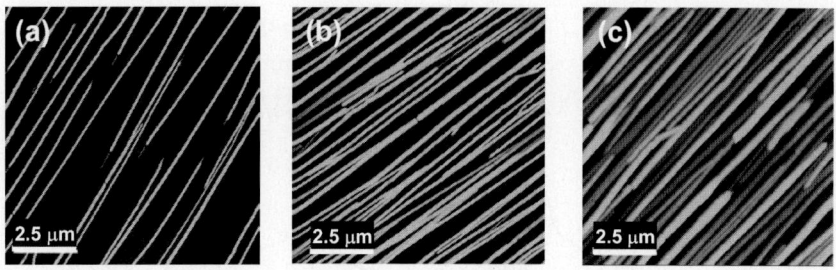

Fig. 10.1. $10 \times 10\,\mu m^2$ AFM topographic images of the surface morphology of p-6P film grown by HWE on (001)-oriented muscovite mica. Growth time (t), data color scale (range), and nanofibers' average base width $\langle b \rangle$ and height $\langle h \rangle$ are (**a**) $t = 5$ min, range = 0–150 nm, $\langle b \rangle \approx 125$ nm, $\langle h \rangle \approx 100$ nm; (**b**) $t = 40$ min, range = 0–220 nm, $\langle b \rangle \approx 210$ nm, $\langle h \rangle \approx 110$ nm; (**c**) $t = 120$ min, range = 0–700 nm, $\langle b \rangle \approx 350$ nm, $\langle h \rangle \approx 290$ nm. Reprinted with permission from [17] and [19]

times longer than 5 min, no remnants of the early crystallites are yet visible, but only linear fibers. AFM images of films obtained with growth times varying from 5 to 120 min are shown in Fig. 10.1. A statistical analysis of the nanofibers' cross-sectional dimensions shows that, when the growth time is varied from 5 to 120 min, the average height $\langle h \rangle$ of the fibers increases from ≈110 to ≈290 nm, while their base width $\langle b \rangle$ increases from ≈210 to ≈350 nm. The more material is deposited, the broader get the distributions of height and width of the fibers. As a matter of fact, closed-packed and highly disordered nanofiber interconnects are realized with long deposition times.

The laser pump beam is (1) focused on the sample to a spot whose size is ∼120 μm, which is much larger than the cross-section of the nanofibers, thus permitting to excite a fairly large number of nanofibers; (2) the emission from the entire spot is collected and spectrally resolved to obtain the response of the nanofibers' ensemble.

No evidence of excitation-induced spectral narrowing or lasing is reported in fibers at early stages of nucleation, in reason of the fact that needles' cross-sectional dimensions are far too small to enable waveguiding of the deep-blue emission of p-6P [12, 13]. Only spontaneous emission is observed in films obtained with deposition times of 5 min or shorter (Fig. 10.1a) for pump fluences of up to $2\,\mathrm{mJ\,cm^{-2}}$ per pulse. From the statistics of the needles' cross-sectional dimensions, and on the basis of Maxwell's equations solved for dielectric waveguides on a substrate support with realistic parameters [12], we estimate that in a film grown with a deposition time of 5 min, the fraction of nanofibers supporting optical waveguiding near 425 nm (where p-6P is highly emissive) is only a few percent.

Figure 10.2 shows ensemble-averaged emission spectra of the film of close-packed nanofibers having $\langle h \rangle \approx 110$ nm (imaged in Fig. 10.1b) for different values of the excitation fluence at room temperature. At low excitation levels, spontaneous emission exhibits a typical vibronic progression with the 0–1

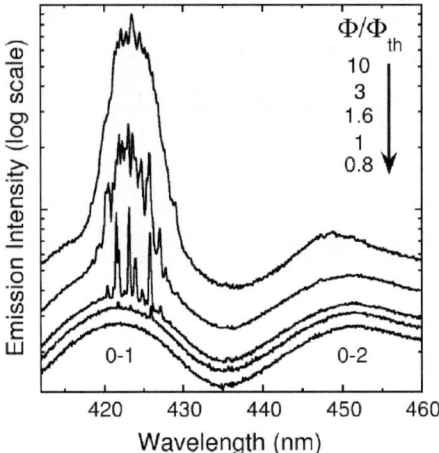

Fig. 10.2. Ensemble-averaged optical emission spectra of close-packed p-6P nanofibers (growth time = 40 min) for different values of the pump fluence normalized to the threshold fluence (ϕ/ϕ_{th}). The spectra are taken at room temperature. Reprinted with permission from [17] and [19]

and 0–2 vibronic bands peaked near 425 and 450 nm, respectively. For excitation fluences exceeding threshold values (ϕ_{th}) that can be as low as $1\,\mu\text{J}\,\text{cm}^{-2}$ per pulse, resolution-limited (laser-like) lines emerge from the spontaneous emission spectrum at the 0–1 peak. Being the film absorption ∼30%, the surface filling factor ∼50%, and further assuming that the conversion efficiency of the absorbed pump energy into (emissive) singlet excitons be equal to unity, a threshold fluence $\phi_{th} = 1\,\mu\text{J}\,\text{cm}^{-2}$ per pulse yields a singlet density $N_{th} \approx 1.2 \times 10^{17}\,\text{cm}^{-3}$, which is over four orders of magnitude smaller than the p-6P molecular density in the crystal. Long-term stability (over a time period of weeks or longer) is verified only for thresholds of the order of $10\text{--}30\,\mu\text{J}\,\text{cm}^{-2}$ per pulse ($N_{th} \sim 1\text{--}3 \times 10^{18}\,\text{cm}^{-3}$). Laser-induced degradation of the p-6P films can in fact occur after intensive irradiation of the samples in ambient conditions.

As expected from the films' anisotropy and epitaxial alignment, the emission features a high degree of linear polarization (perpendicular to the fiber axis) of 7 dB or larger across the entire emission spectrum (Fig. 10.3). Because of the spatial inhomogeneity of the p-6P film(s), fairly large variation in ϕ_{th} (up to a factor of 10) is found as the pump spot is moved across the sample surface. The laser-like emission is also found to be rather isotropic. The spectral pattern of the narrow lines strongly depends on the position of the excitation spot (Fig. 10.4), but no changes are found when the emission is recorded at different times in the same film area at fixed pump fluence, from which one infers that the spikes are not due to experimental artifacts. All these characteristics, though shown here in room temperature measurements, have been assessed to be independent of temperature in the 30–300 K range.

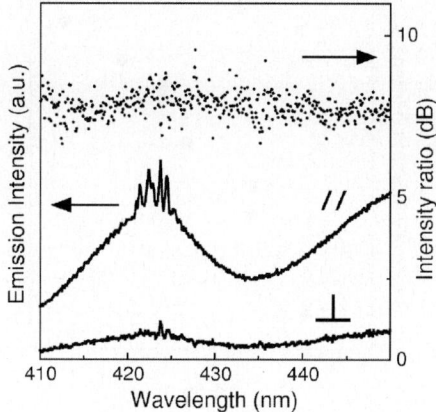

Fig. 10.3. *Solid lines*: Ensemble-averaged emission spectra of close-packed p-6P nanofibers (growth time = 40 min) measured after polarization filtering along the direction parallel ($\|$) and perpendicular ($\perp$) to the long molecular axis of p-6P. *Dots*: Intensity ratio between the two polarization filtered spectra. The spectra are taken at room temperature. Reprinted with permission from [17]

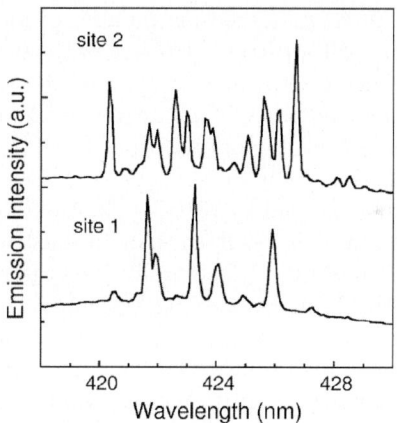

Fig. 10.4. Ensemble-averaged emission spectra of close-packed p-6P nanofibers (growth time = 40 min) in two different positions across the sample surface at the same pump fluence (above threshold). The spectra are taken at room temperature. Reprinted with permission from [17]

The low-threshold laser-like emission is attributed to *coherent* random lasing due to recurrent scattering occurring preferentially along the direction of the needles' axes. It is reasonable to suppose the coexistence of low-loss close-loop paths and open-loop paths with higher losses (for light amplification) within the pumped area; the former are responsible for low-threshold coherent emission, while the latter contribute with nonresonant ASE, hence causing spectral narrowing at higher pump fluences. This picture is in fact

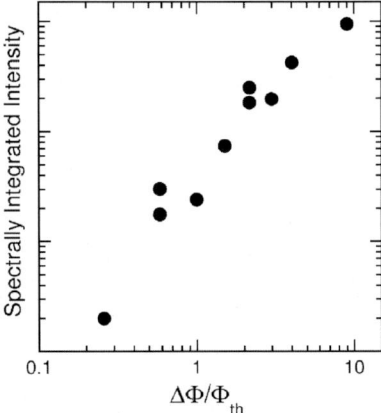

Fig. 10.5. Spectrally integrated, ensemble-averaged nonlinear emission intensity (random lasing plus ASE) vs. normalized pump excess fluence $\Delta\phi/\phi_{th}$ in close-packed p-6P nanofibers (growth time = 40 min). Reprinted with permission from [17]

consistent with the experimental results. As clearly visible in Fig. 10.2, when the pump fluence is raised, an increasing number of narrow peaks appear in the emission spectrum until spectral narrowing of the 0–1 band dominates the system response. The occurrence of spectral narrowing is partly responsible for the strong decrease in visibility of the narrow peaks with increasing pump fluence. Evidence of incipient line narrowing of the 0–2 band is also observed at the highest pump levels.

Further analysis is performed by plotting the emission intensity as a function of the normalized pump excess fluence, defined as $\Delta\phi/\phi_{th} = (\phi-\phi_{th})/\phi_{th}$ (Fig. 10.5). The signal intensity is spectrally integrated over the ASE band after subtraction of the luminescence background. The nonlinear emission intensity grows superlinearly with $\Delta\phi/\phi_{th}$. At low values of $\Delta\phi/\phi_{th}$, the superlinear behavior is explained considering that the number of random modes that reach threshold increases with $\Delta\phi/\phi_{th}$, while its persistence at high $\Delta\phi/\phi_{th}$ values is attributed to the ASE process.

Similar results are obtained in thicker p-6P films (Fig. 10.6). In the film grown with 120 min deposition time (imaged in Fig. 10.1c), one does however observe that (1) coherent features are less pronounced than in the film realized with 40 min growth time; (2) the threshold fluences are typically higher ($\sim$10–100 μJ cm^{-2} per pulse). The larger morphological inhomogeneity, proven by the broader distributions of fiber width and height, is suggested to be responsible for an increased efficiency of out-of-plane light scattering, detrimental to coherent random lasing. Intermediate film growth times (of a few tens of minutes) are thus inferred to yield best lasing performance.

It is also suggested that cross-sectionally thin nanofiber segments, not supporting waveguiding in the deep-blue spectrum of p-6P, always give a spontaneous emission contribution to the ensemble response. This is consistent

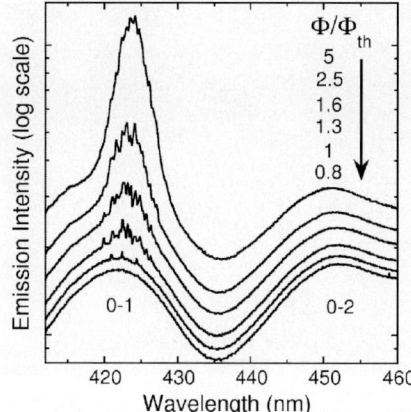

Fig. 10.6. Ensemble-averaged emission spectra of close-packed p-6P nanofibers (growth time = 120 min) for different values of the pump fluence normalized to the threshold fluence (ϕ/ϕ_{th}). The spectra are taken at room temperature. Reprinted with permission from [19]

with the fact that the total luminescence intensity of nanofibers' ensembles does not clamp when the pump fluence is raised above random lasing threshold, as clearly visible both in Figs. 10.2 and 10.6.

10.5 Optical Amplification and Random Laser Action in Single Organic Nanofibers

Ensemble-averaged measurements demonstrate that the optical response of the p-6P films depends sensitively upon the morphological characteristics of the nanofibers, a single fiber being liable to exhibit (1) spontaneous emission only when optical waveguiding is lacking due to reduced fiber width or height; (2) ASE, if waveguiding is supported but disorder-induced coherent feedback is lacking; (3) coherent random lasing, if waveguiding and coherent feedback are both present.

It is the purpose of this Section to illustrate the mechanisms about the origin of coherent feedback in single nanofibers, and to measure net gain and stimulated emission cross-section in fibers where light amplification occurs without feedback.

10.5.1 Coherent Random Lasing in Single Nanofibers

Optical emission measurements are performed on OMBD-grown p-6P films obtained with deposition times in the 2–10 min range [18, 20]. These films feature very long (up to several hundred micrometers) and isolated nanofibers in off-center regions of the substrate area coated with p-6P. Cw epifluorescence

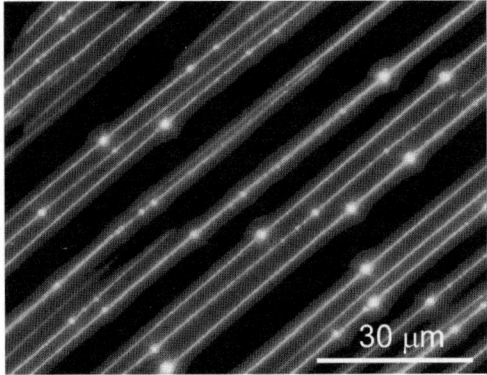

Fig. 10.7. Cw fluorescence micrograph of a p-6P film grown on (001)-oriented muscovite mica by OMBD. Reprinted with permission from [20]

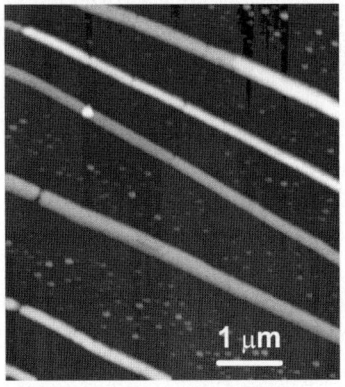

Fig. 10.8. Gray-scale AFM topographic image of the surface morphology of p-6P nanofibers grown on (001)-oriented muscovite mica by OMBD. Black and white levels correspond to fiber heights of 0 and 95 nm, respectively. Reprinted with permission from [18]

micrographs show that the nanofibers are well aligned parallel to one another (Fig. 10.7). The fluorescence emission profiles of individual fibers are very homogenous, except for special locations where the intensity of the emission scattered out of the substrate surface is strongly enhanced, yielding bright spots in the epifluorescence image.

The nanofibers are characterized by AFM topography to get better insight into how the fibers' morphological inhomogeneities determine light scattering and, consequently, coherent feedback for laser action. Figure 10.8 depicts the surface morphology of $\sim$5×5.5 μm^2 film area displaying sparse nanofibers. Typical features of isolated nanofibers on the length scale of a few micrometers are visible. The nanofibers have a base width of 300 nm or larger, which enables waveguiding of the deep-blue p-6P emission. The small islands lying between adjacent nanofibers are remnants of the nucleation process of p-6P into oriented fibers. Occurrence of cracks, typically 50–300 nm wide, implies segmentation of the nanofibers. Such thin breaks occur possibly at the end of the material growth process as a result of a surface thermal gradient while the substrate is cooling [10] or are due to an instability similar to the case of functionalized organic nanofibers on mica [35]. Breaks characterize most nanofibers in most samples albeit there exist break-free isolated nanofibers. Together with special sites where sudden variation in fiber width or height occurs (also evident in Fig. 10.8), cracks are allegedly responsible for (1) strong light scattering into out-of-plane directions; (2) back-scattering (i.e., modal reflection) of the waveguided modes, possibly contributing to the build-up of one-dimensional coherent feedback.

Random lasing from isolated nanofibers starts at pump fluences of the order of $10\,\mu\mathrm{J\,cm^{-2}}$ per pulse. Lasing nanofibers are reported in Fig. 10.9. Figure 10.9a shows a lasing micrograph taken slightly above lasing threshold. It displays both lasing from vertically aligned nanofibers and (weak) spontaneous emission from a set of neighboring nanofibers from a domain with

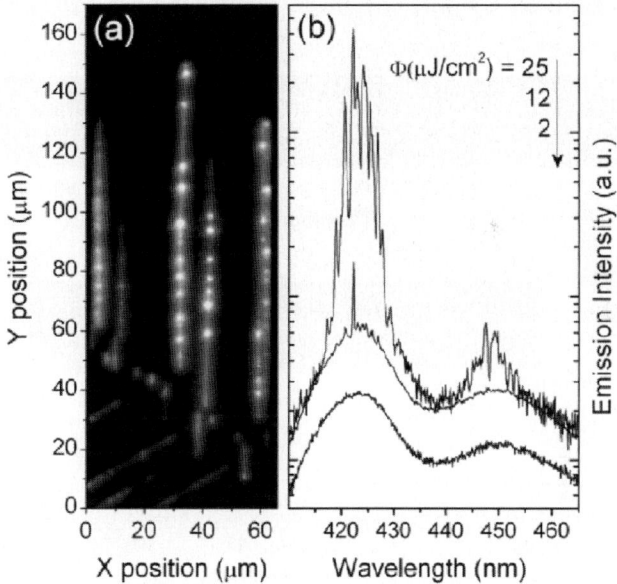

Fig. 10.9. Logarithmic gray-level scale image of the luminescence and coherent emission intensity from nanofibers excited at a pump fluence (ϕ) of $15\,\mu\mathrm{J\,cm^{-2}}$ per pulse. The Y coordinate refers to the position along the vertical direction, which is parallel to the input slit of the detection system. (**b**) Emission spectra of the nanofiber placed at $X \approx 30\,\mu\mathrm{m}$ in panel (**a**), extending vertically from $Y \approx 50\,\mu\mathrm{m}$ to $Y \approx 150\,\mu\mathrm{m}$ for different values of the pump fluence. The emission intensity is spatially integrated over the whole nanofiber length. Reprinted with permission from [18]

different needles' orientation (lower part of the graph). Again, out-of-plane scattering of the lasing emission is highly spotted, confirming that optical waveguiding is interleaved with outcoupling at special sites along the fibers. Spatially integrated emission spectra of the ∼100 μm long nanofiber placed at the center of the field of view in Fig. 10.9a are reported in Fig. 10.9b. When threshold is reached, coherent random lasing appears at the 0–1 peak. Laser action starts also at the 0–2 peak with only slightly higher thresholds. Importantly, the total luminescence intensity sharply saturates for pump fluences larger than the threshold fluence, in agreement with basic laser theory for a single emitter such as a single nanofiber. On the contrary, luminescence clamping cannot be observed in ensemble-averaged measurements on close-packed nanofibers, where not all the excited fibers reach oscillation threshold (Sect. 10.4).

Strict correlation is found between the lasing properties and the morphological characteristics of individual nanofibers. Figure 10.10a shows the AFM topographic image of the same ∼100 μm long nanofiber as depicted in Fig. 10.9. The image highlights the presence of a needle bifurcation on one side of the fiber, hence suggesting the existence of a double-fiber support for optical waveguiding. By zooming in on a smaller fiber region (Fig. 10.10b)

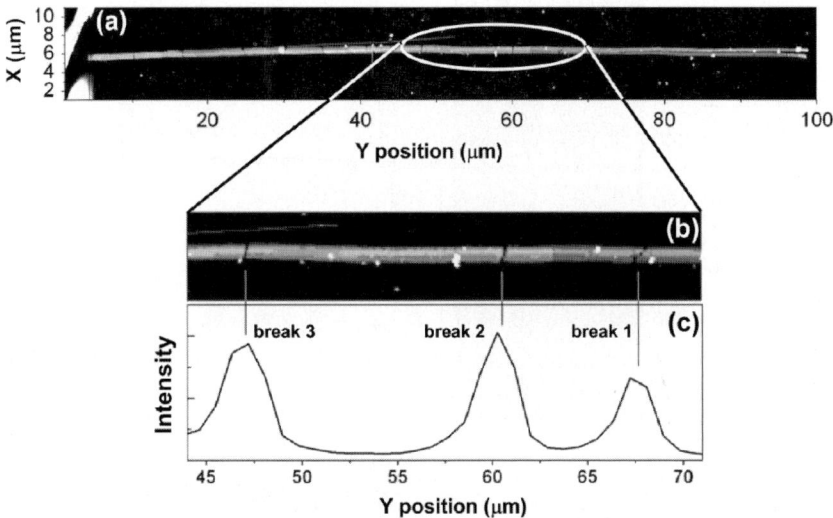

Fig. 10.10. (a) Gray-level scale AFM topographic image of the surface morphology of the isolated nanofiber whose emission spectra are shown in Fig. 10.9b. Black and white levels correspond to fiber heights of 0 and 95 nm, respectively. (b) AFM image of the same nanofiber zoomed in on a smaller region. (c) Spatial profile of the lasing emission intensity zoomed in on the same fiber region. The pump fluence is 15 μJ cm^{-2} per pulse. The vertical markers placed across panels (b) and (c) show the correspondence between the positions of the bright lasing spots and those of the fiber breaks. Reprinted with permission from [18]

and comparing the fiber morphology with the random lasing intensity profile zoomed in on the same region (Fig. 10.10c), it turns out that scattering of the lasing emission guided in the nanofiber occurs at the fiber breaks. In fact, excellent correspondence is found between the positions of the bright lasing spots and those of the fiber breaks. Measurements in other isolated nanofibers give very similar results. Therefore, correlated lasing measurements and AFM topographic measurements indicate that back-reflections of the waveguided light at the fiber break interfaces are the source of the optical feedback responsible for one-dimensional random lasing in isolated nanofibers.

Model Calculations. The experimental results are further supported by computer simulations of the optical spectra of one-dimensionally disordered systems. The theoretical model relates to the propagation of a coherent optical field of variable wavelength through a nanofiber using a standard transfer-matrix approach. Coherent propagation (with arbitrary material gain) accounts for the transmission resonances, which are the relevant channels for lasing [36]; thus, a coherent propagation model highlights the salient features of lasing. Calculations are done for a light field propagating in a multilayered structure at normal incidence, neglecting cross-sectional (modal) effects. In the model structure, several material slabs, simulating fiber segments, are separated by thin air gaps, which in turn stand for fiber breaks. The refractive index step between the material ($n \approx 1.7$ for p-6P [12]) and air causes partial back reflection of the optical field at each material–air interface. Light scattering into directions other than the fiber axis is introduced in the model in the form of linear extinction of the field propagating through the air gaps. Other optical losses such as material self-absorption and roughness scattering are compensated by gain, with the aim to simulate the system response near the lasing instability [23]. The total loss determines the spectral width of the coherent propagation modes.

Simulations are carried out with the structural data, deduced from AFM, of the nanofiber imaged in Fig. 10.10a, in terms of both material slab lengths and air gap widths. The coherent intensity spectrum is calculated at specific break locations and compared to the lasing emission spectrum measured at the same locations (Fig. 10.11). The theoretical spectra are in qualitative agreement with the experimental ones, from which it is inferred that a transfer-matrix model is well suited to describe the basic aspects of random lasing in one-dimensionally disordered systems like single nanofibers. In particular, the model reproduces the average line spacing and accounts for intensity variation of the lines as a function of the position along the nanofiber. It is worth pointing out that solving Maxwell's equations in the *real*, quasi-one-dimensional nanofiber structure is necessary to reproduce the exact positions of the lines. The lack of detailed structural information about a disordered system inevitably results in failure to predict the shapes and positions of its lasing resonances accurately [37, 38].

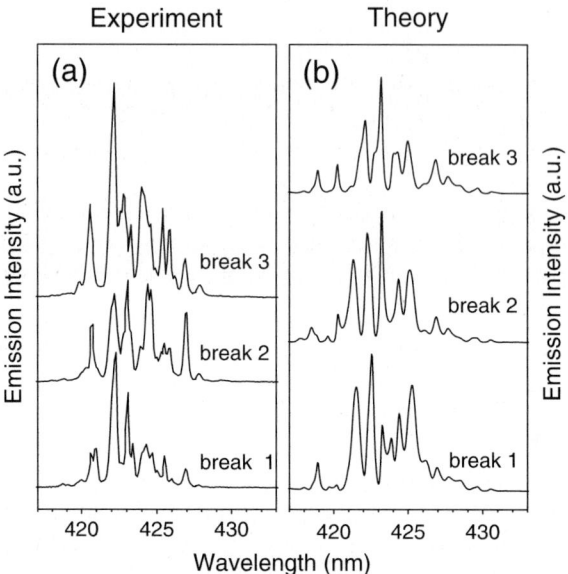

Fig. 10.11. Random lasing spectra measured at the locations of the three fiber breaks shown in Fig. 10.10b. The pump fluence is $25\,\mu\mathrm{J\,cm^{-2}}$ per pulse. (**b**) Optical intensity spectra calculated at the same locations based on the transfer-matrix model explained in Sect. 10.5.1. Reprinted with permission from [18]

10.5.2 Optical Amplification in Single Nanofibers

Light amplification is the process underlying random lasing in organic nanofiber waveguides. Net gain measurements in nanofibers are necessary to better understand the nanofibers' photonic properties, and to estimate the stimulated emission cross-section of the constituent material. To this purpose, one shall consider homogenous (breakless) nanofibers, in which retrieval of net gain is not hindered by random laser action.

Breakless fibers are identified by checking for the absence of intense scattering spots in the optical emission patterns. Figure 10.12 displays the results obtained in such a nanofiber. Micrographs taken at low pump fluences (Fig. 10.12a) show that the luminescence emission pattern is rather homogenous. When the pump fluence is raised above a threshold value, strong increase in scattered intensity is revealed at the fiber end regions (Fig. 10.12b), which indicates that spontaneous emission is amplified along the fiber waveguide, and then outcoupled at the fiber tips. As expected in the case of ASE, the emission spectra exhibit line narrowing for increasing pump fluence. Spectral narrowing is detected at both the 0–1 and 0–2 bands. Gain saturation effects are also observed at high fluences ($>350\,\mu\mathrm{J\,cm^{-2}}$ per pulse) on the 0–1 band. The lack of sharp spectral features implies the absence of coherent feedback within the fiber; hence, it is inferred that the fiber tips are not characterized

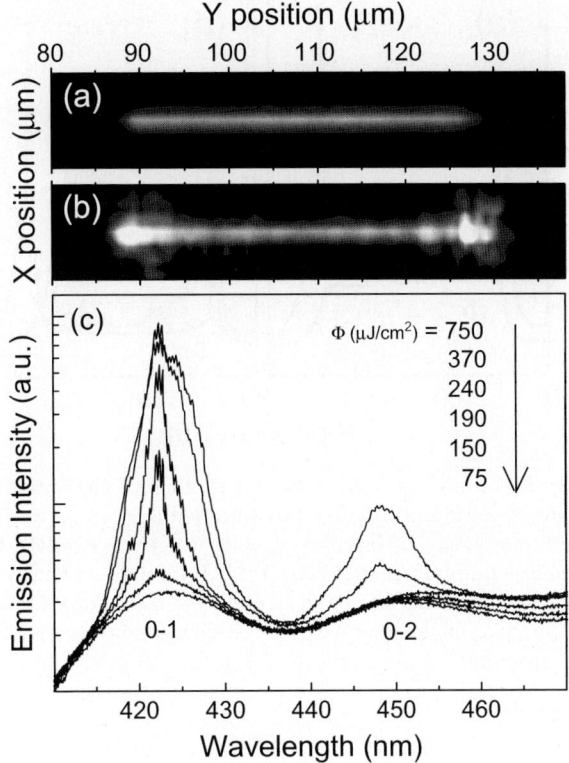

Fig. 10.12. Optical emission micrographs taken with a pump fluence of 75 (**a**) and 370 µJ cm^{-2} per pulse (**b**). (**c**) Emission spectra of the same nanofiber for different values of the excitation fluence. The spectra are spatially integrated over the nanofiber region (from $Y \approx 88\,\mu$m to $Y \approx 128\,\mu$m). Reprinted with permission from [20]

by very well defined facets. Homogenous nanofibers not exhibiting coherent feedback can in fact be treated as single-pass waveguide amplifiers.

Figure 10.13a displays the scattered emission intensity profiles extracted from the micrographs reported in Fig. 10.12a,b. While the emission profile is almost constant across the whole nanofiber at low pump fluences, above the onset of spectral narrowing the scattered intensity is position dependent and increases as the position approaches the end tips. This is consistent with the ASE process as inferred from spectral narrowing. Two independent emission spatial profiles for the 0–1 and 0–2 emission bands are then generated from the emission space-wavelength spectrograms (Fig. 10.13b).

The net gain of the nanofiber-based waveguide is estimated on the basis of the following procedure. ASE yields an output intensity $I(L) \sim (\exp(gL)-1)/g$ [1], where L is the length of the amplifying region and g the net (modal) gain coefficient. Inside the nanofiber (of length L), the total intensity of the

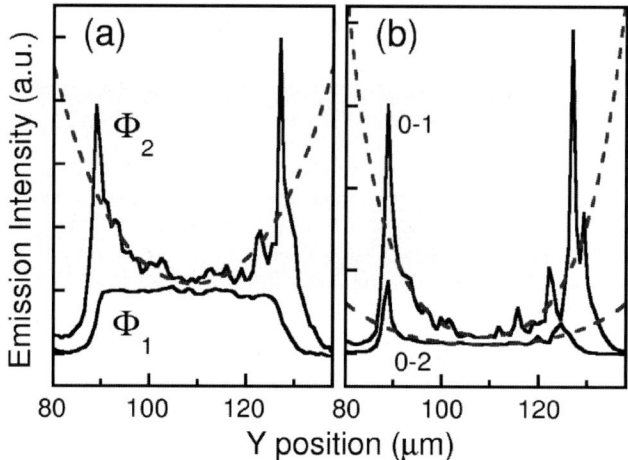

Fig. 10.13. (a) Spectrally integrated spatial profiles of the emission intensity of the nanofiber shown in Fig. 10.12 for two pump fluences: $\phi_1 = 75\,\mu\text{J cm}^{-2}$ and $\phi_2 = 370\,\mu\text{J cm}^{-2}$ per pulse. (b) Intensity profiles spectrally resolved for the 0–1 and 0–2 bands, taken at a pump fluence of $750\,\mu\text{J cm}^{-2}$ per pulse. The red dashed lines in both panels are fits to the data using the formula given for $I_T(d)$ in Sect. 10.5.2. The fit curves are extended outside the nanofiber region for better visibility. Reprinted with permission from [20]

amplified light at a distance d from a fiber tip will thus be $I_T(d) = I(d) + I(L-d)$. Owing to the uniformity of the intensity profiles taken at pump fluences below the ASE onset, it is reasonable to assume that the out-of-plane scattering efficiency is constant across the nanofiber. The same function $I_T(d)$ can thus be used for curve fitting to the measured intensity spatial profiles, using g as a free parameter. Fit curves for $L = 40\,\mu\text{m}$ are shown as the dashed lines in Fig. 10.13a,b. The tip regions are excluded from the fitting procedure since the scattering efficiency is strongly enhanced at the fiber tips. For the profiles relating to the 0–1 and 0–2 bands, best fitting yields the net gain values $g_{0-1} = (1,250 \pm 100)\,\text{cm}^{-1}$ and $g_{0-2} = (750 \pm 100)\,\text{cm}^{-1}$ at the (highest) pump fluence of $750\,\mu\text{J cm}^{-2}$ per pulse. ASE kicks in when the net gain value is comparable to the inverse fiber length ($g \sim L^{-1}$); that explains why in short nanofibers ASE-induced spectral narrowing is characterized by rather high thresholds.

In the limit of linear absorption of the pump energy, the excitation density (N) created by each pulse at the highest pump fluence is estimated to be $\sim 10^{20}\,\text{cm}^{-3}$ (see Sect. 10.4). Further neglecting population/gain time relaxation, the stimulated emission cross-section (σ_{SE}) of p-6P can be estimated from the relation $g = \Gamma g_{\text{mat}} - \alpha$ [39], where $g_{\text{mat}} = \sigma_{\text{SE}} N$ is the material gain coefficient, Γ the confinement factor of the optical intensity inside the nanofiber, α the total propagation loss. Based on Maxwell's equations solved for dielectric waveguides on a substrate support [12], the optical confinement factor in typical p-6P nanofibers deposited on mica is estimated to be ∼1%, while optical waveguiding measurements performed in singly selected organic nanofibers yield $\alpha \sim 100\text{–}300\,\text{cm}^{-1}$ [12,13]. Using $g = 1,250\,\text{cm}^{-1}$, $\varGamma = 0.01$, $N = 10^{20}\,\text{cm}^{-3}$, and $\alpha = 300\,\text{cm}^{-1}$, one obtains $\sigma_\text{SE} \approx 1.5 \times 10^{-15}\,\text{cm}^2$. Such a high σ_SE value is consistent with the high p-6P crystal order in the nanofibers. It is worth mentioning that recent works by other groups reported similar σ_SE values in neat films of oligothiophene dioxide ($6 \times 10^{-16}\,\text{cm}^2$ [40]) and poly (*p*-phenylenevinylene) derivatives ($10^{-15}\,\text{cm}^2$ [41]).

10.6 Potential Applications of Self-assembled Organic Nanofibers

Practical application of self-assembled organic nanofibers in photonics, optoelectronics, and related fields stems from a number of technological achievements, among which are (1) the ability of controlling the material aggregation process into highly crystalline nanostructures with suitable morphological characteristics, so as to achieve the desired photonic response (e.g., well-controlled lasing modes), and (2) the possibility of transferring large numbers of homogeneous nanofibers onto substrates suitable for device realization (such as ITO, Si, and SiO_2), while ensuring a high degree of needle orientation. Concerning the transfer capability, encouraging results have already been obtained with self-assembled organic nanofibers from p-6P [42] and thiophene/*p*-phenylene co-oligomers [43]. Additional requirements depend upon the specific application. In the following, we briefly discuss the potential of organic nanofibers for photonic sensing and electrically pumped lasing.

Nanofiber-based Photonic Sensing. Optical interrogation of miniaturized devices based on suitable random media is currently being envisioned as an effective solution for enabling new encoded marking and remote sensing capabilities for next-generation information technologies [33, 44].

We point out that model calculations of light propagation in a disordered system simulating a self-assembled organic nanofiber (Sect. 10.5.1) demonstrate that random lasing is very sensitive to variation in the nanofibers' structural characteristics, e.g., position, distance, and width of break-induced air gaps. The photonic sensitivity of an individual nanofiber can be studied by introducing a structural perturbation and then monitoring its effect on the coherent random modes. One can probe the effect induced by neutralization of the multiple reflections occurring in a nanofiber within a break-induced air gap, e.g., by filling the gap with a material having the same refractive index as that of the fiber material. The short-dashed line in Fig. 10.14 relates how the intensity spectrum measured at a fiber break location changes when the nearest break (placed at a distance of a few micrometers) is optically neutralized. The structural parameters are the same as those used for the calculations reported in Fig. 10.11b. The present model simulations suggest that *attoliter*

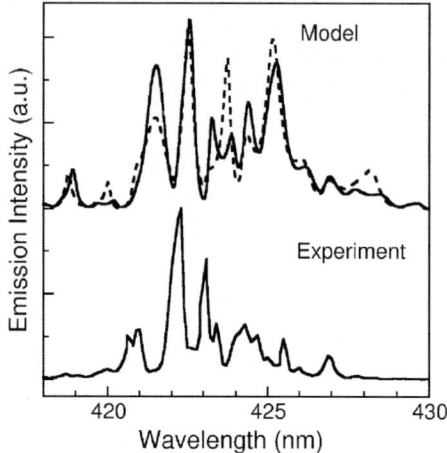

Fig. 10.14. *Lower solid line*: Random lasing spectrum measured at the position of break 1 in the nanofiber imaged by AFM in Fig. 10.10a. The pump fluence is 25 µJ cm^{-2} per pulse. *Upper solid line*: Field intensity spectrum calculated at the same fiber location on the basis of the transfer-matrix model described in Sect. 10.5.1 (see also Fig. 10.11b). *Dashed line*: Field intensity spectrum calculated at the position of break 1 after optical neutralization of the nearest break (break 2). Reprinted with permission from [45]

contamination by an index matching fluid could indeed be sufficient to induce a variation of ∼100% in emission intensity of a gain-narrowed mode, which gives a proof-of-concept demonstration of nanofibers as detectors of ultralow volume changes [45].

We add that surface adsorption of molecular species in nanofibers assembled from suitably functionalized oligomers [35] could generate photonic chemosensing, e.g., by modulation of the effective index of the waveguided modes. Alternatively, chemisorption processes at the optically active surface of the nanofibers can introduce nonradiative deactivation pathways that compete with stimulated emission, thereby generating photonic sensitivity. Operation at the onset of random lasing should also lead to enhanced sensitivity of nanofiber-based sensors owing to the increased slope efficiency of the emission intensity above threshold [46].

Electrically pumped Nanofiber Lasers. It is known that singlet–singlet (S–S) annihilation is the dominant nonradiative loss mechanism in crystalline organic materials [47]. In films of close-packed p-6P nanofibers, the monomolecular recombination and S–S annihilation rate constants have been determined to be 1.8×10^9 s^{-1} and ∼10^7 cm^3 s^{-1}, respectively [17]; hence, the S–S annihilation onset density (N_{SS}) amounts to ∼2×10^{16} cm^{-3}. We mention here that random lasing threshold densities as low as 6×10^{16} cm^{-3} have been reported in films of close-packed nanofibers (Sect. 10.4). Assuming a spin statistical singlet-to-triplet generation ratio of 0.3, and further neglecting nonradiative

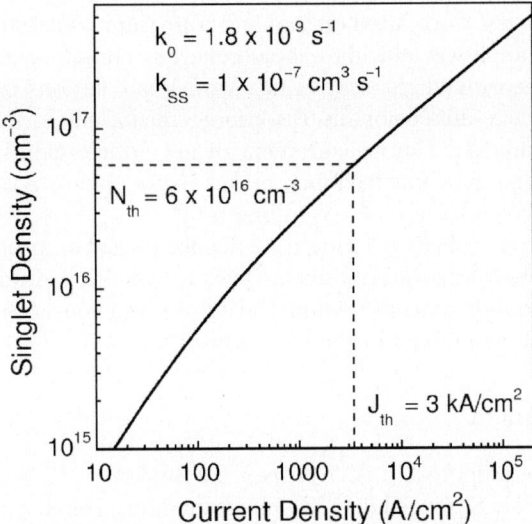

Fig. 10.15. Singlet exciton density vs. dc current density in a p-6P nanofiber. The current density is estimated assuming a spin statistical singlet-to-triplet generation ratio (0.3). The depth of the recombination region is set equal to the nanofiber height of 110 nm. The monomolecular recombination rate constant (k_0) and singlet–singlet annihilation rate constant (k_{SS}) used for the calculation are given in the plot. Density-dependent nonradiative decay processes other than single–singlet annihilation are neglected. The vertical and horizontal dashed lines mark the threshold current density $J_{\text{th}} \approx 3\,\text{kA cm}^{-2}$ and the corresponding threshold singlet density $N_{\text{th}} = 6 \times 10^{16}\,\text{cm}^{-3}$, respectively. Reprinted with permission from [17] and [19]

losses other than S–S annihilation, dc current densities of $3\text{--}5\,\text{kA cm}^{-2}$ would thus be required to reach the minimum lasing threshold densities estimated from the experiments (Fig. 10.15). The question whether such current densities are sustainable in organic nanofibers based on highly crystalline materials such as p-6P calls for further experimental investigations. In fact, deep knowledge about, e.g., the electrode-induced optical losses, the breakdown electric field, the thermal dissipation efficiency, and the effects of ambient conditions on material degradation is necessary to assess the feasibility of electrically pumped nanofiber-based devices in a quantitative way.

10.7 Summary and Conclusions

In summary, we reviewed the experimental data available on optical amplification and random lasing in self-assembled organic nanofibers based on highly crystalline p-6P, epitaxially grown on muscovite mica.

Waveguide ASE and low-threshold coherent random lasing are reported in both ensemble-averaged measurements and spatially resolved measurements in single nanofibers. Correlated optical and topographic measurements carried out in single nanofibers elucidate how coherent emission arises from (one-dimensional) recurrent light scattering. Model calculations of coherent light propagation in one-dimensionally disordered media support the experimental findings. Moreover, the measurement of net optical gain in homogeneous nanofibers lacking coherent feedback makes it possible to retrieve the stimulated emission cross-section of crystalline p-6P.

Implications of coherent random lasing for application of self-assembled nanofibers to photonics and optoelectronics are envisioned and discussed. In particular, a proof-of-concept demonstration is given concerning the potential of single organic nanofibers in photonic sensing.

Acknowledgement

The authors are indebted to A. Andreev, F. Balzer, G. Hlawacek, H. Hoppe, H.-G. Rubahn, H. Sitter, and C. Teichert for film growth and characterization. They also thank V.G. Bordo, R. Frese, and N.S. Sariciftci for fruitful discussions.

References

1. For a review, see, e.g., M.D. McGehee, A.J. Heeger, Adv. Mater. **12**, 1655 (2000); N. Tessler, Adv. Mater. **11**, 363 (1999), and references therein
2. G. Heliotis, R. Xia, G.A. Turnbull, P. Andrew, W.L. Barnes, I.D.W. Samuel, D.D.C. Bradley, Adv. Funct. Mater. **14**, 91 (2004)
3. M. Berggren, A. Dodabalapur, R.E. Slusher, Z. Bao, Nature **389**, 466 (1997)
4. S. Guha, W. Graupner, R. Resel, M. Chandrasekhar, H.R. Chandrasekhar, R. Glaser, G. Leising, Phys. Rev. Lett. **82**, 3625 (1999), and references therein
5. V. Podzorov, E. Menard, A. Borissov, V. Kiryukhin, J.A. Rogers, M.E. Gershenson, Phys. Rev. Lett. **93**, 086602 (2004)
6. H.J. Brouwer, V.V. Krasnikov, T.A. Pham, R.E. Gill, G. Hadziioannou, Appl. Phys. Lett. **73**, 708 (1998)
7. N. Johansson, J. Salbeck, J. Bauer, F. Weissörtel, P. Bröms, A. Andersson, W. Salaneck, Adv. Mater. **10**, 1136 (1998); T. Spehr, A. Siebert, T. Fuhrmann-Lieker, J. Salaneck, T. Rabe, T. Riedl, H.H. Johannes, W. Kowalsky, J. Wang, T. Weimann, P. Hinze, Appl. Phys. Lett. **87**, 161103 (2005)
8. H. Yanagi, T. Morikawa, Appl. Phys. Lett. **75**, 187 (1999)
9. A. Andreev, G. Matt, C.J. Brabec, H. Sitter, D. Badt, H. Seyringer, N.S. Sariciftci, Adv. Mater. **12**, 629 (2000)
10. F. Balzer, H.-G. Rubahn, Appl. Phys. Lett. **79**, 3860 (2001)
11. H. Plank, R. Resel, S. Purger, J. Keckes, A. Thierry, B. Lotz, A. Andreev, N.S. Sariciftci, H. Sitter, Phys. Rev. B **64**, 235423 (2001)
12. F. Balzer, V.G. Bordo, A.C. Simonsen, H.-G. Rubahn, Phys. Rev. B **67**, 115408 (2003)
13. V.S. Volkov, S.I. Bozhevolnyi, V.G. Bordo, H.-G. Rubahn, J. Microsc. **215**, 241 (2004)

14. F. Balzer, K. Al Shamery, R. Neuendorf, H.-G. Rubahn, Chem. Phys. Lett. **368**, 307 (2003)
15. H. Yanagi, A. Yoshiki, Appl. Phys. Lett. **84**, 4783 (2004)
16. H. Yanagi, T. Ohara, T. Morikawa, Adv. Mater. **13**, 1452 (2001)
17. F. Quochi, F. Cordella, R. Orrù, J.E. Communal, P. Verzeroli, A. Mura, G. Bongiovanni, A. Andreev, H. Sitter, N.S. Sariciftci, Appl. Phys. Lett. **84**, 4454 (2004); F. Quochi, A. Andreev, F. Cordella, R. Orrù, A. Mura, G. Bongiovanni, H. Hoppe, H. Sitter, N.S. Sariciftci, J. Lumin. **112**, 321 (2005)
18. F. Quochi, F. Cordella, A. Mura, G. Bongiovanni, F. Balzer, H.-G. Rubahn, J. Phys. Chem. B **109**, 21690 (2005)
19. A. Andreev, F. Quochi, F. Cordella, A. Mura, G. Hlawacek, G. Bongiovanni, H. Sitter, C. Teichert, N.S. Sariciftci, J. Appl. Phys. **99**, 034305 (2006)
20. F. Quochi, F. Cordella, A. Mura, G. Bongiovanni, F. Balzer, H.-G. Rubahn, Appl. Phys. Lett. **88**, 041106 (2006)
21. F. Balzer, J. Beermann, S.I. Bozhevolnyi, A.C. Simonsen, H.-G. Rubahn, Nano Lett. **3**, 1311 (2003)
22. For a topical review, see, e.g., H. Cao, Waves Random Media **13**, R1 (2003); H. Cao, J. Phys. A: Math. Gen. **38**, 10497 (2005)
23. A.L. Burin, M.A. Ratner, H. Cao, R.P.H. Chang, Phys. Rev. Lett. **87**, 215503 (2001)
24. H. Cao, Y.G. Zhao, S.T. Ho, E.W. Seeling, Q.H. Wang, R.P.H. Chang, Phys. Rev. Lett. **82**, 2278 (1999)
25. S.F. Yu, C. Yuen, S.P. Lau, W.I. Park, G.-C. Yi, Appl. Phys. Lett. **84**, 3241 (2004)
26. S.P. Lau, H.Y. Yang, S.F. Yu, H.D. Li, M. Tanemura, T. Okita, H. Hatano, H.H. Hng, Appl. Phys. Lett. **87**, 013104 (2005)
27. B.Q. Sun, M. Gal, Q. Gao, H.H. Tan, C. Jagadish, T. Puzzer, L. Ouyang, J. Zou, J. Appl. Phys. **93**, 5855 (2003)
28. K. Yoshino, S. Tatsuhara, Y. Kawagishi, M. Ozaki, A.A. Zakhidov, Z.V. Vardeny, Appl. Phys. Lett. **74**, 2590 (1999)
29. R.C. Polson, Z.V. Vardeny, Appl. Phys. Lett. **85**, 1289 (2004)
30. S.V. Frolov, Z. Vardeny, K. Yoshino, A. Zakhidov, R.H. Baughman, Phys. Rev. B **59**, R5284 (1999)
31. M. Anni, S. Lattante, R. Cingolani, G. Gigli, G. Barbarella, L. Favaretto, Appl. Phys. Lett. **83**, 2754 (2003); M. Anni, S. Lattante, T. Stomeo, R. Cingolani, G. Gigli, G. Barbarella, L. Favaretto, Phys. Rev. B **70**, 195216 (2004)
32. H. Cao, J.Y. Xu, D.Z. Zhang, S.-H. Chang, S.T. Ho, E.W. Seelig, X. Liu, R.P.H. Chang, Phys. Rev. Lett. **84**, 5584 (2000)
33. D. Wiersma, Nature **406**, 132 (2000)
34. A. Andreev, H. Sitter, C.J. Brabec, P. Hinterdorfer, G. Springholz, N.S. Sariciftci, Synthetic Met. **121**, 1379 (2001)
35. M. Schiek, A. Lützen, R. Koch, K. Al-Shamery, F. Balzer, R. Frese, H.-G. Rubahn, Appl. Phys. Lett. **86**, 153107 (2005)
36. A.L. Burin, M.A. Ratner, H. Cao, S.H. Chang, Phys. Rev. Lett. **88**, 093904 (2002)
37. H. Cao, J.Y. Xu, S.-H. Chang, S.T. Ho, Phys. Rev. E **61**, 1985 (2000)
38. V. Milner, A.Z. Genack, Phys. Rev. Lett. **94**, 073901 (2005)
39. D. Schneider, T. Rabe, T. Riedl, T. Dobbertin, M. Kröger, E. Becker, H.-H. Johannes, W. Kowalsky, T. Weimann, J. Wang, P. Hinze, J. Appl. Phys. **98**, 043104 (2005)

40. D. Pisignano, M. Anni, G. Gigli, R. Cingolani, M. Zavelani-Rossi, G. Lanzani, G. Barbarella, L. Favaretto, Appl. Phys. Lett. **81**, 3534 (2002)
41. S. V. Frolov, M. Ozaki, W. Gellermann, K. Yoshino, Z.V. Vardeny, Phys. Rev. Lett. **78**, 729 (1997)
42. J. Brewer, H.-G. Rubahn, Phys. Stat. Sol. C **2**, 4058 (2005); J. Brewer, C. Maibohm, L. Jozefowski, L. Bagatolli, H.-G. Rubahn, Nanotechnology **16**, 2396 (2005)
43. H. Yanagi, T. Morikawa, S. Hotta, Appl. Phys. Lett. **81**, 1512 (2002)
44. D.S. Wiersma, S. Cavalieri, Nature **414**, 708 (2001)
45. S.B. Petersen, M.T. Neves-Petersen, F. Quochi, F. Cordella, K. Thilsing-Hansen, A. Mura, G. Bongiovanni, H.-G. Rubahn, Proc. SPIE **5937**, 1 (2005)
46. A. Rose, Z. Zhu, C.F. Madigan, T.M. Swager, V. Buloviæ, Nature **434**, 876 (2005)
47. M.A. Baldo, R.J. Holmes, S.R. Forrest, Phys. Rev. B **66**, 035321 (2003)

Part IV

Applications

11

Fabrication and Characterization of Self-Organized Nanostructured Organic Thin Films and Devices

A. Andreev, C. Teichert, B. Singh, and N.S. Sariciftci

11.1 Introduction

Organic electronics (a cross point of organic chemistry, condensed matter physics, materials science, and device physics) [1, 2] has made remarkable technological breakthroughs for the last decade, enabling the realization of viable devices such as organic light-emitting diodes (OLEDs) [3, 4], organic field-effect transistors (OFETs) [5], organic solar cells [6] as well as memory circuits [7, 8]. Some products such as medium resolution emissive OLED displays [9, 10] are already commercially available and others are now in various stages of commercialization. On the other hand, nanoscale technology has clear advantages to be gained from exploiting self-organized growth, as it avoids the need for highly sophisticated patterning of surfaces with nanometer-size objects. Ideally, in organic applications the functional properties can be obtained essentially in one self-assembled molecular layer, so that organic electronics in principle would offer a maximum degree of miniaturization.

π–conjugated polymers combine properties of classical semiconductors with the inherent processing advantages of plastics and therefore play a major role in low cost, large area optoelectronic applications [1–6,9–11]. Unfortunately, polymers are highly disordered in the solid state. Consequently, carrier transport is dominantly influenced by localization resulting in low charge carrier mobilities ($\mu \ll 1\,\mathrm{cm}^2\,\mathrm{V}^{-1}\,\mathrm{s}^{-1}$) [12, 13]. Therefore, an important part of research aims toward significant improvement in the performance of organic devices and deeper understanding of physical processes using small molecular systems, in which highly ordered or even crystalline states can be obtained. Two different approaches towards well ordered small molecular systems are distinguished, namely (1) the growth of free-standing single crystals of high purity and (2) thin film growth by epitaxial vacuum preparation techniques. Unfortunately, bulk organic single crystals showing very high mobilities [14,15] are basically technologically irrelevant because of their poor mechanical stability. Therefore, there is a great interest in the second alternative, which can lead to highly ordered self-organized films with well-defined orientation of the molecules [16]. The last task still remains a challenge because the basic physical mechanisms governing epitaxial growth and self-organization in such films are not fully understood yet. There is a significant number of publications on epitaxy of molecular thin films, which are described in detail in corresponding Chaps. 3, 5, and 7.

Using OFET geometry, device performance was recently significantly improved (mobilities up to 0.1–$1\,\mathrm{cm}^2\,\mathrm{V}^{-1}\,\mathrm{s}^{-1}$), if the degree of the order in *pentacene* and *oligo-thiophene* films increased [5, 12, 17–19]. A clear correlation of the degree of order and the carrier mobility was found also for other thin film structures [20–22]. Higher order, implying enhanced charge carrier transport through the organic layer, is also of major importance in OLEDs, electro-optical modulators, photodetectors, and photovoltaic devices.

The challenging task for the future is to grow self-organized organic layers with anisotropic optical and electronic properties [16, 23]. Such phenomena can not be observed in conjugated polymer thin films, which are usually disordered and isotropic. In addition to improvements in device performance (as a future vision we can mention polarized OLEDs [16, 24]), such investigations are also of considerable fundamental interest providing a deeper insight into the underlying physical effects (anisotropic carrier mobility and diffusion, surface diffusion and sticking coefficients during epitaxy, etc.).

Among the large variety of small molecules interesting for organic electronics we will restrict ourselves to two representative groups: rigid-rod-like and, in less extend, rounded cage-like molecules, respectively. As representative for the general class of rigid-rod-like molecules of high intrinsic anisotropy we choose *oligo-phenylenes* (p-nP, where $n = 4$–6 is the number of repeating phenyl rings). These anisotropic molecules, especially *para*-sexiphenyl (see Fig. 11.1a), show in solid state a typical herringbone structure, which leave the molecular axes parallel for all molecules in the lattice but molecular planes nonparallel [25]. Since the transition dipole moment in such molecules is polarized along their long axis, such a crystallographic feature has interesting physical consequence: polarization of absorbed/emitted light strongly depends on the relative orientation of the long axis of the molecules on the

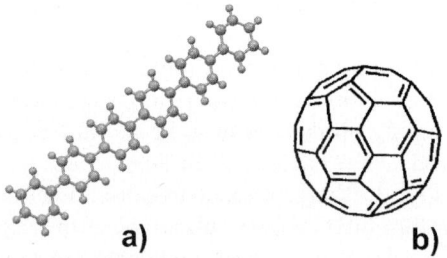

Fig. 11.1. Schematical pictures of para-sexiphenyl $C_{36}H_{26}$ (**a**) and C_{60} (**b**) molecules indicating single and double C–C bonds

substrate [26–28]. Additionally, p-6P is an excellent emitter in the deep blue with very high quantum yield [26–28]. These properties are a prerequisite for desired anisotropic optical and electrical properties of p-6P film if highly ordered growth is achieved.

In contrast to *oligo-phenylenes*, *fullerene* C_{60} represents the class of highly symmetrical molecules (Fig. 11.1b). The closed-cage nearly spherical molecule C_{60} and related *fullerene* molecules (C_{70}, C_{84}, C_{60} derivatives, etc.) have attracted a great deal of interest in recent years because of their unique structure and properties. In particular, upon photoexcitation C_{60} is known as excellent electron acceptor relative to conjugated *polymers* and *oligomers*, capable of taking on as many as six electrons [29]. In the case of donor–acceptor (D–A) blends or bilayers, this leads to ultrafast photoinduced charge transfer with subsequent long-lived charge separated states [11]. Such D–A systems were successfully used in fabricating organic solar cells and photosensitive OFETs [1, 2, 6, 30]. On the other hand, C_{60} is a very good material for highly ordered thin films, showing well-aligned epitaxial growth with the {111} planes of C_{60} parallel to the surface of layered substrates and especially of mica [31–33] (see also Chap. 5).

11.2 Experimental Methods

11.2.1 Organic Materials and Growth Techniques

P-6P crystallizes monoclinic in the space group $P2_1/c$. The lattice constants are $a = 8.091$ Å, $b = 5.568$ Å, and $c = 26.241$ Å with $\beta = 98.17°$ [25]. In a crystal the molecules are planar [34]. The diameter of a single phenyl ring is 4.3 Å and periodic distance between phenyl subunits is $C_s0 = 4.32$ Å [25]. Molecular structures of p-4P and p-5P are rather similar with $C_s = 4.41$ Å and $C_s = 4.39$ Å, correspondingly [35]. C_{60} molecules crystallize into fcc-structure with a lattice constant of 14.17 Å and a nearest-neighbor C_{60}–C_{60} distance of 10.02 Å [36]. Both p-6P and C_{60} are commercially available with a relatively high purity grade ($\sim$99%). More details about pre-handling of molecular materials and their crystallography can be found in the other chapters (Chaps. 5–7) of this book.

Two different vacuum evaporation techniques were mainly used in order to grow molecular structures discussed in this chapter, namely Organic Molecular Beam Epitaxy (OMBE) and Hot Wall Epitaxy (HWE) [16, 37]. As substrate material single crystalline mica(001), KCl(001), Al(111), and Au(111) were used. For device applications precleaned and patterned transparent highly conductive indium–tin–oxide (ITO) substrate – now standard substrate in the field of organic electronics – was used. For details of the growth and substrate preparation the readers are referred to tutorial chapter and Chaps. 3, 7 (OMBE), and 5 (HWE and their variants).

11.2.2 OFET: Device Fabrication

OFETs can be fabricated with various device geometries as depicted in Fig. 11.2. The most commonly used device geometry is bottom gate (Fig. 11.2a,b) with top contact party because of borrowing the concept of the very successful thin-film silicon transistor using thermally grown Si/SiO_2 oxide as gate dielectric. Recently, it has been shown that solution-processed organic dielectrics (dielectric constant up to 18) are promising for high performance OFETs [38–40]. As depicted in Fig. 11.2c,d organic dielectrics are especially suitable for the top-gated structure of OFET, since it does not destroy the underlying organic semiconductors. Top-gate bottom contact structure devices allow patterning the bottom source–drain electrodes on top of any flexible or rigid substrate first before building the rest of the device. Top-gate top contact structure devices allow growing organic semiconductor films on top of any flexible or rigid substrate.

In this study, we have chosen *divinyltetramethyldisiloxane-bis(benzocyclobutane)* (BCB, also known as *Cyclotene*^{TM}) as gate dielectric. Previous studies showed that BCB is a promising dielectric material possessing a large dielectric breakdown voltage, a nearly temperature independent thermal expansion coefficient and the rather high glass transition temperature in excess of 350°C [41]. A scheme of the device geometry used and the molecular structure of the dielectric are shown in Fig. 11.3. The device is fabricated on ITO/glass substrate.

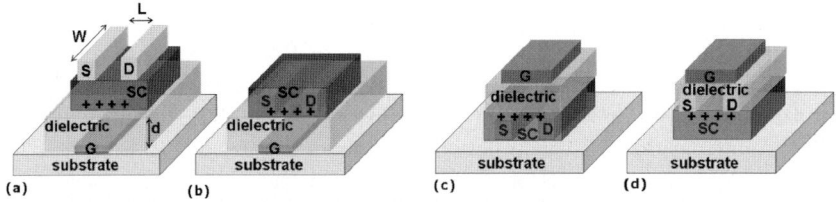

Fig. 11.2. Schematic of the bottom-gate organic field-effect transistors (OFETs) with top contact (**a**) and bottom contact (**b**) structures. Top-gate/bottom contact (**c**) and top-gate/top contact (**d**) OFETs are also shown

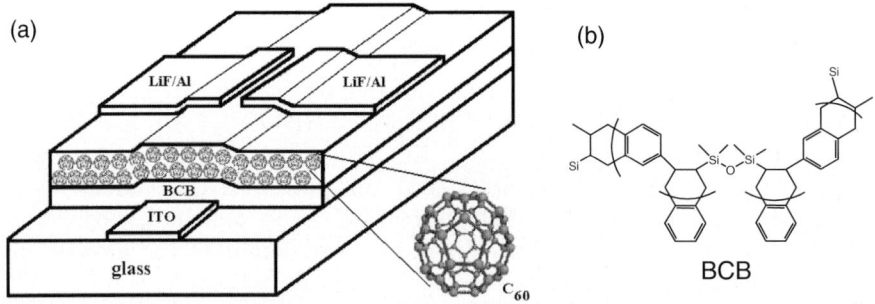

Fig. 11.3. (**a**) Schematic of the top contact C_{60} OFET device structure used in this study and (**b**) chemical structure of BCB

BCB (used as received from *Dow Chemicals*) was spin coated at 1,500 rpm for 1 min from a 30% molar solution in *mesitylene*, yielding films with a thickness of ~1.7 μm. The first round of curing of the BCB was done by thermally crosslinking the polymer for 30 min at 250°C in an argon atmosphere. After crosslinking the BCB, the film is insoluble and practically free of residual organic solvents. In the second round, the BCB dielectric is preheated in-situ inside the hot wall epitaxy system under high vacuum conditions for 20 min. All C_{60} films were grown by HWE on top of the dielectric at a constant substrate temperature of 130°C at a deposition rate of 0.8 Å s^{-1}. Device transportation from the HWE system to the glove box for metal evaporation and further electrical characterization was carried out under an argon environment. The top source and drain electrodes (LiF/Al 0.6/60 nm) were evaporated under vacuum (2×10^{-6} mbar) through a shadow mask. Other metals such as Au, Cr/Au, Ca/Al, and Mg/Al were also investigated. The channel length L of the device is 20–25 μm and the channel width is $W = 1.4$ mm, giving a W/L ratio of 70–56. For dielectric constant and capacitance–voltage measurements, separate devices with Metal–Insulator–Semiconductor (MIS) structures were fabricated under identical conditions and measured. From the measurement of the BCB film thickness, $d \approx 1.7$ μm, and the dielectric constant of BCB, $\varepsilon_{BCB} \approx 2.6$, a dielectric capacitance $C_{BCB} \approx 1.3$ nF cm^{-2} was calculated. The electrical characterization of the devices was carried out in an argon environment inside a glove box. An *Agilent E5273A* instrument was employed for the steady state current–voltage measurements. All measurements were performed with an integration time of 2 s in steps of 2 V.

11.2.3 Electrical Characterization Using an OFET (Operating Principle)

To demonstrate the operating principle of the OFETs, a simplified energy level diagram for Fermi level of source–drain metal electrode and HOMO–LUMO (Highest Occupied Molecular Orbital–Lowest Unoccupied Molecular Orbital) levels of a semiconductor are shown in Fig. 11.4. If there is no gate voltage applied (Fig. 11.4a), the organic semiconductor that is intrinsically undoped will not show any charge carriers. Direct injection from the source/drain electrodes is the only way to create a current in the organic semiconductor. Such currents will be relatively small due to high resistance of the organic semiconductors and the large distance between the source and drain electrodes. When a negative gate voltage is applied (Fig. 11.4b), positive charges are induced at the organic semiconductors adjacent to the gate dielectric (p-type conducting channel is formed). If the Fermi level of source/drain metal is close to the HOMO level of the organic semiconductor, then positive charges can be extracted by the electrodes by applying a voltage V_{ds} between drain and source. Such organic semiconductors with ability to conduct only positive charge carriers are called p-type semiconductors.

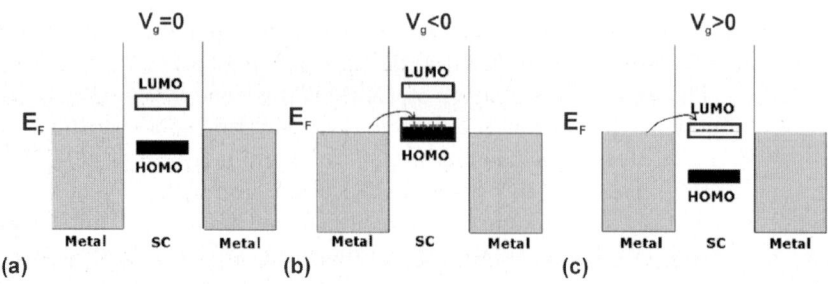

Fig. 11.4. Illustration of the working principle of an OFET with respect to applied V_g. (**a**) No charges are injected when $V_g = 0$ V. (**b**) When a negative voltage is applied to the gate, positive charges are induced at the organic semiconductor/organic insulator interface (at the channel). (**c**) When a positive voltage is applied to the gate, negative charges are induced at the organic semiconductor/organic insulator interface

When a positive voltage is applied to the gate (Fig. 11.4c), negative charges are induced at the semiconductor adjacent to the dielectric interface (n-type conducting channel is formed). If the Fermi level of source/drain metal is far away from the LUMO level, so that electron injection/extraction is very unlikely, then low I_{ds} is expected due to high contact barriers. If the Fermi level of source/drain metal is close to the LUMO level of the organic semiconductor, then negative charges can be injected and extracted by the electrodes by applying a voltage V_{ds} between drain and source. Such organic semiconductors with ability to conduct only negative charge carriers are n-type semiconductors. In some organic semiconductors, both electrons and holes can be injected and transported – an effect known as *ambipolar transistor*.

Output characteristics of a typical OFET based on organic semiconductor *poly(2,5-thienylene vinylene)* (PTV) are shown in Fig. 11.5a in order to describe typical OFET characteristics and the methods used to calculate the mobility μ and I_{on}/I_{off} ratio [42, 43]. Figure 11.5b shows a typical plot of drain current I_{ds} vs. drain voltage V_{ds} at various gate voltages V_g, which corresponds to a device using PTV film of thickness 50 nm, 500 nm spin-coated *polymethyl-methacrylate*, PMMA as the gate insulator on top of heavily doped Si substrate as gate electrode, and gold source and drain electrodes in top contact geometry as depicted in Fig. 11.5a. At low V_{ds}, I_{ds} increases linearly with V_{ds} (linear regime) and is approximately determined from

$$I_{ds} = \frac{WC_i}{L} \mu (V_g - V_T) V_{ds}, \quad (11.1)$$

where L is the channel length, W is the channel width, C_i is the capacitance per unit area of the insulating layer, V_T is the threshold voltage, and μ is the field-effect mobility. The latter can be calculated in the linear regime from the transconductance,

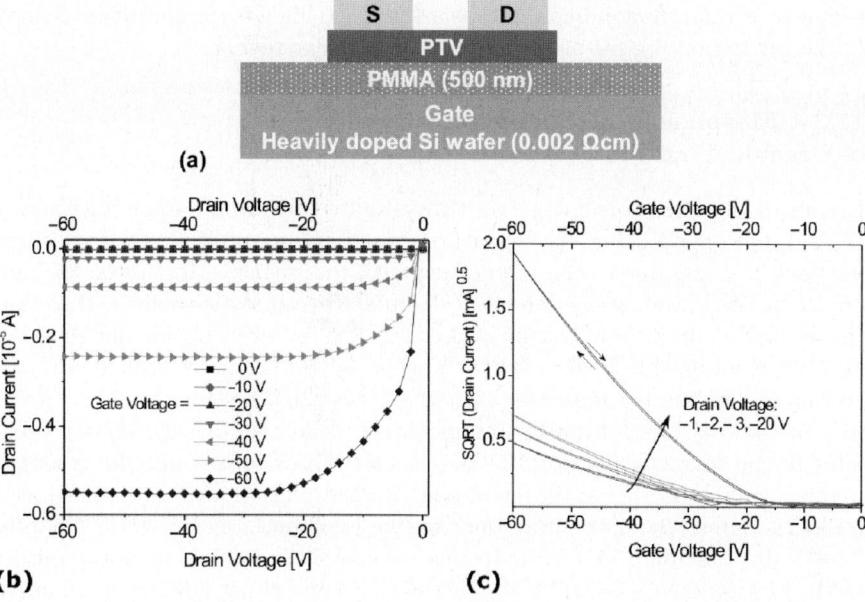

Fig. 11.5. (a) Scheme of a p-channel PTV OFET fabricated and measured in author's laboratory. (b) Output characteristics (drain–source voltage, V_{ds} vs. drain–source current, I_{ds}) of this transistor. (c) Transfer characteristics ($\sqrt{I_{ds}}$ vs. gate voltage V_g for different V_{ds}) with arrow indicating forward and reverse direction of sweeping of V_g

$$|g_m|_{\text{lin.}} = \left(\frac{\delta I_{ds}}{\delta V_g}\right)_{V_{ds}=\text{const.}} = \frac{W}{L}\int_0^{V_{ds}} \mu C_i \, dV = \frac{WC_i}{L}\mu V_{ds}, \quad (11.2)$$

by plotting I_{ds} vs. V_g at low V_{ds} and equating the value of the slope of this plot to g_m.

Figure 11.5c, which corresponds to Fig. 11.5b, shows such a plot, and the calculated mobility value is $7 \times 10^{-4}\,\text{cm}^2\,\text{V}^{-1}\,\text{s}^{-1}$ at $V_{ds} = -2\,\text{V}$. The value of V_{ds} is chosen so that it lies in the linear part of the I_{ds} vs. V_{ds} curve. The current modulation (the ratio of the current in the accumulation mode over the current in the depletion mode, also referred to as I_{on}/I_{off}) for the device of Fig. 11.5a is slightly above 10^3 when V_g is scanned from 0–60 V (Fig. 11.5c). For V_{ds} more negative than V_g, I_{ds} tends to saturate (saturation regime) owing to the pinch-off of the accumulation layer, and this regime is modeled by

$$I_{ds} = \frac{\mu W C_i}{2L}(V_g - V_T)^2. \quad (11.3)$$

In the saturation regime, μ can be calculated from the slope of the plot of $\sqrt{I_{ds}}$ vs. V_g as shown in Fig. 11.5c. For the same device as in Fig. 11.5a, the mobility calculated in the saturation regime was $1 \times 10^{-3}\,\text{cm}^2\,\text{V}^{-1}\,\text{s}^{-1}$. The difference between calculated mobility values is assigned to higher charge carrier density in the saturation regime as compared to the linear regime.

11.2.4 Morphological Characterization of Organic Thin Films and Devices

Details of the film morphology were investigated by atomic force microscopy (AFM). On approaching the surface, the oscillation amplitude decreases and its phase changes due to the interaction with the surface. Most often the amplitude is then used as the detection signal and scanning is performed so that the oscillation amplitude is kept at predefined setpoint value smaller than the oscillation amplitude in free space. The phase shift can be also monitored. In this mode, the tip interacts with the surface most strongly in the lower part of the trajectory. Finally, in noncontact AFM (NC-AFM), the tip oscillates in the attractive region of the Van der Waals forces and the resonant frequency shift of the cantilever measured by a phase lock loop is used as a feedback signal, providing true noninvasive imaging (see Ref. 41 in Chapter 1.5). Note, that high and even atomic resolution is possible in both contact mode AFM and NC-AFM. For further details concerning different variants of the AFM the reader is referred to review paper [39] in Chapter 1.5. A *Digital Instruments MultiMode IIIa* scanning probe microscope with an *AS-130(J)* scanner and a phase extender was used. The phase extender allows to image topography and phase shift simultaneously and thus enabling the differentiation between materials with different viscoelastic properties [44] as discussed in Tutorial chapter. The measurements were done under ambient conditions mostly using tapping mode to eliminate the risk of surface damage due to lateral forces between tip and surface. Most of the presented images have been obtained with Pt-coated Si-tips and slow scan speeds. Some of the high resolution images were recorded using high density carbon (HDC) tips [45]. These tips consist of a carbon whisker grown in scanning electron microscope on conventional Si-tip. These probes have a tip radius of less than 5 nm and an opening angel of less than 10°. The resulting high aspect ratio allows to precisely image crystallites with nearly vertical side walls. For details concerning working principles of the AFM and its measurement modes reader is referred to Tutorial chapter.

11.2.5 Optical and Structural Characterization of Organic Thin Films and Devices

To get more detailed information on the alignment of the molecules relative to the substrate and to each others, several optical techniques were employed. All preliminary measurements were performed at room temperature. Polarized absorption spectra were recorded with an UV–Vis HP-spectrophotometer at normal incidence. Polarized photoluminescence spectra were measured on a *Hitachi F-4010 Fluorescence Spectrometer* at normal incidence. Infrared reflection measurements were performed using a Bruker IFS-66 FTIR spectrometer. Further, steady-state photoluminescence (PL) and time resolved PL measurements at low temperatures were performed in order to reveal possible trapping states for neutral excitations (*excitons*) in highly ordered nanostructures. The conventional steady-state PL was measured at temperatures ranging from 4.2 to 300 K using an optical helium cryostat. Time-resolved PL measurements were performed with a monochromator and a 0.1 ns gating registration system containing a photomultiplier tube and a strobe-oscilloscope. A nitrogen laser with a pulse duration of 8 ns and a repetition rate of 100 Hz was used for optical excitation at 337.1 nm. For the experimental details see, for instance, [46].

X-ray diffraction (XRD) was used to evaluate the overall crystalline quality of the films as well as molecular orientation within the films. Standard XRD measurements were performed using Cr-K_α radiation and a secondary graphite monochromator. Detailed XRD investigations were performed at the F3 station at the Cornell High Energy Synchrotron Source (CHESS). For details the reader is referred to the Chap. 6.

11.3 Anisotropy of Self-Organized Organic Thin Films

11.3.1 Anisotropic Epitaxial Growth of p-6P on Mica(001)

As a first example of anisotropically self-organized epitaxial growth we will discuss the case of p-6P molecules deposited onto mica(001). As it was described earlier, the p-6P represents nicely the general class of rigid-rod-like molecules of high intrinsic anisotropy and, therefore, anisotropical growth can be expected, if deposition conditions are properly chosen. A detailed discussion of the obtained results will be given below, but at first same historical review should be given.

Thin films of p-6P were traditionally grown by physical vapor deposition [27,28,47,48]. It was found that p-6P films deposited on isotropic substrates at high substrate temperatures exhibit preferred growth in (001) orientation with the molecules aligned roughly perpendicular to the substrate [27, 28, 47–49]. At low substrate temperatures a preferred growth in (11$\bar{2}$) or/and in (20$\bar{3}$) orientations develop and the layers consist dominantly of molecules, whose long axes are inclined preferentially parallel to the substrate [47,48]. Generally, thicker p-6P films grown by vapor deposition consisted of randomly oriented islands or grains, whose shape and structure depend on the purity of the source material and on growth conditions.

Next step in increasing the degree of anisotropy and order was made by MBE on GaAs(001) substrates [50]. Anisotropic growth was achieved with quite large p-6P islands (several micrometers long, about 1 µm wide), which were oriented into two perpendicular twin directions [100] and [010], corresponding to substrate symmetry. The structural analysis of these islands was made by electron diffraction and indicated that the growth was well defined, with p-6P molecules tilted of 0° and 50° relative to the surface normal [51]. These reports demonstrated the influence of a suitable growth technique on the degree of anisotropy and order in p-6P thin films.

The first really large-scale ordered anisotropic structures of p-6P with a strongly expressed preferential direction were obtained by HWE on mica substrates [16]. Freshly cleaved mica ($K_2Al_4[Si_6Al_2O_{20}](OH)_4$) – a sheet silicate consisting of consecutive layers of K, Al, SiO_4, and OH) – was selected as convenient substrate material because of well defined surface with very small roughness and weak surface bonds similar to van der Waals bonds of p-6P. The p-6P molecules were evaporated at 240°C and substrate temperature was varied from 70 to 170°C. The HWE growth parameters result on mica in highly ordered anisotropic structures of the deposited layers, independent on the growth temperature [16, 52, 53].

Figure 11.6a shows an atomic force microscopy (AFM) image of the surface morphology of a typical thicker p-6P film grown by HWE on mica at 130°C. The structure consists of very long high aspect ratio needle-shaped *fibers*, which could be resolved into closed chains of three-dimensional islands (crystallites) [54]. In more detail the results of corresponding growth experiments are described in the corresponding Chap. 5. The straight chains run parallel to each other and are separated by relatively flat areas. Their perfect orientation is mediated by the surface geometry of the substrate (probably via dipole field-induced dipole interaction p-6P-substrate [55]) and is not influenced by surface steps [16, 56]. AFM line profile measurements show that these chains can reach width, height, and length of up to 800 nm, 300 nm, and much more than 100 μm (submillimeter), respectively. Hence, the chains are

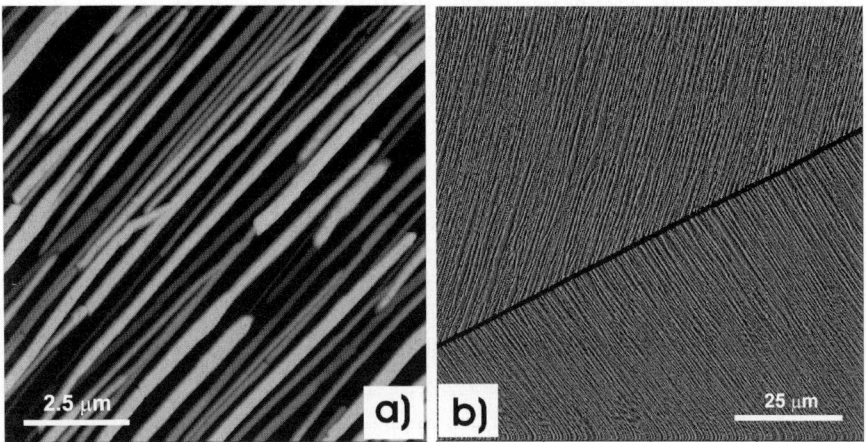

Fig. 11.6. (a) AFM image revealing the chain morphology of p-6P films – z-scale is 700 nm; (b) $100 \times 100\,\mu m^2$ image (*deflection images*) characterizing the morphology on flat terraces, separated by a big cleavage step of the mica

strongly anisotropic and their axes align according to one preferential direction. On perfectly cleaved mica this preferential direction is not even changed for the whole film surface of ~20 mm², as found by many light-microscope studies. On the other hand, it was shown that on mica terraces, separated by big cleavage steps, the preferential direction of the chains can change by 120° from terrace to terrace, while their morphology stays identical [16,56]. A typical example is shown in Fig. 11.6b. The turnaround of the *chains* by 120° seems to be based on an interaction between the deposited p-6P molecules and the different layers of mica [55,56].

To get more detailed information about the alignment of p-6P molecules relative to the substrate and to the chains, several samples were investigated using different optical techniques [16]. In Fig. 11.7 the polarized UV–visible absorption spectrum of a p-6P film is shown for different angles of the polarizer with respect to the chains. The plotted curves show a well pronounced maximum around 366 nm for the polarizer perpendicular to the chains, while no significant absorption can be observed for the excitation beam polarized parallel to the chains. Moreover, the maximum centered at ≈366 nm becomes tilted when the angle between the chains direction and electric field vector of the exciting light is close to 0°. Finally, one observes at 366 nm a dichroic absorption ratio between parallel and perpendicular polarization of more than 11. It is important to mention here that the optical absorbancy of p-6P films depends strongly on the orientation of the molecules relative to the substrate/incident beam: the absorption parallel to long molecule axis is characterized by a peak at about 380 nm and perpendicular to that by an absorption band around 280 nm [27]. Therefore, it was concluded that the dominant peak at ≈366 nm,

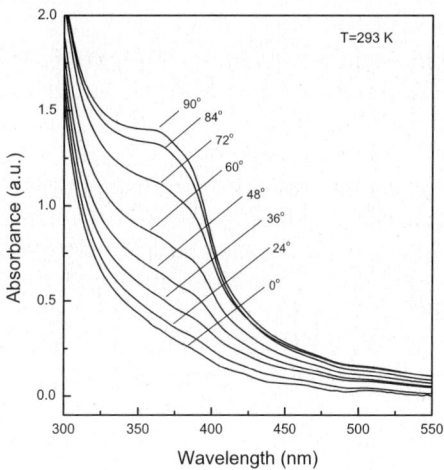

Fig. 11.7. Angular resolved polarized UV–vis absorption spectra of a p-6P film (normal incidence). Angle of field relative to needles: 0° – field parallel to the *chains* direction, 90° – field perpendicular to the *chains* direction

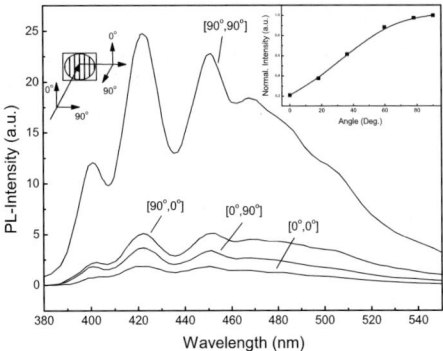

Fig. 11.8. Polarized PL spectra of a p-6P film at room temperature for excitation at 350 nm. The left insert shows a schematic representation of measuring geometry. The right insert shows the angular dependence of the emission for excitation polarized at 90°

observed for the polarization perpendicular to the chains is due to the π–$\pi^\star$ absorption of molecules, which are slightly tilted towards the substrate surface with their long axes almost perpendicular to the *chain* axis [27].

One of the applications of such anisotropic p-6P structures can be polarized LEDs and lasers. Obviously, a necessary condition for polarized electroluminescence is polarized photoluminescence. To determine the polarization degree of the emitted light, photoluminescence (PL) measurements were performed in a two polarizer geometry for the pump polarization and probe polarization (Fig. 11.8, left insert). Typical photoluminescence spectra are shown in Fig. 11.8. For all four permutations of polarizations a well known PL spectrum (upper spectrum) with three pronounced bands [57, 58] were observed, showing a fine vibronic structure [57] (for a detailed discussion see Sect. 11.4). The maximum of emission is observed again if the excitation acts perpendicular to the needle's direction, consistent with UV–vis absorption data. The dichroic ratio for 90°–90° emission compared to 0° − 0° is ≈14. The strong emission bands have the electric field vector component perpendicular to the film surface, indicating that the p-6P molecules are aligned not absolutely flat on the substrate, but are tilted slightly out of plane.

IR spectroscopy provides information on the molecular basis and presents therefore a powerful tool to study the alignment of single molecules in oriented, thin organic films [59]. For p-6P films grown by HWE angular dependences for three well known vibrational bands were analyzed: the band at 1,479 cm^{-1} assigned earlier to the ring stretching vibration along the molecular axis [57], and two bands at 760 and 815 cm^{-1}, which are assigned to the C–H out-of-plane bending vibration on mono- and parasubstituted phenyl rings [59]. The results of these investigations are shown in Fig. 11.9. The polarization dependent IR reflection spectra confirm the alignment of the p-6P molecules perpendicular to the needles. Indeed, the intensity of the C–H bands (vibration

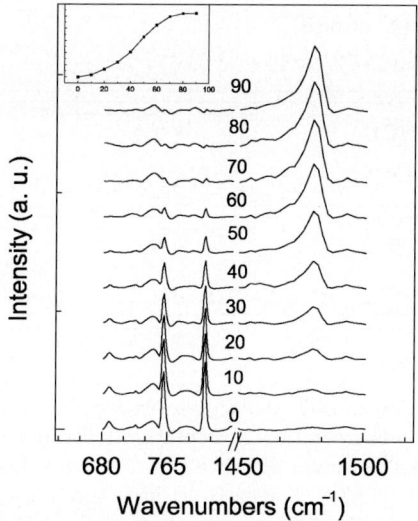

Fig. 11.9. Angular resolved polarized IR reflection spectra of a p-6P film. Angle of field relative to needles: 0° – field parallel to the chains direction, 90° – field perpendicular to the chains direction. The insert shows the angular dependence of the 1,479 cm^{-1} vibration

polarized perpendicular to long axis of molecules) can be tuned between 0 and 100% by changing the angle between the needles and the electric field vector. The ring stretching band (vibration polarized along the molecular axis) demonstrates the opposite dependence on the polarization. Moreover, the intensities of bands show a strong cosine-type dependence on the angle (see insert in Fig. 11.9). This behavior can be only understood if the p-6P molecules are mostly arranged with their long axes perpendicular to the chains, showing only a small inclination towards the substrate surface. This is again in a good agreement with UV–vis absorption and PL data.

To get more detailed information about the orientation of the p-6P molecules relative to substrate surface and each other, profound structural investigations (TED, TEM, and XRD) were performed [52–54]. It was shown that the crystalline structure (so called β-phase [25]) of the films remains nearly identical in the whole range of the film thicknesses investigated. Three different epitaxial relationships between the mica substrate and p-6P crystallites were observed at the early as well as at an advanced growth stage: $(11\bar{1})_{PSP}\|(001)_{mica}$ & $[1\bar{2}\bar{1}]_{PSP}\|[\bar{3}40]_{mica}$ (orientation (A)); $(\bar{1}\bar{1}1)_{PSP}\|(001)_{mica}$ & $[\bar{1}10]_{PSP}\|[\bar{3}40]_{mica}$ (orientation (B)), and $(11\bar{2})_{PSP}\|(001)_{mica}$ & $[\bar{2}0\bar{1}]_{PSP}\|[\bar{3}10]_{mica}$ (orientation (C)) [53]. Note that in all three cases the alignment of the long molecular axes is approximately the same. The molecular orientation relative to the substrate as extracted from these data is very similar to that obtained from optical measurements. The high

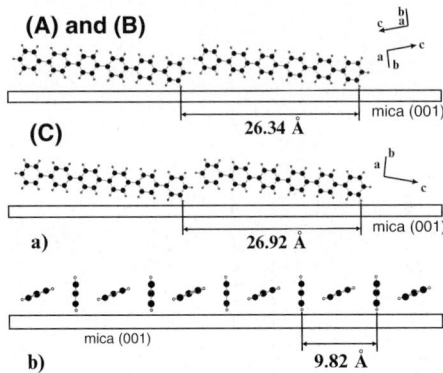

Fig. 11.10. (a) Side view of p-6P molecules relative to the mica (001) substrate as they are oriented within their epitaxial orientations. Top part: orientations ((A) and (B)) with the same intermolecular distance of $a = 26.34$ Å. Bottom part: orientation (C) with another $a = 26.92$ Å. For each of these three orientations, the unit cell vectors of epitaxially grown p-6P are given at the right side; (b) parallel view

degree of epitaxial alignment is also evidenced by rocking curves with an FWHM as low as 0.062° [60].

Based on performed optical and structural investigations one was able to determine the exact orientation of p-6P molecules on the mica surface as schematically depicted in Fig. 11.10. As demonstrated in Fig. 11.10a the long molecular axes are nearly perpendicular to the chains and tilted by 5° with respect to the surface. Only the lowest aromatic rings of the molecules are in contact with the substrate surface, as visible in Fig. 11.10a. This fact can be also clearly seen in Fig. 11.10b, which shows the parallel view of the p-6P molecules. The long molecular axes of all molecules within the unit cell are oriented parallel to each other, whereas adjacent molecular planes are tilted about 66°, which form the typical herringbone structure of p-6P.

Recently, similar long-range ordered anisotropic structures of p-nP ($n=4$–6) consisting of needle-like crystallites were obtained on mica substrates by OMBE [55, 61]. In particular, for the p-6P molecules it was shown that the growth of needle-like structures (fibers) occurs only in a rather narrow surface temperature range $\Delta T \approx 25°$ at a small adsorption rate of $0.025 \, \text{nm s}^{-1}$ [55]. This observation was used to locally influence needle growth via laser-induced surface heating [62]. It was demonstrated that the appropriate combination of substrate prehandling and laser heating allows to fabricate isolated fibers, microrings, arrays of separated aggregates, or dense arrays of parallel oriented nanofibers of p-6P [62–64]. Profound optical and structural investigations were performed [63], showing that such structures are highly ordered and therefore suitable for optoelectronic organic devices – see corresponding Chaps. 3 and 9.

Although the growth of lying p-6P molecules discussed above is promising for anisotropic OLEDs, the standing molecular orientation (long molecule

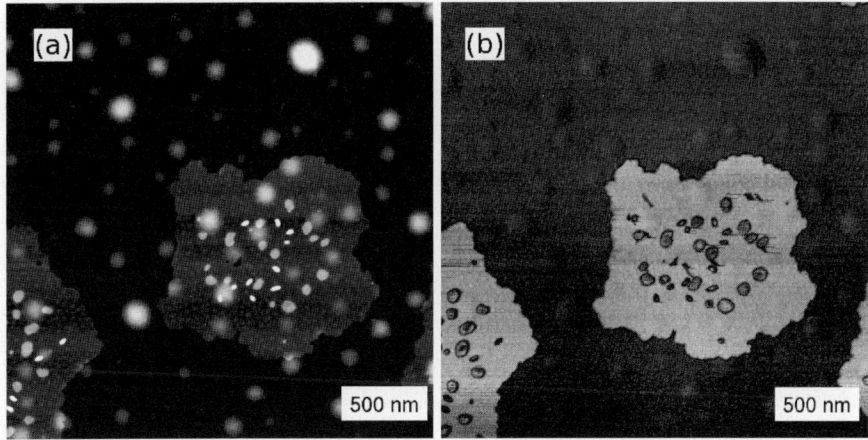

Fig. 11.11. p-6P islands formed by upright standing molecules on mica(001) [61]. (a) Height image of a 1.3 nm thick film showing two islands (z-scale: 10 nm); (b) corresponding phase image (z-scale: 20°)

axis is perpendicular to surface) is also desired, for example, for OFETs [26]. This orientation is quite unusual for p-6P grown on mica(001). Nevertheless, recently it was shown that exposure to ambient conditions just prior to p-6P deposition allows the growth of films formed from upright standing molecules [61]. In detail, cleaved mica surface was exposed to ambient conditions for 2 weeks and then preheated at an increased temperature of 500°C. After a preheating step, p-6P films were deposited by OMBE at very low deposition rate of only 0.85 Å min^{-1} [61]. Figure 11.11a shows the morphology of a 1.3 nm thick p-6P film grown at room temperature. A $\sim$1 μm^2 large island with the characteristic height of 2.5 nm can be seen. From this height one can deduce that the molecules are standing upright, since the length of a p-6P molecule is 2.6 nm [25]. In addition to big islands some smaller round structures can be seen in Fig. 11.11a. The simultaneously recorded phase image (Fig. 11.11b) reveals that only the large 2.5 nm high structures are p-6P, as only they show a significant phase contrast due to their different viscoelastic properties [44]. Therefore, the round structures showing no phase contrast are attributed to subsurface water bubbles in the mica. These bubbles can be formed due to the recombination of OH– groups found in the mica unit cell to H_2O at high temperature during the preheating step.

11.3.2 Anisotropic Epitaxial Growth of p-6P on KCl(001)

Other nice example of anisotropically self-organized growth gives the epitaxial system p-nP ($n = 5, 6$) on KCl(001). Freshly cleaved cubic KCl(001) with its well-defined (001)-oriented cleavage surface feature many advantages for the epitaxy of the p-6P. On the one hand, KCl with the lattice constant

$a = 0.629$ nm and ionic lattice spacing $d_{110} = 0.444$ nm provide a rather small lattice mismatch with p-6P ($\approx 3\%$ along KCl [110] direction) [35], which is a prerequisite for commensurable thin films growth. Moreover, KCl surface shows the ionic type of the bonding, which is stronger than van der Waals bonds on the mica surface. On the other hand, single crystalline KCl(001) has been proven as a versatile model substrate for many other organic compounds [65–67].

Needle-like structures of p-6P on the cleaved KCl(001) surface were first obtained by Yanagi and coworkers in 1997–1999 using mask-shadowing vapor-deposition technique [35, 68], just before anisotropic p-6P growth on mica was discovered [16]. p-6P molecules were deposited onto KCl(001) surface kept at 150°C through a mesh mask having round holes, which faced the KCl substrate with an intervening space. As was shown by fluorescence microscopy and AFM, the molecules that passed through the hole were not only deposited on the round surface exposed to the molecular beam, but also grew in linear needle-like crystallites (lying epitaxial orientation of the molecules [68]) with a submillimeter length and a width of several hundreds of nanometers in the shadowed region of the surface. All crystallites were oriented orthogonally along KCl [110] directions, in accordance with surface symmetry of the KCl substrate. It was also shown that flat layered crystals with standing orientation can grow on KCl(001) at the same time as needles [35].

Grazing-incidence X-ray diffraction was used in order to determine the details of the complex polymorphism of such p-6P films. It was demonstrated that two of the known five orientational phases feature preferential in-plane alignments along the [110] in-plane direction of the substrate: p-6P($11\bar{1}$)||KCl(001) with [$1\bar{1}0$]||KCl[110] as well as p-6P(001)||KCl(001) with [$1\bar{1}0$]||KCl[110] [69]. Later structural and morphological investigations have shown that the first one can be ascribed to lying molecules comprising needles, while the last one corresponds to standing molecules comprising layered crystals [70, 71].

Anisotropic electroluminescence cells were fabricated using p-6P films with standing or lying molecule orientation by its wet-transferring from KCl surface onto indium tin oxide (ITO) coated glass with following deposition of an electron transport layer (*2-(4-biphenyl)-5-(4-tert-butylphenyl)-1,3,4-oxadiasole*) and Al contacts [72]. It was demonstrated that the carrier transport efficiency of those cells depends on molecule orientation in the active p-6P layer: the cells with lying p-6P molecules exhibited much better voltage-luminance characteristics compared to the cells with standing orientation. These results were explained in terms of the higher efficiency of the hole injection and transport for the parallel stacks of lying p-6P molecules [72]. The authors have also emphasized that the lying p-6P orientation is an attractive configuration for polarized light emitting devices.

Polarized fluorescence of p-6P was used to investigate the needle-like crystallites [35, 68] in detail. Figure 11.12a shows fluorescence micrograph of the needle-like crystallites under excitation at 465 nm. The images were measured

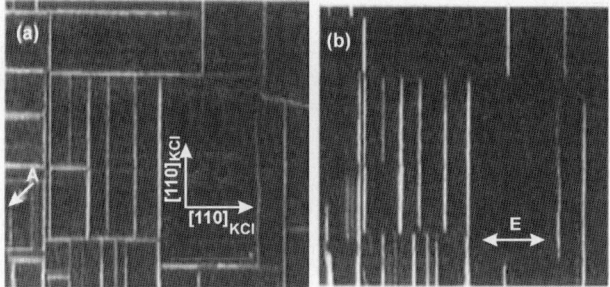

Fig. 11.12. Fluorescence micrograph p-6P crystallites grown at 150°C. The fluorescence image (a) was taken without polarizer. The image (b) was taken from the same area as (a) with horizontal polarization of the electric field. Reprinted with permission from [35]

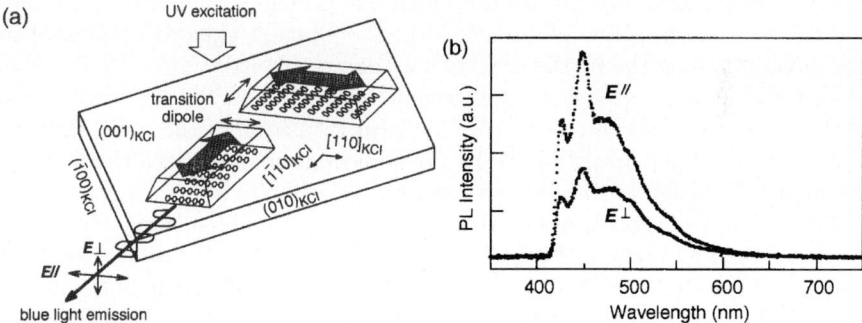

Fig. 11.13. (a) Schematic representation for the orientation of p-6P molecules within the needle-like crystals and the configuration of the polarized emission from the crystals. (b) Luminescence spectra taken under the cross-sectional observation of the needle-like crystals with the polarizations of $E_\parallel$ and $E_\perp$, respectively, shown in (a). Reprinted with permission from [68]

through a rotating analyzer to define the polarized emission [68]. It was shown that only the needles oriented vertically gave a fluorescence and the ones oriented horizontally totally vanished, if the electric field of the analyzer was horizontal (Fig. 11.12b). On the other hand, the fluorescence image turned completely around when the electric field direction of the analyzer was rotated vertically. This perfect selectivity in the orthogonally polarized fluorescence images demonstrated that the lying p-6P molecules in the needle-like crystallites biaxially orient on KCl(001) surface as schematically shown in Fig. 11.13a. The molecular axis is aligned perpendicular to the needle axis of the crystal that orients along KCl [110] directions. Therefore, all the HOMO–LUMO transition dipoles are also arranged orthogonally along the same directions to result in such a perfect polarization. To prove this observation, the polarized photoluminescence was studied in the cross-section as schematically shown in Fig. 11.13a [68]. Corresponding spectra taken at both polarizations $E_\parallel$ and $E_\perp$ are shown in Fig. 11.13b. One can see that PL intensity is much higher for $E_\parallel$ than for $E_\perp$, as it is expected from spatial distribution of the light emitted from the orthogonally oriented needles. On the other hand, the dichroic ratio here is much smaller than in the case of uniaxially oriented p-6P chains on mica(001) [16].

To achieve epitaxial needle-like growth on the whole substrate surface, a modification of the HWE technique was used by the same authors [68, 73]. Epitaxial crystals growing as orthogonal oriented long needles were obtained, with a length of more than 100 μm, a width and height in the ranges of 300–800 and 80–150 nm, respectively, as was measured by AFM. These needles were found as suitable for self-waveguiding amplification of blue light emission from the p-6P. They were optically pumped with a frequency-tripled Nd:YAG pulsed laser ($\lambda_{ex} = 355$ nm, pulse width 5 ns, repetition rate 10 Hz). The excitation laser was horizontally focused through a cylindrical lens onto the KCl surface in a line of 1 mm width; the emitted light was collected in the direction along the needle axis with a charge-coupled device (CCD) spectrometer. Figure 11.14a shows the output intensity and the line width as a function of pulse energy for the (0–1) optical transition of the p-6P. Demonstrated spectral narrowing above the threshold was characterized as amplified stimulated emission (ASE) by self-waveguiding [73].

The degree of anisotropy of this gain-narrowed emission was also investigated [73]. Figure 11.14b shows the angular dependencies of the emission outputs at the (0–1) band ($\lambda = 425$ nm), where Φ_{ex} and Φ_{em} denote the angles of the polarization plane of the excitation laser pulse and the emitted light, respectively, with respect to the molecular axis in the needle crystal. The Φ_{ex}

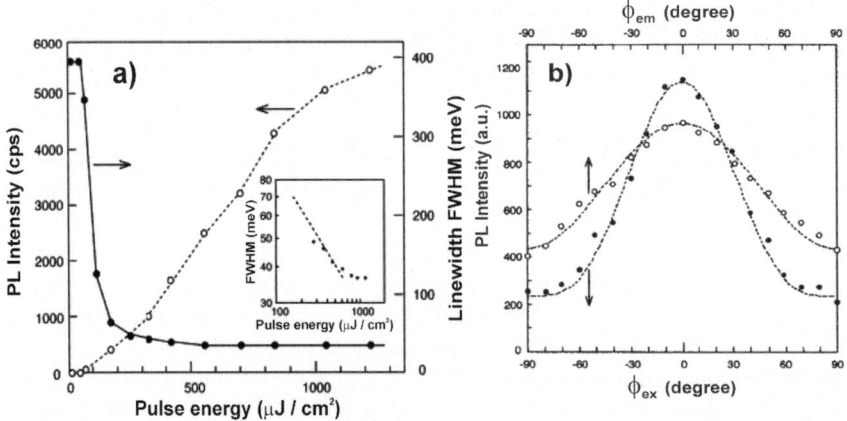

Fig. 11.14. (a) Photoluminescence intensity and full width at half maximum (FWHM) of the emission band at $\lambda = 425$ nm as a function of excitation pulse energy. The inset shows a logarithmic plot of the line width vs. excitation energy. (b) Dependence of the output intensity on Φ_{ex} and Φ_{em}. Reprinted with permission from [73]

and Φ_{em} dependent outputs were collected without polarization of the emitted and excited light, respectively. One can see that the Φ_{em}-dependent emission outputs can be well fitted using $\cos^2 \Phi_{\mathrm{em}}$, which is characteristic of uniaxially aligned transition dipoles of p-6P along the KCl $\langle 110 \rangle$ direction [74]. In contrast, the curve of Φ_{ex}-dependent emission outputs can be fitted only by a higher power (n) function, $\cos^n \Phi_{\mathrm{ex}}$ ($n \sim 3.7$ in Fig. 11.14b). This Φ_{ex}-dependence with higher n demonstrates again that the self-waveguided gain is based on cascading dipole interactions, i.e., stimulated emission or ASE. The lower dichroic ratio (ratio of the output at $\Phi = 0°$ to that at $\Phi = \pm 90°$) and higher background at $\Phi = \pm 90°$ were ascribed to the fact that the polarization of the emitted light is disrupted by scattering at the KCl steps and orthogonally crossed needles [73].

Recently, high anisotropic p-6P films were grown on KCl(001) by *standard* HWE process, using which p-6P films on mica were grown [75, 76]. Using tapping mode AFM and XRD, it was demonstrated that previously found [35] needle-like crystallites and layered islands coexist at later growth stages, while the needle-like crystallites present the initial growth stage of such films. The more interesting latter growth stage was investigated in detail [75, 76]. Figure 11.15 shows representative high-resolution topographic and phase images of a p-6P film in this stage with corresponding cross-sections. The topographic image together with the cross-section (Fig. 11.15, left) reveal that all terraces found

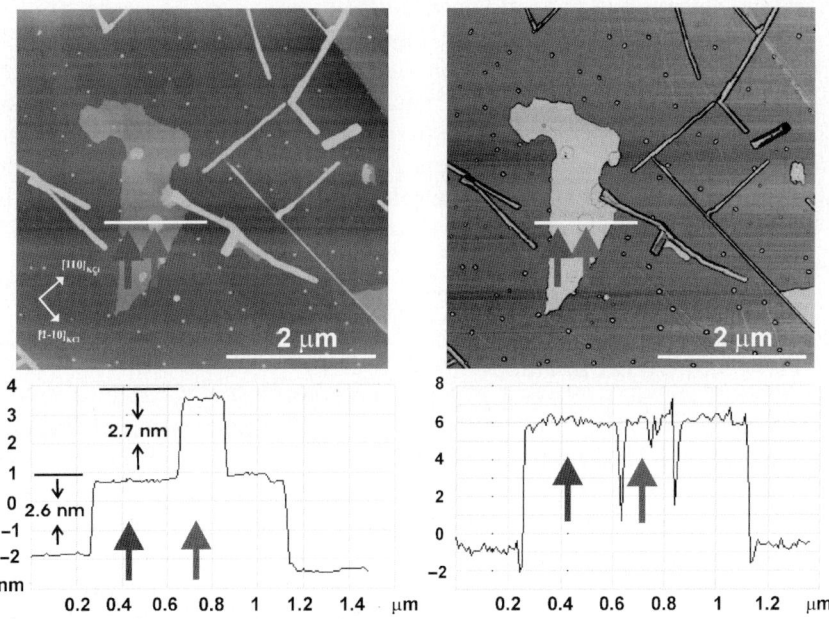

Fig. 11.15. Topographic (*left*) and phase (*right*) images of a p-6P film grown on KCl(001) by HWE and corresponding cross-sections (*bottom*). Growth time is 1 min and growth temperature is 90°C. z-scales are 30 nm and 30°, respectively

are in average about 2.6 nm high, which corresponds to one monolayer of standing p-6P molecules. Thus, it was assumed that the islands are composed of upright molecules. By comparing the topographic image with the phase image (Fig. 11.15, right) it was also found that there is a strong phase shift between the substrate and the first layer in the islands as well as between substrate and the needles, but no phase shift between the 1st and the 2nd layer in the islands. Therefore, it was concluded that the islands start to grow directly on the bare KCl(001) surface [75]. Moreover, it was mentioned that this conclusion drawn from AFM results is well supported by cross-correlation of the data obtained earlier by X-ray diffraction [77], which also indicated the absence of p-6P wetting layer on KCl(001).

Molecular orientations in p-6P films grown on KCl(001) by HWE have been proven by X-ray diffraction (XRD) and transmission electron microscopy measurements [77]. In the line with previous data [69], these investigations revealed at least four different epitaxial orientations parallel to the surface of the substrate: $(11\bar{1})$, $(11\bar{2})$, and $(20\bar{3})$ reflections due to lying molecules, whereas (006) reflection was due to standing molecules. One ascribes the last orientation to layered mounds, with the long axes of the molecules approximately perpendicular to the substrate. Correspondingly, the needles comprise other three epitaxial orientations, where the long molecular axes are aligned approximately parallel to KCl(001) surface. It is important to mention here that recent XRD and TEM/TED investigations demonstrated the true single crystalline nature of islands and needles [78].

p-nP ($n = 4$–6) films were recently grown on KCl(001) substrates also by OMBE [63]. Using optical absorption and LEED it was shown that for small depositions rates of less than $0.2\,\text{Å}\,\text{s}^{-1}$ and the substrate temperature at about 370 K p-nP molecules are oriented perpendicular to the substrate surface, forming different orientational domains [79]. In contrast, for higher deposition rates ($2\,\text{Å}\,\text{s}^{-1}$) the molecules were parallel to the surface. Anisotropic needle-shaped aggregates oriented in two orthogonal directions were also detected by fluorescence microscopy.

11.3.3 Anisotropic Epitaxial Growth of p-6P and p-4P on TiO$_2$ and Metal Surfaces

As was described earlier, the recent advances in hot-wall epitaxy (HWE) and organic molecular beam epitaxy (OMBE) allowed to grow highly ordered thin films of p-6P and dielectric substrates like mica or KCl with high degree of anisotropy at the macroscopic scale [31–35]. Nevertheless, for many optoelectronic applications a conductive substrate is needed to utilize an electrical contact to the film. For this purpose the TiO$_2$(110) substrate was recently proposed and successfully used by authors of [80]. Moreover, detailed investigations using STM and AFM have revealed that p-6P growth by OMBE on TiO$_2$(110) is highly anisotropic, but much more complex compared to those on mica(001) or KCl(001) substrates [80,81]. For details the reader is referred here to corresponding Chap. 7.

Various attempts were also made to achieve anisotropic growth of *oligophenylenes* on metallic surfaces. This was motivated by a different surface bonding type (covalent) of metals compared to mica (van der Waals) or KCl (ionic) as well as by the relevance of metal contacts for device applications. OMBE was used to deposit p-4P and p-6P compounds from a home-made quartz glass Knudsen cell onto Al(111) and Au(111) substrates at room temperature. The deposition rate was between 0.2 and 0.72 nm min^{-1}. The Au and Al single crystal were prepared with a standard Ar$^+$-ion sputtering cleaning technique with subsequent annealing. The total amount of the material was directly measured during deposition with a water-cooled quartz microbalance [81,82].

First, we will discuss the results obtained on crystalline Al(111) substrates [81]. For a 35 nm-thick p-6P film it was shown that the surface is covered with small $100 \times 200\,\text{nm}^2$ islands. The height of these islands is 60 nm. The islands form small linear arrangements consisting of up to 5–6 entities. These chains typically form pairs that include an angle of roughly 60°. Two sets (six directions in each) rotated by 30° with respect to each other could be identified. This is in good agreement with the 12 alignment groups for p-6P growth at room temperature on Al(111) reported in [83]. It was also reported that for this particular case the aromatic plane of the p-6P molecule is parallel to the substrate surface, in contrast to many other cases discussed in this chapter, like on mica(001) [52], where the molecules are tilted with respect to the surface.

Next step was made with p-4P films grown on gold substrates [82, 84]. Figure 11.16a summarizes the AFM results obtained for the 200 nm thick p-4P film grown on Au(111). This film shows four different morphological features

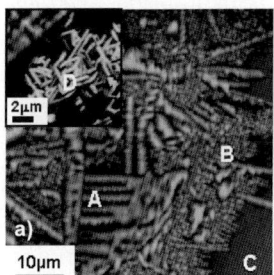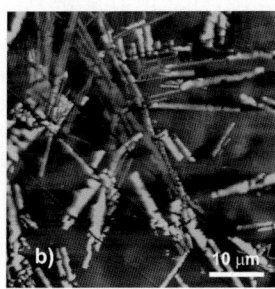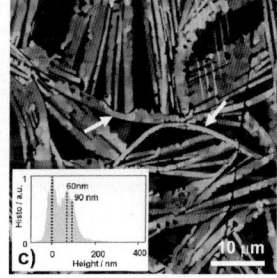

Fig. 11.16. High-resolution AFM images of p-4P films grown at 300 K on Au(111). (a) 200 nm film (z-scale is 600 nm). The inset shows a $10 \times 10\,\mu\text{m}^2$ image (z-scale is 1 µm) of an additional structural feature; (b) 30 nm film (z-scale is 800 nm); (c) 30 nm film with 0.15 ML carbon precoverage (z-scale is 400 nm). The insert in (c) represents the corresponding height histogram. Also in (c), two curved p-4P chains are indicated by arrows

with lateral extensions in the range of 10 μm that exist in close vicinity. The area denoted by A consists of parallel chains of tetragonal islands with a base edge length of about 1 μm. Their height is in the range of 200–250 nm. Within a single chain the islands are all identically oriented. Again, at least six different chain directions could be identified. Area B features significantly smaller, plate-like crystallites. The structural features A and B can either abut against each other or they can be separated by regions C free of three-dimensional islands. A fourth structural feature exists, which is presented with higher resolution in the inset of Fig. 11.16a. This area D exhibits a loose group of individual crystallites with a length ranging from 0.8 to 2 μm. The cross-sectional analysis revealed that these islands are almost as high (250–550 nm) and as wide (300–500 nm). In other words, these individual islands are rods with a nearly quadratic cross-section.

More thinner p-4P films were also deposited on Au(111) at the same conditions in order to clarify the details of the growth [82]. Figure 11.16b shows an AFM image of a 30 nm thick film. One can see that p-4P forms isolated needle-like chains of different lengths and orientations with a height of up to 300 nm and a length between 5 and 40 μm. However, the lateral chain width seems to be nearly constant throughout a single p-4P chain. Several individual chains can also agglomerate to form fanlike bunches. At least six dominant chain directions could be identified, which seems to be clearly related to the fcc(111) surface geometry. Thus, one can conclude that the Au(111) substrate mediates the chain formation.

XRD and LEED investigations [84] have shown that despite the inhomogeneous island morphology of p-4P films shown in Fig. 11.16a,b, the molecular arrangement within these islands is highly regular with respect to the Au(111) surface, i.e., the p-4P crystallites grow epitaxially. The dominating crystalline orientation of the p-4P bulk is those with (211) net plane of p-4P parallel to the Au(111) substrate. It turns out that both the long axes and the short axes of the p-4P molecules are lying parallel to the (211) plane. In addition, the measurements using the X-ray diffraction pole figure technique have shown that the long axes of the molecules are azimuthally aligned along high-symmetry directions of the Au substrate, i.e., the Au$\langle 1\bar{1}0 \rangle$ or the Au$\langle 11\bar{2} \rangle$ directions, respectively [85].

The influence of a predeposited surfactant (carbon) was also investigated for Au(111) substrates to provide a deeper insight into the formation of p-4P chains. Carbon precoverage was prepared by means of X-ray induced dissociation of p-4P molecules adsorbed on the Au surface prior to the growth. Figure 11.16c shows the obtained effect for a 30 nm p-4P film. One can see that after depositing of 0.15 ML of the carbon the order and anisotropy in the film are decreased. Many needles are not straight any more, but curved with continuously changing edge orientation as visible in Fig. 11.16c. Additionally, the height histogram (insert in Fig. 11.16c) reveals a nonuniform height distribution for this film. The dominant heights are 60, 90, and 120 nm. These results were explained in [82] as following. With increasing carbon coverage, the coupling of the p-4P molecules to the substrate – which is responsible for the locking of the chain orientation to certain preferred directions – is getting obviously weaker. As a result one can observe the bended needles. In accordance with that curved manifestations of p-6P phenyl aggregates have also been observed on specially prepared [62–64] or contaminated [86] muscovite surfaces.

Summarizing, these investigations have shown that single crystalline TiO_2 and metal substrates present good template for the growth of a wide variety of anisotropic film morphologies, which can be interesting for device applications. On the other hand, especially on metal substrates the degree of the anisotropy and order achieved is smaller compared to dielectric substrates like mica or KCl. This can be preliminary explained by the fact that strong covalent bonds on the metal surfaces can inhibit anisotropic surface diffusion of the molecules and therefore decrease the degree of anisotropy.

11.4 Luminescent and Lasing Properties of Anisotropic Organic Thin Films

One of the applications of the anisotropic molecular structures can be polarized LEDs and lasers. Obviously, a necessary condition for a polarized electroluminescence is an *efficient polarized* photoluminescence. The results discussed in previous sections show clearly that the *polarized* photoluminescence can be achieved if anisotropic needle-like structures can be fabricated. Therefore, the second important issue is the *efficiency* and the *properties* of such polarized emission. On the other hand, it is well recognized that closer packing of the molecules in molecular thin films favors the formation of molecular aggregates (for example, dimer-like structural defects) as was shown for *oligo-phenylenevinylene* [87] and α-*sexithienyl* (T_6) [88, 89]. Such aggregates, being essentially defects even in otherwise well ordered films, might alter considerably the luminescent properties as well as affect both exciton energy and charge carrier transport [90,91]. The interest in aggregates in anisotropic p-6P structures was motivated also by the fact that recent results on isotropic polycrystalline p-6P films [92, 93] pointed to the molecular character of the light emission, with no clear evidence of intermolecular excited states. The photoluminescence (PL) and time-resolved PL techniques were therefore recently employed in order to reveal possible formation of aggregates (as trapping states for *excitons*) in highly ordered anisotropic p-6P structures [94].

p-6P films have been grown by HWE on mica(001) as described in the Sect. 11.3.1 of this chapter, resulting in highly ordered anisotropic morphology as shown in Fig. 11.6. Substrate temperature was varied from 78 to 170°C. The obtained results in [94] are shown in Fig. 11.17 presenting steady-state PL spectra measured at temperatures ranging from 5 K up to room temperature. One can see that at 5 K the PL spectra have a well known structure (see also Fig. 11.8) consisting of the main band at 392 nm (3.16 eV) followed

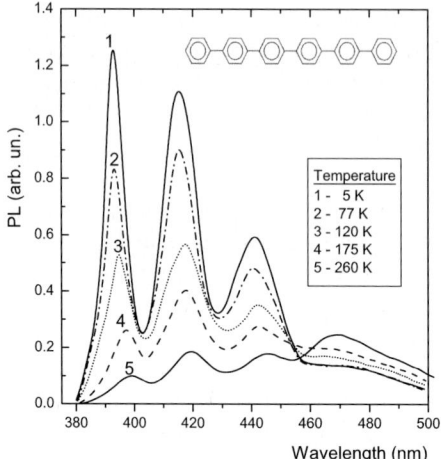

Fig. 11.17. Temperature-dependent steady-state PL spectra of a p-6P film (curves 1–5 correspond to $T = 5$, 77, 120, 175, and 260 K, respectively) grown at substrate temperature 78°C

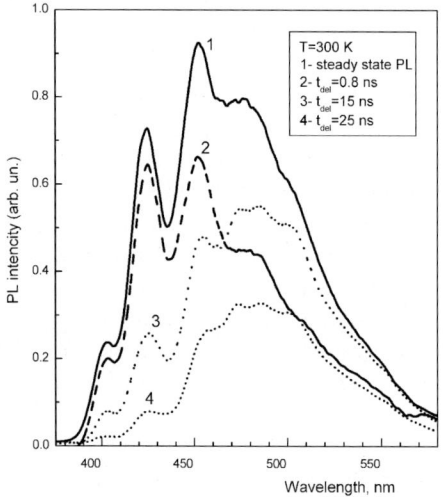

Fig. 11.18. Room temperature time-resolved PL spectra of a p-6P film monitored with the delay time $t_{\mathrm{del}} = 0.8$, 15, and 25 ns (*curves 2–4*, respectively). *Curve 1* – steady-state PL

by a vibronic progression at 415 nm, 442 nm, and about 467 nm (Fig. 11.17, curve 1). Such a PL spectrum has been ascribed in literature [92, 93] to the fluorescence due to the radiative decay of the intrinsic bulk excitons with intrachain character (hereafter the exciton spectrum). Apparently, there is an additional very broad structureless band centered around 480 nm, which overlaps with the exciton spectrum. The temperature dependence of the exciton spectrum and the lower energetic broad band is opposite – while the exciton bands show a rather strong decrease with elevating temperature, the intensity of the broad band tends to increase with increasing temperature (Fig. 11.17). It was therefore assumed that the additional broad structureless band results from emission from low energy *aggregate* states in p-6P films (hereafter *aggregate* band).

A straightforward evidence for the presence of two different emissive centers in p-6P films came from time-resolved PL measurements. Figure 11.18 shows PL spectra monitored at room temperature with the delay time $t_{\text{del}} = 0$ ns, 0.8, 15, and 25 ns after the excitation. As one may see the delayed PL spectra ($t_{\text{del}} = 15$ and 25 ns – curves 3 and 4, respectively) are almost completely dominated by a very broad aggregate band overlapping with just weak traces of the exciton PL structure. Moreover, the aggregate band decays considerably slower featuring a life-time estimated as 4 ns, which is about an order of magnitude larger than the apparent lifetime of 400 ps of singlet excitons reported for p-6P films [93]. This longer lifetime provides greater opportunity to excitation to find nonradiative pathways, thus decreasing the quantum yield of the photoluminescence. It should be mentioned that the 480 nm aggregate band was much less pronounced at 77 K and almost not detectable in the delayed PL spectra monitored at 5 K.

Furthermore, one tried [94] to correlate the PL spectra of p-6P films with systematically varying growth conditions. Figure 11.19 presents room-temperature PL spectra obtained for a set of the films grown at different substrate temperatures T_{Sub}, namely 78, 90, 110, 130, and 150°C. As one can see the concentration of aggregates is notably increased when T_{Sub} reaches 130 and 150°C. This fact was explained with temperature facilitated formation of aggregates during the growth due to enhanced diffusion of p-6P molecules.

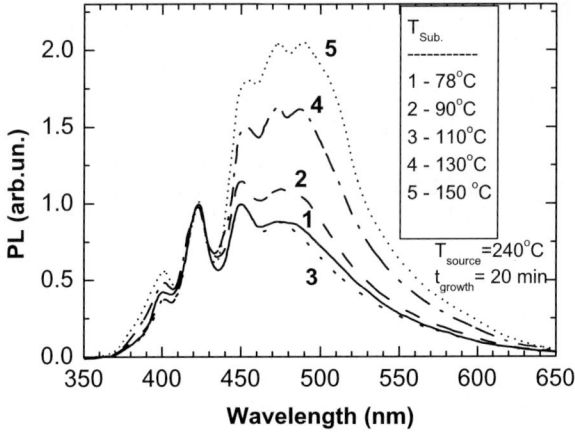

Fig. 11.19. Steady-state PL spectra monitored at ambient temperature for the films grown at different substrate temperatures $T_{\text{Sub}} = 78$, 90, 110, 130, and 150°C. All curves are normalized to the intensity of the exciton band at 420 nm

Summarizing these findings, the authors of [94] pointed out that all obtained PL results are fully consistent with aggregate nature of broad red-shifted emission band. The aggregates are likely *dimer-like* structural defects formed during the film growth, in which intermolecular interaction is enhanced. This explanation is strongly supported by two facts: (1) deliberate ex-situ thermal annealing of studied p-6P films to the temperatures T_{Sub} in ambient atmosphere had no effect on their PL spectra; and (2) the relative weight of the aggregate component depended (see Fig. 11.19) considerably on HWE growth conditions. Moreover, one might assume that in contrast to *small grain* isotropic polycrystalline films, an extended length of nanochain parts in highly ordered p-6P films should facilitate efficient energy migration over a large distance along the needles that, in turn, should result in larger probability to encounter an aggregate trap even though the latter might exist in very low concentration. Since the aggregates in p-6P films create rather deep traps, one can neglect retrapping. In contrast, exciton migration is expected to be rather limited in polycrystalline films. Efficient Förster energy transfer from *isolated* to aggregated chromophores occurs on a much faster timescale than the exciton lifetime, which means that even the presence of a small number of aggregated sites in a film can dominate the emissive properties. This can explain the considerable decrease of the overall PL intensity accompanied by relative enhancement of aggregate emission in PL spectra upon increasing temperature. Although the concentration of aggregates in investigated films is probably very low so they are not apparently visible in PL at 5 K, aggregate emission relatively increases at higher temperatures due to an increased exciton diffusion, which has a thermally activated character [95, 96] and hence a larger probability to encounter an aggregate trap.

The next important step in exploration of exceptional optical properties of highly ordered anisotropic p-6P structures was made by F. Quochi and coauthors [97, 98]. Based on results of [52, 53, 95, 96], they achieved for the first time the random lasing from self-organized p-6P nanochains grown by HWE. These exceptional results are discussed extensively in corresponding Chap. 10.

11.5 Devices Based on Organic Thin Films

11.5.1 OFETs Based on C_{60} Thin Films Grown by HWE

Effort to increase charge carrier mobility in OFETs with n-type organic materials have been difficult due to several physical reasons: rapid degradation under ambient condition and sensitivity of the electron transport properties to purity of the crystal [99]. Among the n-type organic semiconductors, to mention a few, C_{60} [100], *methanofullerenes* [101], *fluorinated phthalocyanines* [102], *naphthalenes*, and *N-substituted naphthalene 1,4,5,8-tetracarboxylic diimide* [103] or *N,N'dialkil-3,4,9,10-perylene tetracarboxylic diimide* (PTCDI,

PTCDI-C_5 and PTCDI-C_8H, correspondingly) [104] show highest obtained mobilities of up to $0.6\,\mathrm{cm^2\,V^{-1}\,s^{-1}}$ [105]. Most of these devices are fabricated on untreated inorganic SiO_2 or Al_2O_3 dielectrics. Because of the very different physical nature of organic active layer and inorganic dialectics, this procedure results often in highly disordered films, leading to a poor performance [106]. Although the van der Waals-like interactions between organic molecules and inorganic substrates are rather weak, the crystallographic phases, the orientation, and the morphology of the resulting organic films critically depends on the interface and growth kinetics. On the other hand, the first few monolayers closer to the dielectric interface determine the charge transport in field-effect devices [107]. These interface related phenomena can be investigated, in particular, using organic active layers grown on the top of an organic insulator, by studying their structure-performance relationships. In this section, we discuss corresponding results obtained for OFETs based on C_{60} films grown by HWE on top of organic dielectrics (BCB, see Sect. 11.2.2 and Fig. 11.3) [38].

Preheating procedure (in-situ thermal cleaning of the substrate surface just before growth) has been chosen in [38] as a factor controlling the properties of the interface between dielectric and active layer. At the first, topography of the bare BCB surfaces was inspected using AFM resulting in a surface roughness below 5 nm [101]. Further, the morphology of the C_{60} layers was investigated in dependence on preheating conditions. The morphologies of two 5 nm thick C_{60} films are shown in Figs. 11.20a,b, demonstrating a striking difference of the morphology, if the growth proceeds on preheated substrates. One can see that on preheated dielectric at 250°C, the resulting C_{60} film shows rather nanocrystalline-like structure, while amorphous-like C_{60} films are found on top of nontreated substrates. In contrast, thicker C_{60} films (100–300 nm) show practically the same morphology as demonstrated in Figs. 11.20c,d. Apparently, interaction of C_{60} with BCB results on preheated dielectric in large crystalline structure of C_{60}. As the film thickness becomes higher, surface does not influence the growing any more, which leads to the same morphologies irrespective of the initial growth condition.

One have started comparable electrical characterizations with the measurements of transistor characteristics for devices fabricated on nonpreheated surfaces. n-channel OFETs with a clear saturation of the drain–source current (I_{ds}) for drain–source voltages (V_{ds}), exceeding the applied positive gate voltage (V_{gs}), were routinely obtained. I_{ds} were in the range of sub-mA. The transfer characteristic ($I_{ds}(V_{gs})$) in the saturated region with a $V_{ds} = 60\,\mathrm{V}$ featured a negligible hysteresis. I_{ds} increased quadratically with V_{gs}, which was fitted to the well known equation (11.3) [108].

In contrast, OFETs fabricated on preheated at 250°C substrates have been found to have a high intrinsic conductivity. Figure 11.21a shows an n-channel transistor with its saturated regime. I_{ds} here has a V_{gs} independent slope, which is the signature of high intrinsic conductivity. Origin of this is the increased mobility, but not doping. Doping can be excluded as the C_{60} films were grown in a clean HWE system on precleaned BCB dielectric layers. Moreover,

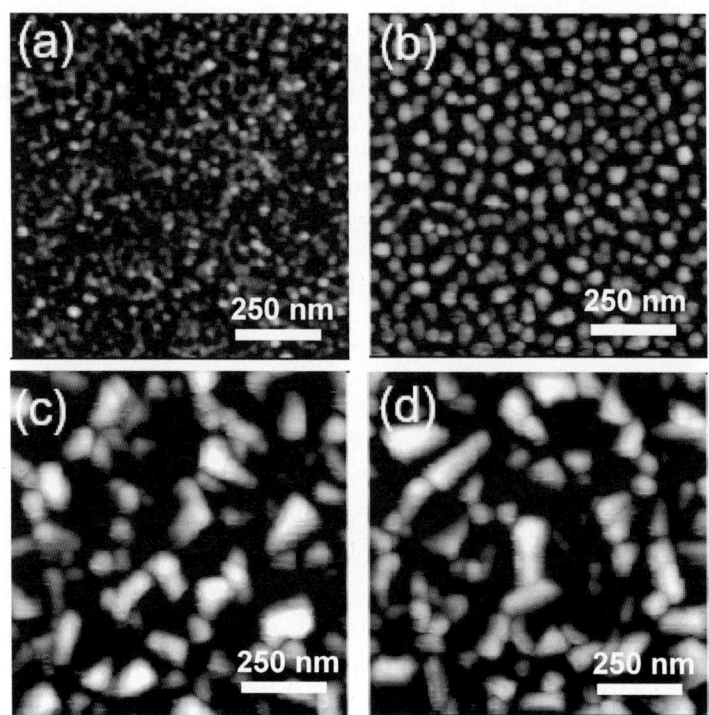

Fig. 11.20. AFM topography images of the C_{60} films grown on BCB substrates: (**a**) $d = 5$ nm – without preheating; (**b**) $d = 5$ nm – with preheating at 250°C; (**c**) $d = 300$ nm – without preheating; and (**d**) $d = 300$ nm – with preheating at 250°C. z-scale is 25 nm (**a**) and 40 nm (**b–d**)

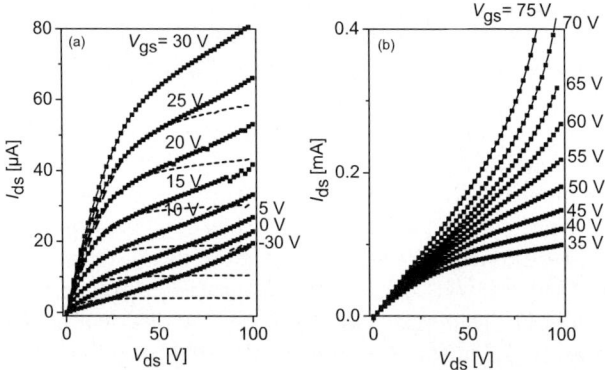

Fig. 11.21. Output characteristics of C_{60} OFETs on preheated at 250°C dielectric BCB. (**a**) $I_{ds}(V_{ds})$ curves (*symbols*) where the bulk current has been subtracted from all the respective curves (*dashed curves*); (**b**) $I_{ds}(V_{ds})$ curves at high applied V_{gs}

in the OFETs fabricated on preheated substrates one observed an increase in I_{ds} at least by three times at the same V_{gs} values, as compared to those on nontreated substrates. Therefore, all the curves were replotted by subtracting the bulk current $I_{ds}(V_{ds})$ for $V_{gs} = 0\,\text{V}$ from the respective curves. In the corrected curves (dash curves) one can see a well defined pinch-off $I_{ds}(V_{ds})$ for an applied V_{gs}. Applying a large V_{gs} resulted in a nonlinear $I_{ds}(V_{ds})$ curve as shown in Fig. 11.21b. This observation can be due to a voltage dependency of the mobility. Such nonlinearity is theoretically outlined in [109,110]. Therein $\mu(V_{gs})$ is attributed to trap-filling effects and/or space-charge limited currents of the drain current itself beside the gate induced space charge. However, at low V_{gs} space charge limited current and trap-filling are negligible and not observable in the output characteristics. Although in [111–113] it has been mentioned that LiF/Al makes ohmic contact to C_{60}, there exists a low contact resistance since a weak deviation from linearity of $I_{ds}(V_{ds})$ at low V_{ds} was observed. A very similar phenomena occur in the case of reduced channel lengths [114]. However, in the present study, contact resistance is not taken into account for calculating of the μ_e. On the other hand, it can also be mentioned that such a large magnitude of the current density is observed only in semiconductors with low trap densities. These traps can be filled with large carrier injection/accumulation. This phenomena leads to a trap filled regime, a behavior which is normally observed very rarely [115,116].

Finally, one can compare the transfer characteristics of the OFETs fabricated under different conditions. As shown in Fig. 11.22, the slope $\Delta\sqrt{I_{ds}}/\Delta V_{gs}$ for the OFETs based on preheated at 300°C dielectrics is much higher than the corresponding one for nontreated dielectrics. Theoretical fit curves (continuous lines) for the experimental data using (11.1) are also plotted in Fig. 11.22. These curves agree reasonably well with the experimental data, suggesting a constant μ_e as function of V_{gs} in the range of the voltages investigated. Accordingly, μ_e calculated using (11.1) increased from 0.5 to $6\,\text{cm}^2\,\text{V}^{-1}\,\text{s}^{-1}$ as a result of preheating of the dielectric layer. Note, that previous reports on C_{60} OFETs grown on SiO_2 have shown a highest obtained μ_e of $0.5\,\text{cm}^2\,\text{V}^{-1}\,\text{s}^{-1}$ [100]. The increase in μ_e could be attributed preliminary to the transition (while going from the unheated to the preheated dielectric) from disordered interface made of dendritic grains of different orientations with a rather high concentration of defects, mainly located at grain boundaries, to a well ordered C_{60} film as depicted in Figs. 11.20a,b.

As discussed in detail by Kalb et al. [117], OFET mobility is not dependent on the gate voltage, if active layer fabricated from high mobility material – as was also observed in the referred study. Additionally, the contact resistance becomes more important for extracting parameters from the current–voltage curves. In order to check this, other contacts such as Au, Cr/Au, Mg/Al, and Ca/Al were also fabricated. Among all these contacts, better current injection was obtained with LiF/Al. Hence the argument of a better electron injection with a lower work function electrode does not hold in the present study. Previous studies have shown pinning of Fermi-level at the interface of metal-C_{60}.

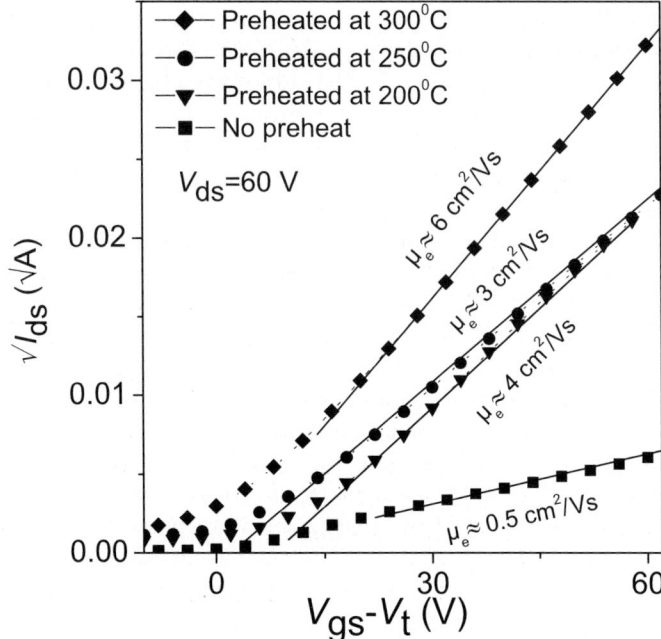

Fig. 11.22. $\sqrt{I_{ds}}$ vs. V_{gs} plots (*symbols*) for C_{60} OFETs fabricated at different growth conditions along with theoretical fit curves (*continuous lines*). Original data points are taken without subtracting the bulk current. The obtained electron mobilities μ_e are indicated along with the respective curves

LiF is proposed to be lowering the barrier for electron injection as can be seen from organic solar cells studies [111–113]. Therefore, the observation of a structure-performance relationship seems to be notable. Most important parameter is the morphology of the first few monolayer of C_{60} as shown in Fig. 11.20 – C_{60} films with large crystallites features a higher current and it is proposed that such films have also a lower trap density. This results in non-linear transport phenomena (Fig. 11.21) because of the trap filling effects at large gate voltages.

11.5.2 Anisotropic Current–Voltage Characteristics of p-6P Chains on Mica

Although organic materials in general display low charge carrier mobilities, small molecule crystals can exhibit considerable high mobilities at room temperature, if local impurities and extended defects, such a grain boundaries, are minimized [14, 15]. Organic thin-film transistors have shown, for example, mobilities as high as $3\,\mathrm{cm^2\,V^{-1}\,s^{-1}}$ [118], and with single crystalline films room temperature mobilities of $15\,\mathrm{cm^2\,V^{-1}\,s^{-1}}$ were obtained [14, 15]. However, the fabrication of OFETs with fragile molecular systems may disrupt the molecular order due to the preparation of a complex layer and electrode arrangement. For this reason studies using the Space Charge Limited Current method (space charge limited current, SCLC) are important. With this method, which requires only two electrodes, above mentioned experimental difficulties can be eliminated. This section is devoted therefore to recently obtained results concerning current–voltage (I–V) characteristics of well ordered nanochains of p-6P grown on mica [119].

p-6P films were grown by HWE on mica(001) substrate at 90°C as described in Sect. 11.3.1. Typical chain-like structure of p-6P was formed at this growth conditions with about 200 nm in the width and 60 μm in the length. To study the anisotropy in the transport properties, Ag electrodes (gap-type geometry as schematically shown in Fig. 11.23) were evaporated on the top of the films with electrode edges either perpendicular to the chains direction (Fig. 11.24a) or parallel to the chains direction (Fig. 11.24b). The electrode gap lengths L were varied from 20 to 100 μm, which allows the study of the current dependence on the gap length. The width of the electrode contacts were kept constant at 1,000 μm.

The steady state current-voltage characteristics of a nano-chain film with electrode edges perpendicular to the chains direction (current flowing parallel

Electrodegap L >> thickness of the semiconductor t

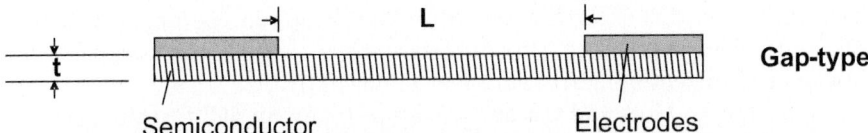

Fig. 11.23. Electrode configuration of gap-type geometry (side view)

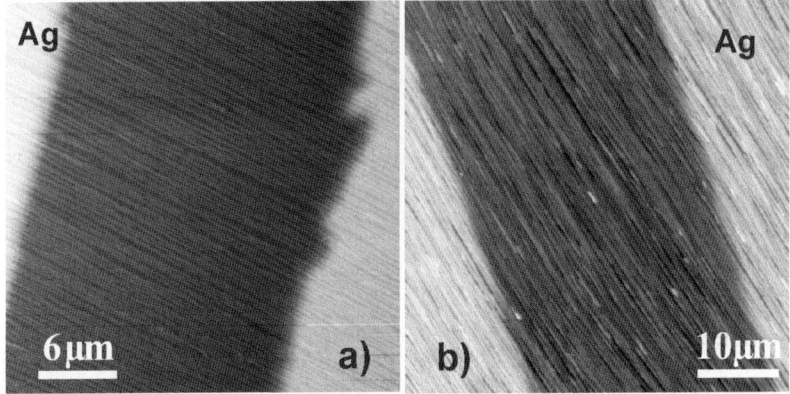

Fig. 11.24. AFM pictures of the devices with Ag-electrodes deposited with the electrode edges perpendicular to the chains direction (**a**) and parallel to the chains direction (**b**)

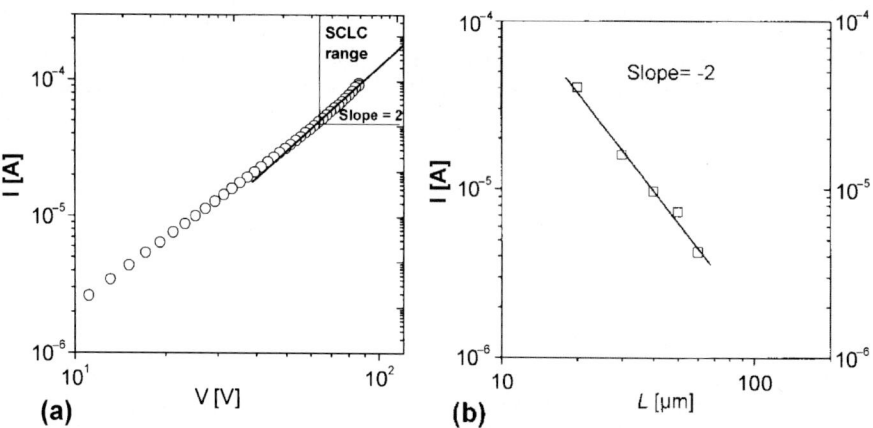

Fig. 11.25. (a) Double logarithmic plot of current-voltage characteristics with electrode edges perpendicular to the chain direction, demonstrating SCLC region for voltage >50 V, (b) I vs. L, for $V = 60$ V (from the SCLC region) giving a slope of -2

to the cains direction) at room temperature are shown in Fig. 11.25(a) in a double logarithmic plot. The slope changes from lower values to 2 at voltages higher than 50 V, indicating a transition to a SCLC behaviour. No measurable current was found for arrangements with the electrode edges parallel to the chains direction (current flowing almost perpendicular to the chains direction). The reason for that is attributed to the anisotropy in the film morphology due to the chain structure. Probably, free-space between two parallel chains hinders the current flow between them significantly.

According to Mott and Gurney–theory [120], for an SCLC regime between two electrodes in sandwich structure with an electrode gap length L (equivalent to the film thickness), the current density J is inversely proportional to the third power of L,

$$J = \frac{9}{8} \varepsilon \varepsilon_0 \, \mu_{\text{eff}} \frac{V^2}{L^3} \, . \tag{11.4}$$

In (11.4), lateral effects are not taken into account. Although often used, (11.4) is not valid for gap-type geometry [121] and its usage leads to an overestimation of the calculated charge carrier mobility [122]. Therefore, a theory was developed for a gap-type arrangement shown in Fig. 11.23, where lateral effects were taken into account [123]. The space-charge-limited current here is inversely proportional to the square of L per unit width, W,

$$\frac{I}{W} = \frac{2}{\pi} \varepsilon \varepsilon_0 \, \mu_{\text{eff}} \frac{V^2}{L^2} \, . \tag{11.5}$$

To verify the validity of this formalism, gap length dependency of SCLC current was studied as shown in Fig. 11.25(b) in double logarithmic plot. From the plot of $I(L)$ a slope of -2 is obtained, which further confirms the validity of equation (11.5).

Using equation (11.5), an effective mobility $\mu_{\text{eff}} = 14\,\text{cm}^2/\text{Vs}$ for arrangements with electrode edges perpendicular to the chain direction is calculated. The usage of equation (11.4) would overestimate μ_{eff} by a factor of more that 50. For films with electrode edges perpendicular to the chain directions there is no current even for a large applied voltage which is consistent with a large free-space between the nano-needles.

11.6 Conclusions

Using a wide variety of characterization techniques it could be shown, that by choosing the right combination of substrate type, surface geometry, and growth conditions, it is possible to grow *oligophenyls*, and especially *para-sexiphenyl*, in a variety of anisotropic morphologies. In the case of p-6P grown on mica or KCl substrates, two molecular orientations with respect to substrate surface are generally found: molecules lying almost flat on the surface forming long anisotropic needle-like crystallites and standing upright molecules forming large-area isotropic islands. Self-organization process, however, that lead to a characteristic morphology is very complex and a variety of substrate and molecule properties have to be considered.

It was also shown that particular growth morphology of organic nanoaggregates can lead to special anisotropic physical properties like luminescence and lasing [16, 94, 98], waveguiding [63], and a very high mobility ($\mu_{\text{eff}} = 14\,\text{cm}^2\,\text{V}^{-1}\,\text{s}^{-1}$) along the long axis of needle-like chains [119]. This opens the field for a variety of interesting applications ranging from polarized OLEDs and low threshold organic lasers to high-mobility OFETs.

It was demonstrated finally that highly ordered growth is an essential precondition to obtain high carriers mobility in organic thin films. In particular, considerable high mobilities up to 6–$14\,\text{cm}^2\,\text{V}^{-1}\,\text{s}^{-1}$ were obtained in highly ordered films of C_{60} and p-6P grown by HWE.

Acknowledgments

The authors gratefully acknowledge the financial support by the Austrian Science Foundation within the project cluster *Highly Ordered Organic Epilayers* (especially, FWF projects P-15155, P-15625, P-15626, P-15627) and within the National Research Network *Interface Controlled and Functionalized Organic Films* (especially, FWF projects S9706, S9707, S9708). Part of this work was performed within the Christian Doppler society dedicated laboratory on plastic solar cells funded by the Austrian ministry of economic affairs and Konarka Austria GmbH.

The authors also thank the groups that provided excellent samples for morphological and optical investigations; by name H. Sitter, Institute for Semiconductor and Solid State Physics, Johannes Kepler University Linz; M.G. Ramsey, G. Koller, J. Ivanco, Institute of Physics, University Graz; and

A. Winkler, S. Müllegger, Institute of Solid State Physics, Graz University of Technology. Crystallographic investigations were performed in the group of R. Resel, Institute of Solid State Physics, Graz University of Technology, in close collaboration with B. Lotz and A. Thierry, Institute Charles Sadron, Strasbourg. Experimental support for the XRD studies using synchrotron radiation is greatly acknowledged to D. Smilgies, Cornell High Energy Synchrotron Source, Cornell University, Ithaca, USA. Finally, PhD students and PostDoc's are acknowledged for their individual contributions, by name G. Hlawacek at the Institute of Physics, University Leoben; A. Montaigne Ramil, at Institute for Semiconductor and Solid State Physics, Johannes Kepler University Linz and T. Haber at the Institute of Solid State Physics, Graz University of Technology.

References

1. D. De Leeuw, Physics World p. 31 (1999)
2. A. Heeger, Curr. Appl. Phys. **1**(4–5), 247 (2001)
3. J. Burroughes, D. Bradley, A. Brown, R. Marks, K. Mackay, R. Friend, P. Burns, A. Holmes, Nature **347**(6293), 539 (1990)
4. C. Tang, S. VanSlyke, Appl. Phys. Lett. **51**, 913 (1987)
5. C. Dimitrakopoulos, S. Purushothaman, J. Kymissis, A. Callegari, J. Shaw, Science **283**(5403), 822 (1999)
6. C. Brabec, N. Sariciftci, J. Hummelen, Adv. Funt. Mater. **11**, 15 (2001)
7. W. Clemens, et al., J. Mater. Res. **19**, 1963 (2004)
8. B. Crone, et al., Nature **403**, 521 (2000)
9. R. Friend, et al., Nature **397**, 121 (1999)
10. URL http://www.philips.com
11. N.S. Sariciftci, A.J. Heeger, in *Handbook of Organic Conductive Molecules and Polymers* (Wiley, New York, 1997), pp. 413–455
12. D. Dimitrakopoulos, D. Mascaro, IBM J. Res. Dev. **45**(1), 11 (2001)
13. H. Sirringhaus, et al., Science **280**, 1741 (1998)
14. R. de Boer, et al., Phys. Stat. Sol. A **201**(6), 1302 (2004)
15. V.C. Sundar, Science **303**, 1644 (2004)
16. A. Andreev, G. Matt, C. Brabec, H. Sitter, D. Badt, H. Seyringer, N. Sariciftci, Adv. Mater. **12**(9), 629 (2000)
17. A. Dodabalapur, L. Torsi, H. Katz, Science **268**(5208), 270 (1995)
18. G. Horowitz, R. Hajlaoui, F. Kouki, EPJ Appl. Phys. **1**(3), 361 (1998)
19. G. Horowitz, X. Peng, D. Fichou, F. Garnier, J. Appl. Phys. **67**(1), 528 (1990)
20. D.J. Gundlach, Y.Y. Lin, T.N. Jackson, D.G. Schlom, Appl. Phys. Lett. **71**(26), 3853 (1997)
21. N. Karl, K.H. Kraft, J. Marktanner, M. Munch, F. Schatz, R. Stehle, H.M. Uhde, J. Vac. Sci. Technol. A **17**, 2318 (1999)
22. H. Katz, A. Lovinger, J. Johnson, C. Kloc, T. Siegrist, W. Li, Y.Y. Lin, A. Dodabalapur, Nature **404**(6777), 478 (2000)
23. H.E. Katz, A. Dodabalapur, Z. Bao, *Handbook of Oligo and Polythiophenes* (Wiley-VCH, Weinheim, 1999), p. 459ff
24. M. Era, T. Tsutsui, S. Saito, Appl. Phys. Lett. **67**(17), 2436 (1995)

25. K.N. Baker, A.V. Fratini, T. Resch, W.W. Adams, E.P. Socciand, B.L. Farmer, Polymer **34**(8), 1571 (1993)
26. G. Leising, S. Tasch, W. Graupner, *Handbook of Conducting Polymers*, 2nd edn. (Marcel Dekker, New York, 1997), pp. 847–880
27. A. Niko, F. Meghdadi, C. Ambrosch-Draxl, P. Vogl, G. Leising, Synthetic Met. **76**(1–3), 177 (1996)
28. E. Zojer, N. Koch, P. Puschnig, F. Meghdadi, A. Niko, R. Resel, C. Ambrosch-Draxl, M. Knupfer, J. Fink, J. Bredas, G. Leising, Phys. Rev. B **61**(24), 16538 (2000)
29. P. Allemand, et al., J. Am. Chem. Soc. **113**, 1050 (1991)
30. N. Marjanovic, T. Singh, G. Dennler, S. Günes, H. Neugebauer, N. Sariciftci, R. Schwödiauer, S. Bauer, Org. Electron. **7**(4), 188 (2006)
31. W. Krakow, et al., Appl. Phys. A **56**, 185 (1993)
32. D. Stifter, H. Sitter, J. Cryst. Growth **156**(1–2), 79 (1995)
33. K. Tanigaki, et al., Appl. Phys. Lett. **63**, 2351 (1993)
34. S. Guha, W. Graupner, R. Resel, M. Chandrasekhar, H. Chandrasekhar, R. Glaser, G. Leising, Phys. Rev. Lett. **82**(18), 3625 (1999)
35. T. Mikami, H. Yanagi, Appl. Phys. Lett. **73**(5), 563 (1998)
36. M. Dresselhaus, G. Dresselhaus, P. Eklund, *Science of Fullerenes and Carbon Nanotubes* (Academic Press, London, 1996), p. 171
37. A. Lopez-Otero, Thin Solid Films **49**(1), 3 (1978)
38. A. Montaigne Ramil, T. Singh, N. Haber, N. Marjanovic, S. Gunes, A. Andreev, G. Matt, R. Resel, H. Sitter, S. Sariciftci, J. Cryst. Growth **288**(1), 123 (2006)
39. T. Singh, et al., in *Organic Thin-Film Electronics*, vol. 871E, ed. by A. Arias, N. Tessler, L. Burgi, J. Emerson (Warrebdale, PA, 2005). *Mater. Res. Soc. Symp. Proc.*, p. I 4.9.1
40. X. Peng, et al., Appl. Phys. Lett. **57**, 2013 (1990)
41. R. Schwödiauer, et al., Appl. Phys. Lett. **75**, 3998 (1999)
42. G. Horowitz, R. Hajlaoui, R. Bourguiga, M. Hajlaoui, Synthetic Met. **101**(1), 401 (1999)
43. G. Horowitz, R. Hajlaoui, H. Bouchriha, R. Bourguiga, M. Hajlaoui, Adv. Mater. **10**(12), 923 (1998)
44. S.N. Magonov, V. Elings, M.H. Whangbo, Surf. Sci. **375**, L385 (1997)
45. T. Fujii, M. Suzuki, M. Miyashita, M. Yamaguchi, T. Onuki, H. Nakamura, T. Matsubara, H. Yamada, M. Nakayama, J. Vac. Sci. Technol. B **9**, 666 (1991)
46. Y.P. Piryatinskii, O.V. Yaroshchuk, Opt. Spectrosc. **89**, 860 (2000)
47. R. Resel, N. Koch, F. Meghdadi, G. Leising, W. Unzog, K. Reichmann, Thin Solid Films **305**(1–2), 232 (1997)
48. R. Resel, G. Leising, Surf. Sci. **409**(2), 302 (1998)
49. R. Resel, Thin Solid Films **433**(1–2 special issue), 1 (2003)
50. B. Müller, T. Kuhlmann, K. Lischka, H. Schwer, R. Resel, G. Leising, Surf. Sci. **418**(1), 256 (1998)
51. K. Erlacher, R. Resel, S. Hampel, T. Kuhlmann, K. Lischka, B. Müller, A. Thierry, B. Lotz, G. Leising, Surf. Sci. **437**(1), 191 (1999)
52. H. Plank, R. Resel, S. Purger, J. Keckes, A. Thierry, B. Lotz, A. Andreev, N. Sariciftci, H. Sitter, Phys. Rev. B **64**(23), 2354231 (2001)
53. H. Plank, R. Resel, H. Sitter, A. Andreev, N. Sariciftci, G. Hlawacek, C. Teichert, A. Thierry, B. Lotz, Thin Solid Films **443**(1–2), 108 (2003)

54. C. Teichert, G. Hlawacek, A. Andreev, H. Sitter, P. Frank, A. Winkler, N. Sariciftci, Appl. Phys. A **82**(4 special issue), 665 (2006)
55. F. Balzer, H.G. Rubahn, Appl. Phys. Lett. **79**(23), 3860 (2001)
56. H. Plank, R. Resel, A. Andreev, N. Sariciftci, H. Sitter, J. Cryst. Growth **237–239**(1–4 III), 2076 (2002)
57. G. Louarn, et al., Synthetic Met. **55–57**, 4762 (1993)
58. L. Athouel, et al., Synthetic Met. **84**, 287 (1997)
59. E. Thulstrup, *Elementary Polarization Spectroscopy* (VCH Publishers, New York, 1989)
60. A. Andreev, H. Sitter, R. Resel, D.M. Smilgies, H. Hoppe, G. Matt, N. Sariciftci, D. Meissner, D. Lysacek, L. Valek, Mol. Cryst. Liq. Cryst. Sci. Technol. A **385**(Part II), (2002)
61. G. Hlawacek, Q. Shen, C. Teichert, R. Resel, D.M. Smilgies, Surf. Sci. **601**, 2584 (2007)
62. F. Balzer, H.G. Rubahn, Nano Lett. **2**(7), 747 (2002)
63. F. Balzer, H.G. Rubahn, Adv. Funct. Mater. **15**(1), 17 (2005)
64. F. Balzer, J. Beermann, S. Bozhevolnyi, A. Simonsen, H.G. Rubahn, Nano Lett. **3**(9), 1311 (2003)
65. H. Yanagi, S. Doumi, T. Sasaki, H. Tada, J. Appl. Phys. **80**(9), 4990 (1996)
66. Y. Toda, H. Yanagi, Appl. Phys. Lett. **69**(16), 2315 (1996)
67. D.M. Smilgies, N. Boudet, B. Struth, Y. Yamada, H. Yanagi, J. Cryst. Growth **220**(1–2), 88 (2000)
68. H. Yanagi, T. Morikawa, Appl. Phys. Lett. **75**(2), 187 (1999)
69. D.M. Smilgies, N. Boudet, H. Yanagi, Appl. Surf. Sci. **189**(1–2), 24 (2002)
70. E. Kintzel Jr., D.M. Smilgies, J. Skofronick, S. Safron, D. Van Winkle, J. Vac. Sci. Technol. A **22**(1), 107 (2004)
71. E. Kintzel Jr., D.M. Smilgies, J. Skofronick, S. Safron, D. Van Winkle, J. Cryst. Growth **289**(1), 345 (2006)
72. H. Yanagi, S. Okamoto, Appl. Phys. Lett. **71**(18), 2563 (1997)
73. H. Yanagi, T. Ohara, T. Morikawa, Adv. Mater. **13**(19), 1452 (2001)
74. Y. Yoshida, et al., Adv. Mater. **12**, 1587 (2000)
75. A.Y. Andreev, et al., J. Vac. Sci. Technol. A **24**(4), 1660 (2006)
76. A. Andreev, T. Haber, D.M. Smilgies, R. Resel, H. Sitter, N. Sariciftci, L. Valek, J. Cryst. Growth **275**(1–2), e2037 (2005)
77. T. Haber, A. Andreev, A. Thierry, H. Sitter, M. Oehzelt, R. Resel, J. Cryst. Growth **284**(1–2), 209 (2005)
78. T. Haber, M. Oehzlt, R. Resel, A. Andreev, A. Thierry, H. Sitter, D. Smilgies, B. Schaffer, W. Grogger, R. Resel, J. Nanosci. Nanotechnol. **6**(3), 698 (2006)
79. F. Balzer, H.G. Rubahn, Surf. Sci. **548**(1–3), 170 (2004)
80. G. Koller, S. Berkebile, J. Krenn, G. Tzvetkov, G. Hlawacek, O. Lengyel, F. Netzer, C. Teichert, R. Resel, M. Ramsey, Adv. Mater. **16**(23–24), 2159 (2004)
81. G. Hlawacek, C. Teichert, A. Andreev, H. Sitter, S. Berkebile, G. Koller, M. Ramsey, R. Resel, Phys. Stat. Sol. A **202**(12), 2376 (2005)
82. G. Hlawacek, C. Teichert, S. Müllegger, R. Resel, A. Winkler, Synthetic Met. **146**(3), 383 (2004)
83. R. Resel, I. Salzmann, G. Hlawacek, C. Teichert, B. Koppelhuber, B. Winter, J. Krenn, J. Ivanco, M. Ramsey, Org. Electron. **5**(1–3), 45 (2004)
84. S. Müllegger, I. Salzmann, R. Resel, G. Hlawacek, C. Teichert, A. Winkler, J. Chem. Phys. **121**(5), 2272 (2004)

85. S. Müllegger, I. Salzmann, R. Resel, A. Winkler, Appl. Phys. Lett. **83**(22), 4536 (2003)
86. H. Sitter, private communication
87. H.J. Egelhaaf, et al., Synthetic Met. **83**, 221 (1996)
88. P. Mei, et al., J. Appl. Phys. **88**, 5158 (2000)
89. M. Muccini, et al., Adv. Mater. **13**, 355 (2001)
90. A. Kadashchuk, Y. Skryshevski, Y. Piryatinski, A. Vakhnin, E. Emelianova, V. Arkhipov, H. Bässler, J. Shinar, J. Appl. Phys. **91**(8), 5016 (2002)
91. A. Kadashchuk, A. Vakhnin, Y. Skryshevski, V. Arkhipov, E. Emelianova, H. Bässler, Chem. Phys. **291**(3), 243 (2003)
92. A. Piaggi, G. Lanzani, G. Bongiovanni, M. Loi, A. Mura, W. Graupner, F. Meghdadi, G. Leising, Opt. Mater. **9**(1–4), 489 (1998)
93. A. Piaggi, G. Lanzani, G. Bongiovanni, A. Mura, W. Graupner, F. Meghdadi, G. Leising, M. Nisoli, Phys. Rev. B **56**(16), 10133 (1997)
94. A. Kadashchuk, A. Andreev, H. Sitter, N. Sariciftci, Y. Skryshevski, Y. Piryatinski, I. Blonsky, D. Meissner, Adv. Funct. Mater. **14**(10), 970 (2004)
95. H. Bässler, *Semiconducting Polymers: Chemistry, Physics and Engineering* (Wiley-VCH, Weinheim, 2000), p. 365
96. H. Bässler, *Disordered Effect on Relaxational Processes* (Springer, Berlin Heidelberg New York, 1994), p. 585
97. A. Andreev, F. Quochi, F. Cordella, A. Mura, G. Bongiovanni, H. Sitter, G. Hlawacek, C. Teichert, N. Sariciftci, J. Appl. Phys. **99**(3), (2006)
98. F. Quochi, F. Cordella, R. Orrù, J. Communal, P. Verzeroli, A. Mura, G. Bongiovanni, A. Andreev, H. Sitter, N. Sariciftci, Appl. Phys. Lett. **84**(22), 4454 (2004)
99. R. Kepler, Phys. Rev. **119**, 1226 (1960)
100. S. Kobayashi, T. Takenobu, S. Mori, A. Fujiwara, Y. Iwasa, Appl. Phys. Lett. **82**, 4581 (2003)
101. T.B. Singh, N. Marjanović, P. Stadler, M. Auinger, G.J. Matt, S. Günes, N.S. Sariciftci, R. Schwödiauer, S. Bauer, J. Appl. Phys. **97**, 083714 (2005)
102. Z. Bao, A.J. Lovinger, J. Brown, J. Am. Chem. Soc. **120**, 207 (1998)
103. H. Katz, J. Johnson, A. Lovinger, W. Li, J. Am. Chem. Soc. **122**, 7787 (2000)
104. R. Chesterfield, J.C. McKeen, C.R. Newman, C.D. Frisbie, P.C. Ewbank, K.R. Mann, L.L. Miller, J. Appl. Phys. **95**, 6396 (2004)
105. P.R.L. Malenfant, C.D. Dimitrakopoulos, J.D. Gelorme, L.L. Kosbar, T.O. Graham, Appl. Phys. Lett. **80**, 2517 (2002)
106. G. Horowitz, P. Lang, M. Mottaghi, H. Aubin, Adv. Funct. Mater. **14**(11), 1069 (2004)
107. F. Dinelli, M. Murgia, P. Levy, M. Cavallini, F. Biscarini, D.M. de Leeuw, Phys. Rev. **92**, 116802 (2004)
108. M. Sze, *Physics of Semiconductor Devices* (Wiley-VCH, New York, 1981)
109. G. Horowitz, P. Delannoy, J. Appl. Phys. **70**, 469 (1991)
110. M. Koehler, I. Biaggio, Phys. Rev. B. **70**, 045314 (2004)
111. D. Mihailetchi, J.K.J. van Duren, P.W.M. Blom, J.C. Hummelen, R.A.J. Janssen, J.M. Kroon, M.T. Rispens, W.J.H. Verhees, M.M. Wienk, Adv. Funct. Mater. **13**, 43 (2003)
112. G.J. Matt, N.S. Sariciftci, T. Fromherz, Appl. Phys. Lett. **84**, 1570 (2004)
113. C.J. Brabec, A. Cravino, D. Meissner, N.S. Sariciftci, T. Fromherz, M.T. Rispens, L. Sanchez, J.C. Hummelen, Adv. Funct. Mater. **11**, 374 (2001)

114. E.J. Meijer, G.H. Gelinck, E. Van Veenendaal, B.H. Huisman, D.M. De Leeuw, T.M. Klapwijk, Appl. Phys. Lett. **82**, 4576 (2003)
115. A. Rose, Phys. Rev. **97**, 1538 (1955)
116. M.A. Lampart, Phys. Rev. **103**, 1648 (1956)
117. W. Kalb, P. Lang, M. Mottaghi, H. Aubin, G. Horowitz, M. Wuttig, Synthetic Met. **146**(3), 279 (2004)
118. T.W. Kelly, L.D. Boardman, T.D. Dunbar, D.V. Muyres, M.J. Pellerite, T.P. Smith, J. Phys. Chem. B **107**, 5877 (2003)
119. T.B. Singh, G. Hernandez-Sosa, H. Neugebauer, A. Andreev, H. Sitter, N.S. Sariciftci, physica status solidy (b) **243**, 3329 (2006)
120. N.F. Mott, R.W. Gurney, *Electronic Processes in Ionic Crystals*, 2nd edn., (Clarendon Press, Oxford, 1948), p. 172
121. M. Polke, J. Stuke, E. Vinaricky, Phys. Stat. Sol. **3**, 1885 (1963)
122. O.D. Jurcheschu, J. Baas, T.T.M. Palstra, Appl. Phys. Lett. **16**, 3061 (2004)
123. J.A. Geurst, Phys. Stat. Sol. **15**, 107 (1966)

12

Device-Oriented Studies on Electrical, Optical, and Mechanical Properties of Individual Organic Nanofibers

J. Kjelstrup-Hansen, P. Bøggild, H.H. Henrichsen, J. Brewer, and H.-G. Rubahn

12.1 Introduction

Organic nanofibers are promising candidates for future nanophotonic and nanoelectronic devices due to their optical, electrical, chemical, and morphological properties. *Para*-hexaphenylene (p6P) as well as diverse functionalized quaterphenylene molecules such as p-methyloxylated p-quaterphenylene (MOP4) [1] or p-chlorinated p-quaterphenylene (CLP4) [2] form well-aligned needles or "nanofibers" upon vacuum epitaxy on muscovite mica substrates [3,4]. In this chapter we will use the words "nanofiber" and "needle" synonymously. The nanofibers consist of large single crystalline areas of molecules oriented nearly parallel to the substrate surface with typical dimensions of a few hundred nanometers in width, a few ten nanometers in height, and several hundred micrometers in length. Partly due to their crystalline order and morphology the nanofibers show photonic functionalities such as waveguiding [5–7], lasing [8] as well as nonlinear optical activity [9,10].

For device applications the nanofibers have to fulfill several prerequisites. First, the optical, electrical, and mechanical properties of individual nanofibers have to be well known. Second, methods have to be developed to position individual nanofibers in the device, i.e., to detach them from their growth substrates, transfer them, and arrange them onto the desired device platforms. Third, they have to be connected electrically or optically. Note that an important advantage of *organic* nanofibers is that they are build from molecular blocks that can be functionalized and thus modified with respect to their optoelectronic properties. The drawback is that the nanofibers are rather fragile and thus difficult to handle mechanically.

In this article we review optical, electrical, and mechanical properties of individual organic nanofibers. A parallel chapter reviews in a more general fashion nanooptics of surface-bound aggregates.

As for the photonic aspect it is noted that nanofibers have previously been examined mainly via surface-based optical methods such as epifluorescence microscopy (Fig. 12.1) or scanning near field optical microscopy. The fact that

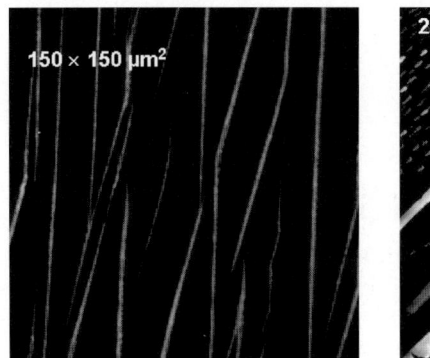 

Fig. 12.1. Epifluorescence (left-hand side) and atomic force microscopy (right-hand side) images of methyloxylated *para*-quaterphenylene (MOP4) nanofibers. Typical fiber heights are of the order of 100 nm

the needles were lying down on the supporting surface made it difficult to view them along the longitudinal axes, making it practically impossible to measure the true angular intensity distribution of emitted light. The angular distribution of fluorescence intensity is of special interest considering that the needles have been shown to act as waveguides and that they exhibit lasing properties [8]. Here we present direct measurements of luminescence angular distributions from individual free-floating or embedded nanofibers.

Constructing nanofiber-based devices requires in addition the ability to integrate the nanofibers with surrounding electrical circuitry. Since the fabrication of the p6P nanofibers themselves takes place via dipole-assisted self-assembly [3] on a special growth substrate, an in situ growth approach, such as demonstrated with carbon nanotubes [11], appears less viable. Recently Briseno and coworkers [12] did demonstrate an alternative method by fabricating large organic single-crystals on a patterned self-assembled monolayer template, but it is not yet clear if the method is compatible with subsequent conventional microfabrication processing steps. So far there is also no experimental proof that growth of nanofibers could become possible on arbitrarily prepatterned substrates.

At present, the most straightforward approach is to keep nanofiber fabrication and device construction as two separate, consecutive steps. This, however, requires either the transfer from the growth substrate and the subsequent accurate positioning of the nanofiber on a prefabricated circuit (i.e., electrodes first), or the transfer to a pristine substrate, followed by additional device fabrication steps (i.e., nanofibers first).

Recently the transfer of either individual nanofibers or oriented arrays of nanofibers from the original growth substrate to arbitrary other substrates was demonstrated. Nanofibers could also be embedded in fluids such as water, which is exemplified in the following paragraphs. As for "mass transfer" of arrays of nanofibers, more details are disclosed in a recent patent [13][1].

12.2 Toward Photonic Devices: The Optical Properties of Isolated Nanofibers

In this part we show that we can view nanofibers at different angles from side- to front-view by investigating freely floating needles in water. This along with measurements of individual nanofibers on or inside micropipettes seen at different angles from top- to side-view, allows the determination of the three-dimensional luminescence angular intensity distribution from the nanofibers. By transferring floating needles to a sucrose solution it has also been possible to obtain three-dimensional images of suspended needles using two-photon laser scanning microscopy.

12.2.1 Preparation and Optical Detection

The needles were transferred from the muscovite mica growth substrate to water by a soft, nondestructive lift-off process. On examination in a fluorescence microscope a large multitude of free floating aggregates of varying lengths were seen. The needles were then fixed in solution by first preparing a solution of 5% sucrose and 95% water in a hot water bath and then applying a few drops of the solution to a microscope slide, where the drops were mixed with a drop of water containing the nanoaggregates.

A Zeiss two-photon laser scanning microscope (LSM) was used to obtain three-dimensional images of individual needles of p-hexaphenylene (p6P) and methyloxylated p-quaterphenylene (MOP4) and to study the fluorescence and second harmonic intensities of single needles as a function of two-photon excitation wavelength. The excitation laser was a Spectra-Physics Mai Tai Femtosecond Ti:sapphire laser with an operating range from 720 to 900 nm.

Glass micropipettes were forged using a micropipette-puller. The tip diameter of the pipettes was about 10 µm. The pipettes were filled with water containing nanoaggregates and left to dry leaving a number of nanofibers deposited on the inside (Fig. 12.2). The pipettes were then placed in a glycerin bath in a holder which enabled them to be rotated around their long axes in a controlled manner.

The glycerin was used to match the index of refraction to the glass of the pipettes thus removing imaging problems caused by refraction of light from the pipette. The holder and pipette with needles were placed under an epifluorescence microscope. By rotating the pipette it was possible to rotate the nanofibers *inside* the pipette and thus to measure the luminescence of

---

[1] Details about practical aspects of nanofiber handling are available via the homepage http://www.nanofiber.dk of the company Nanofiber A/S.

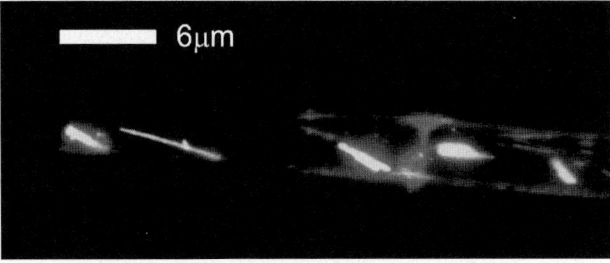

Fig. 12.2. Epifluorescence image of pipette with needles on the inside, submerged in glycerin

the fibers as a function of angle ϕ using a CCD camera mounted on the microscope.

Fibers were also applied to the *outside* surface of the pipettes. The pipettes were then placed in a holder which enabled them to be rotated around their long axes and measurements were made of the angular distribution of emitted luminescence again using an epifluorescence microscope and a CCD camera but without submerging the pipettes in glycerin.

The intensity distribution of emitted light along the long needle axis was determined quantitatively by a series of images of two rotating needles connected so that they formed a "T" [14].

12.2.2 Nanofiber Tomography and Angular Light Emission

Figure 12.3 shows a reconstructed three-dimensional projection of a two-photon laser scanning measurement of two p6P needles in water, one of which having the shape of a "Y." Apparently, one of the arms of the Y is still attached to the straight part (see inset). The images have been corrected for distortions along the focal direction, which were induced by a slight motion of the needles during scanning in z-direction. This correction was performed by adding small fluorescing polystyrene balls of known diameter (1 µm) to the solution and recording simultaneously images of the needles and the balls. Subsequently the known dimensions of the balls were used for a numerical correction.

The need for reconstruction of distortions can be overcome by suspending floating nanoaggregates in a sucrose solution. Figure 12.4a shows four views at different angles of a MOP4 needle in such a solution. The small arm protruding from the needle has a measured width of 430 nm. Thus we estimate the optical resolution to be of the order of 400 nm. Figure 12.4b is an atomic force microscopy (AFM) image of a similar MOP4 nanofiber supported on mica. Although the LSM image does not contain as much detail as the AFM image it reveals the front sides as well as those sides of the needles which have during growth been oriented toward the substrate. Careful inspection of the

12 Individual Nanofiber Devices   305

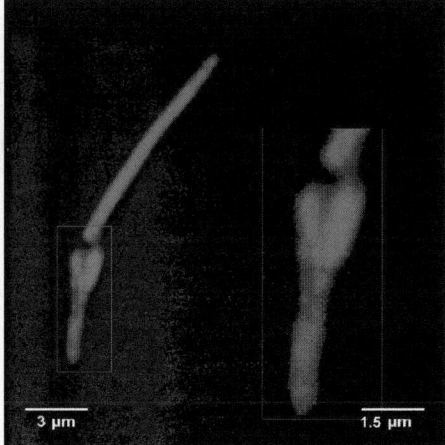

Fig. 12.3. Two-photon laser scanning microscope (LSM) image of *para*-hexaphenylene nanofibers, free floating in water. The *inset* is a magnification of the Y-shaped structure

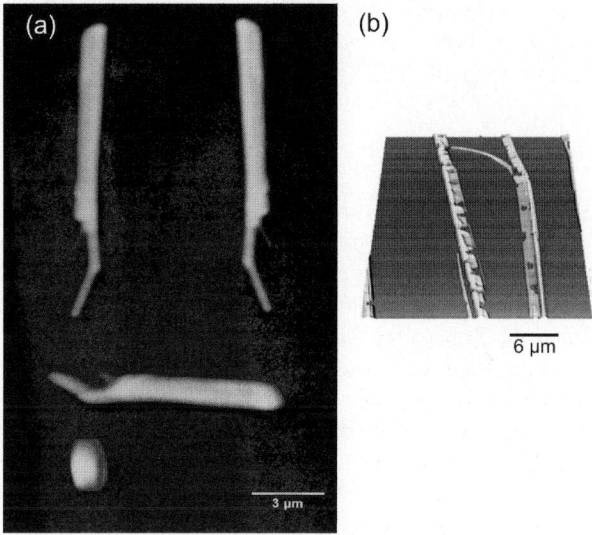

Fig. 12.4. (a) LSM images of a MOP4 nanofiber embedded in sucrose: views for different rotation angles, namely: 0° around x-axis (top left-hand side), 180° around x-axis (top right), 0° around y-axis (lower left-hand-side), 90° around z-axis (lower right-hand-side). (b) Atomic force microscopy image of a similar MOP4 nanofiber grown on a mica substrate. The height of the fiber is around 100 nm

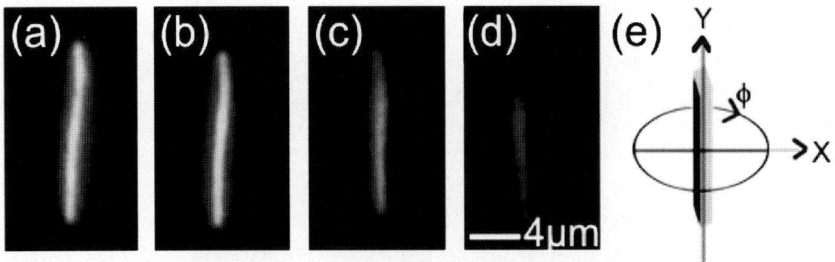

Fig. 12.5. Figures (a)–(d) nanofiber seen at four different angles of ϕ. (e) Schematic showing the angle ϕ relative to the nanofiber

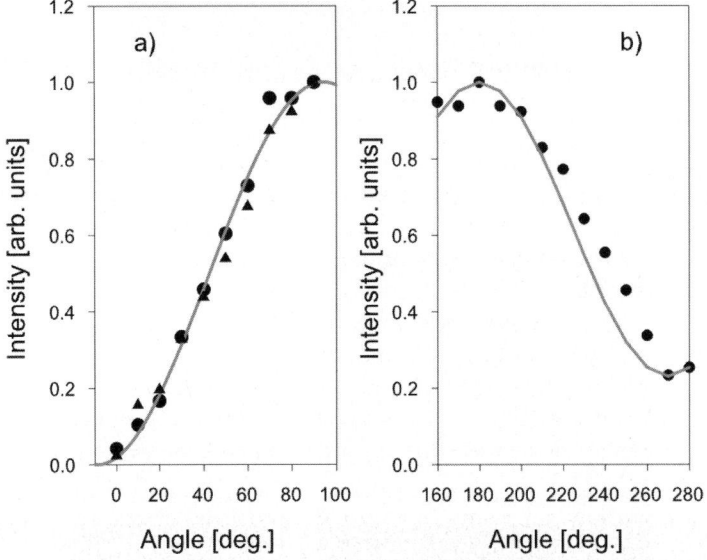

Fig. 12.6. Angular distribution of emitted light from a p6P nanofiber when rotated around the long axis. Figure (a) shows the angular distribution of luminescence found from a needle inside a pipette. The grey lines are best fits of $\cos^2(\phi)$ to the data. Figure (b) shows the angular distribution of light found from a needle on the outside of a pipette. The black circles signify the maximum intensity while the grey line is the best fit of $\cos^2(\phi)$ to the data

3D image in fact shows that the substrate-oriented face of the needle is more flat as compared to the side facing the vacuum.

Figure 12.5 shows four images of a nanofiber on a glass pipette seen from four different angles ϕ, where $\phi = 90°$ refers to the view from the wide side of the nanofibers, i.e., the upper or lower side for surface-bound nanofibers (Fig. 12.5b).

Figure 12.6 depicts the angular distribution of emitted luminescence from nanofibers both on and inside the pipettes. The intensity from the bright side of the fibers is 5–20 times stronger than the light emitted from the dark side of the fibers. Note that due to their single crystalline nature the fibers consist of an array of well-aligned dipole emitters. Therefore it is expected that the emission in the far field will be proportional to $\cos^2(\phi)$. In Fig. 12.6 the data is fitted with $\cos^2(\phi)$, providing indeed good agreement between data and fit.

The difference in the absolute increase in intensity for fibers seen from the bottom (Fig. 12.6a) and from the top (Fig. 12.6b) could be explained by the fibers having different levels of crystalline perfection. A fiber with well-ordered dipoles will have a larger emission anisotropy compared to a less well-ordered fiber.

Freely floating nanofibers in water seen in a fluorescence microscope have a pronounced spatial anisotropy in light emission also with respect to the angle Θ which is counted with respect to the longitudinal axis (Fig. 12.8) [14]. Although the fibers luminesce also out of their broadsides, the tips radiate a more intense cone of light.

Figure 12.7a shows three images of two nanofibers in solution forming a "T". By recording a series of images of the rotating "T" it was possible to measure the angular distribution of emitted luminescence from a nanofiber so that the intensity from the tip could be compared to that resulting from the broadside. Figure 12.7b shows a drawing of the nanofiber "T."

The measured angular distribution of emitted luminescence from a nanofiber can be seen in Fig. 12.8. A large increase of emitted light intensity is seen toward the tip of the nanofibers. This increase is partially due to the fact that the individual emitters are anisotropically oriented. However, a simple estimate on the basis of geometrical optics reveals for the given

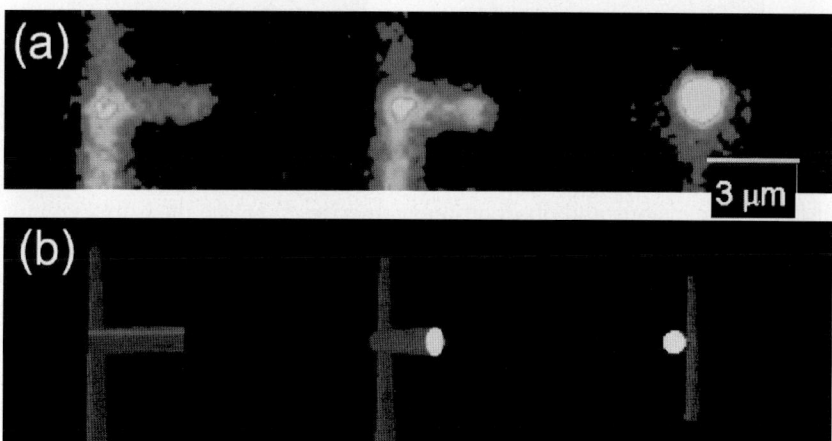

Fig. 12.7. (a) Different viewing angles of fluorescence from two p6P nanofibers. As shown in the sketch in (b) a short fiber is rotating around a nearly space-fixed long fiber. The images show the fiber at ($\Theta = 0°$), ($\Theta = 45°$), and ($\Theta = 86°$)

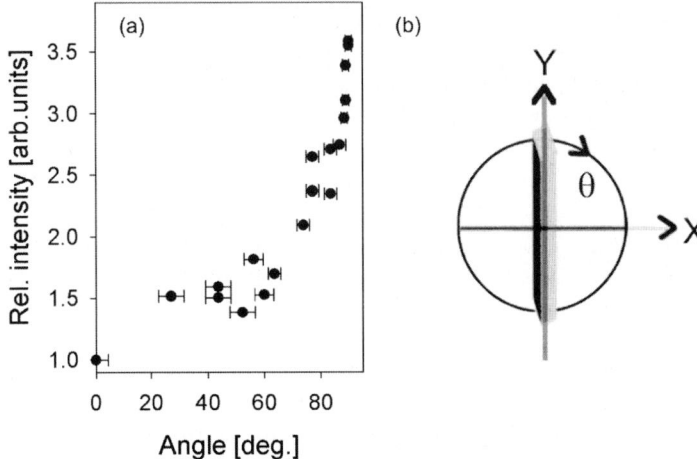

Fig. 12.8. (a) Angular distribution of emitted luminescence from a nanofiber rotated so that the intensity from the tip can be compared to that of the broadside. (b) Schematic showing the angle θ relative to the nanofiber

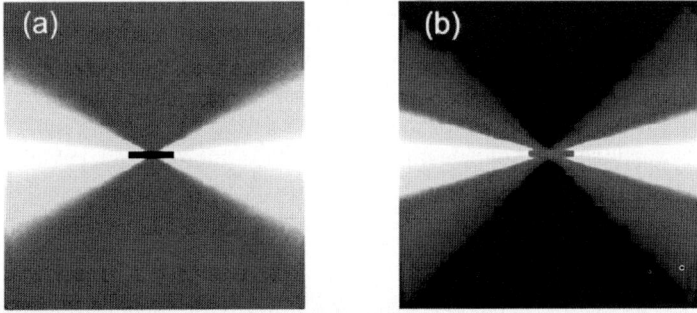

Fig. 12.9. (a) Calculated angular distribution of light from a needle with the nonemitting side up. (b) Calculated angular distribution of light from a needle with the emitting side upward

nanofiber dimensions and observation geometries that the measured enhancement is significantly larger as compared to an anisotropy effect. Therefore it is concluded that waveguiding along the nanofibers also plays an important role. Waveguiding as well as lasing in organic nanofibers has been demonstrated directly with other methods and is discussed in other chapters of this book.

Figure 12.9 shows the resulting angular distribution of light from a single nanofiber, calculated on the basis of intensity measurements along (ϕ) and perpendicular (Θ) to the long nanofiber axes. As seen, the light is emitted in a flat cone from the tips of the nanofibers. It is also well polarized in-plane due to the crystalline orientation of the individual emitters.

12.3 Studies on Electrical Properties

Among the interesting aspects of organic semiconductors are their electrical and electrooptical properties. The first organic field-effect transistor (OFET) was demonstrated in 1986 [15], based on an organic thin film. More ordered films have been used to demonstrate pentacene transistors with hole mobilities up to $1.5\,\mathrm{cm^2\,Vs^{-1}}$ [16]. Recently, FET structures based on organic single crystals have been reported [17, 18]. One method relies on fabricating source and drain contacts on top of a single crystal by metal evaporation as demonstrated by Podzorov and coworkers [17]. This method has allowed the fabrication of a rubrene FET with a hole mobility up to $1\,\mathrm{cm^2\,Vs^{-1}}$. An alternative technique employs an elastomeric stamp, on which an electrode pattern is prefabricated before the stamp is laminated against the crystal surface [18]. This rubrene FET showed a mobility up to $15\,\mathrm{cm^2\,Vs^{-1}}$. For an extensive review see [19].

Equally appealing are the electrooptical properties. In 1986, Tang and VanSlyke demonstrated the first organic electroluminescent diode (OLED) [20] based on a sandwich structure of several thin film layers. Much attention has been devoted to improving different properties, for instance by controlling the molecular ordering in the film to provide OLEDs with polarized electroluminescence [21].

Combining the electroluminescent properties of organic materials with micro and nanofabrication techniques led to fabrication of nanoscale OLEDs [22, 23]. Both studies employed electron beam lithography and reactive ion etching to create nanoscale holes in a thin insulating layer, which had been deposited on top of a transparent anode. The devices were then fabricated by applying the appropriate organic layers and a top cathode. At driving field strengths of $\simeq 10^6\,\mathrm{V\,cm^{-1}}$, electroluminescence could be observed from holes down to $\simeq 60\,\mathrm{nm}$ diameter.

The method proposed here for fabricating nanoscale organic devices is to use self-assembled nanofibers as components. This avoids the need for fabrication of nanoscale structures by lithography, but on the other hand it requires additional steps of transfer and positioning. As an additional benefit, this method can result in a nanoscale *polarized* light source.

12.3.1 Charge Injection and Transport

When considering the electronic structure of organic semiconductors, it is useful to adopt a representation similar to the band picture of inorganic semiconductors [24]. However, there are significant differences between the electronic structure of organic and inorganic semiconductors. In the latter case, the carriers are delocalized over the entire crystal and broad continuous energy bands are formed. In organic crystals the carriers are localized on single molecules and conduction occurs by incoherent hopping between neighboring molecules. The intermolecular interaction is small and the energetic structure of the

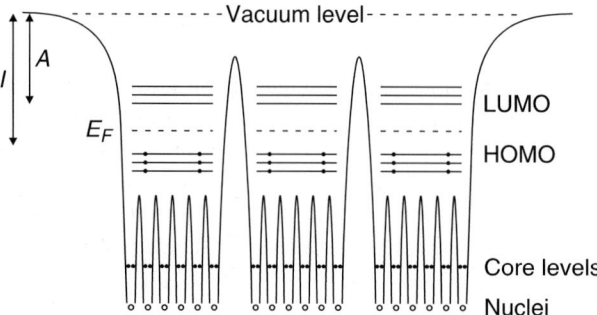

Fig. 12.10. Conceptual energy level diagram of an organic semiconductor comprised of three molecules. The ionization energy I and electron affinity A are determined by the position of the HOMO and LUMO levels with respect to the vacuum level

organic crystal is therefore to a large extent determined by the energy levels of the individual molecules. The two energy levels that govern the charge transport properties are the highest occupied molecular orbital (HOMO) and lowest unoccupied molecular orbital (LUMO) as shown in Fig. 12.10. These levels support hole and electron currents, respectively. It is possible to conduct either hole or electron currents exclusively, but also both simultaneously [25].

The current through an organic crystal can be limited either by energy barriers at the injecting contact or by the crystal bulk. This leads to two different conduction regimes: injection limited current (ILC) or bulk limited current. However, an intermediate situation where both factors influence the current can also occur. In the bulk limited regime, the contact resistance is considered ohmic and small compared to the bulk resistance. In this case, the conduction can either be ohmic or space-charge limited (SCL). At low voltage, the current rises in proportion to the applied voltage. At higher voltage levels, the amount of injected charge can rise to a level where the organic material can no longer support an ohmic current flow. When this happens a surplus of charge accumulates, i.e. a space charge is formed. Due to electrostatic repulsion this limits further injection – hence the name space-charge limited conduction. In this regime the current is limited to increase with the square of the voltage as described by the Mott–Gurney law [25]. If both types of carriers are injected the space charge becomes less significant, and the current may increase faster with voltage [25, 26].

In the injection limited regime the contact resistance is significantly larger than the bulk resistance and the energy barrier at the contact controls the injection. The exact injection mechanism depends on barrier height and is typically described by tunneling phenomena through and thermal emission over the barrier. Therefore, no general theory can describe all situations. However, examples exist on specific systems of both numerical [27], Monte Carlo [28], and analytical [29] calculations. In a mixed mode, the conduction can be injection limited at low bias and become SCL at higher voltages.

The common way of effectively injecting one type of charge carrier is to use contact materials with appropriate work functions that are close to either the HOMO or the LUMO level. This is a simplified approach though, since most metal/organic semiconductor interfaces contain dipoles, which can alter the injection barrier [24, 30]. Nevertheless, it is possible to obtain majority carriers of either kind by appropriate choice of contact material [26] which is useful for OFET devices.

If two different contact materials are used, high-injection rates of both holes and electrons can lead to efficient formation of exciton pairs. The excitons may decay radiatively, which leads to emission of photons with an energy approximately equal to the band gap. This is the principle of electroluminescence and the working principle of OLEDs. In an FET structure the ratio between the injected holes and electrons can be controlled by tuning the energy levels with a nearby gate voltage [31]. This can optimize the light output and be used to switch the OLED on and off.

The conductivity of the organic semiconductor is strongly dependent on the hole and electron mobilities, thus high mobilities give lower driving voltages. However, in very short devices high mobilities increase the probability of carriers reaching the opposite electrode without recombination [32]. This has a negative effect on device efficiency. It can be avoided by using additional layers that transport either holes or electrons while blocking the opposite carrier type. This increases device efficiency and automatically balances hole and electron currents, at the expense of a more complicated device structure [32].

12.3.2 Experiments on Single-Nanofiber Devices

Realizing devices based on individual nanofibers requires the development of techniques for appropriate handling and contacting. This necessitates the transfer of just a few nanofibers from the growth to the device substrate. This can be accomplished by releasing the nanofibers from the growth substrate in a liquid and drop-casting the nanofiber dispersion to the device substrate [14, 33]. A micropipette system facilitates the transfer of very small volumes of the dispersion and thereby only very few nanofibers onto the device substrates of choice.

The device substrate is fabricated by conventional microfabrication techniques. A 1-µm thick silicon dioxide layer is grown on top of a silicon wafer by thermal oxidation. UV lithography and reactive ion etching is used to form a set of electrode supports in the silicon dioxide layer. The nanofiber dispersion is deposited on top of the silicon dioxide electrode supports resulting in the configuration shown schematically in Fig. 12.11a. The device is fabricated by first positioning a silicon nanowire across a nanofiber using a custom-built micromanipulation system [34] as shown in Fig. 12.11b. This silicon nanowire acts as a shadow mask during the subsequent deposition of metal contacts by thermal evaporation. Finally the nanowire shadow mask is removed and bonding wires are ultrasonically attached to make contact to external circuitry;

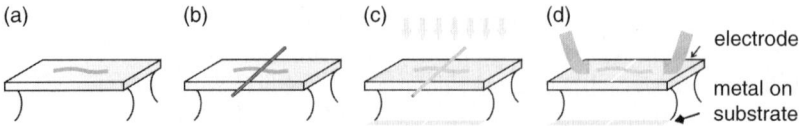

Fig. 12.11. Schematic drawing of the fabrication of an electrical device. (**a**) One or a few organic nanofibers are deposited on top of an insulating electrode support structure. (**b**) A rigid silicon nanowire is positioned across, thereby acting as a shadow mask. (**c**) A thin metallic layer is deposited by thermal evaporation. Notice how the undercut beneath the electrode support prevents electrical contact between the electrode metal, and the metal that unavoidably ends up on the remaining substrate. (**d**) Removal of the shadow mask leaves two electrodes only connected through the nanofiber. Finally the device is wire bonded and ready for test

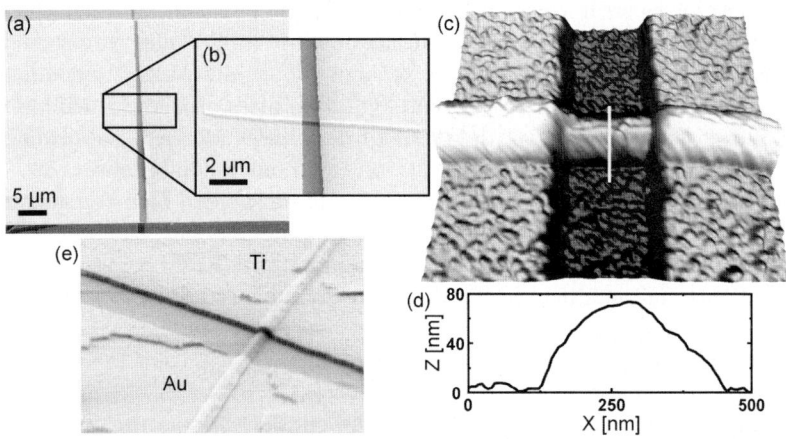

Fig. 12.12. (**a**) SEM image of a finished nanofiber device. (**b**) SEM close-up of the same nanofiber. (**c**) 1.9 μm (width) by 1.6 μm (depth) AFM image of similar device. (**d**) Cross-sectional profile measured along the line indicated in (**c**). (**e**) SEM image showing a device with two different electrode materials, recorded with the sample tilted by 60°. The uncovered part of the nanowire is 820 nm

see Fig. 12.11. The finished devices were inspected with a scanning electron microscope (SEM) and an atomic force microscope (AFM). Figure 12.12 shows examples of different devices.

This shadow masking technique can be extended to provide devices with different electrode materials [34] as is needed in electroluminescent devices. After the silicon nanowire is placed, a silicon flake is added adjacent to the silicon nanowire thus partly covering the nanofiber under investigation. When the first electrode material has been deposited, the silicon flake can be removed by micromanipulation, and the second electrode material is then deposited. Finally the silicon nanowire is removed and bonding wires are attached.

Figure 12.12e shows a device with different cathode and anode materials fabricated this way.

A significant advantage with the nanowire shadow masking is that it does not contaminate the sample, as photo or electron beam lithography with resist could do. Furthermore organic materials would be bleached by the UV light used in conventional photolithography and the chemicals used to process the resist may affect many types of organic materials. The silicon nanowire shadow masking technique is a relatively simple way of fabricating submicron prototype devices with very sharp electrode edges, see Fig. 12.12.

The charge injection and transport in individual nanofibers can be investigated by measuring the current–voltage characteristics. Figure 12.13a shows examples of two such characteristics where Au and Al has been used as contact material, respectively.

The apparent parabolic dependence of the last part of the J–V characteristics could indicate SCL behavior. In order to investigate this in more detail, the measured J–V characteristics were analyzed with the Mott–Gurney theory [25]. This provides an intrinsic upper limit to the current flow. Contact effects, defects, traps, or any other influencing factors will always cause the

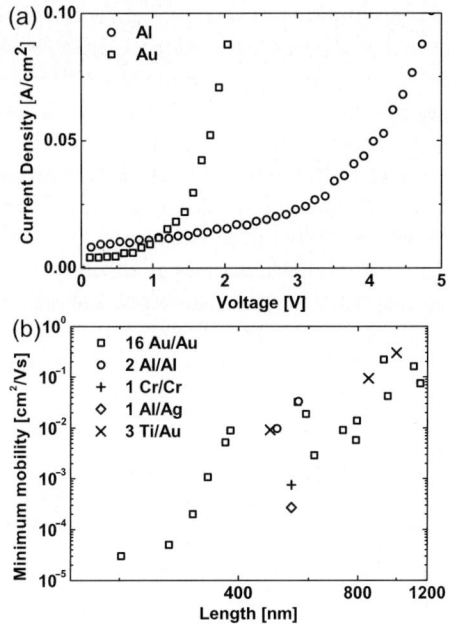

Fig. 12.13. (a) Current density vs. voltage for p6P devices with Au–Au and Al–Al contacts, respectively. The current density has been found as the measured current divided by the cross-sectional area as found from SEM and AFM. (b) *Minimum mobility* values estimated as described in the text for 23 different devices. Syntax: [cathode]/[anode]

current to be smaller than the prediction of the Mott–Gurney theory [35]. Thus, an estimate of the *minimum* mobility $\mu_{\min}$ can be extracted from the J–V measurements [35],

$$\mu_{\min} = \frac{8JL^3}{9\epsilon_r\epsilon_0 V^2} .\qquad(12.1)$$

The exact appearance of the J–V characteristics measured on different samples vary somewhat with Fig. 12.13a showing typical examples. Similar to the method used by de Boer et al. [35], the current measured at the maximum bias voltage was used in the calculation of $\mu_{\min}$.

The extracted minimum values of mobility show a significant spread over four orders of magnitude between 3×10^{-5} and 3×10^{-1} cm^2 Vs^{-1}. This corresponds to the large spread observed in a similar study on tetracene crystals [35]. In order to investigate the cause of this large spread, the minimum mobility values have been plotted vs. device length in Fig. 12.13b. If the current had indeed been space-charge limited, no length dependence of the extracted mobility values is expected as mobility is an intrinsic material property. The observed dependence therefore clearly indicates that the SCL regime is not reached and that the contact barriers have a significant impact on the observed characteristics, i.e., the current is injection limited. It can be noted that the geometry of the nanofiber devices differs somewhat from the parallel-plate geometry assumed in the general Mott–Gurney theory [25]. However, using the theory for the in-plane SCL current in thin layers, which predicts the current to depend inversely on the device length squared [36], would lead to the same conclusion regarding injection-limited current.

It should also be noted that hole and electron mobilities in general are different. In Fig. 12.13b all extracted mobility values have been plotted on the same graph irrespective of the assumed carrier type, which is expected to differ between the different contact metals as discussed shortly. However, no clear dependence on contact material (and thus carrier type) is seen, which indicates that the calculated values could be significantly below their actual mobility values. The extracted values thus only indicate a minimum mobility of 3×10^{-1} cm^2 Vs^{-1}. This is in reasonable agreement with a previous report of a hole mobility of 10^{-1} cm^2 Vs^{-1} measured on a p6P thin film in an FET configuration [37], however, in which contact barriers were also believed to reduce the extracted mobility below its intrinsic value.

As seen in the current–voltage characteristics in Fig. 12.13a, the onset voltage depends on contact material [38]. This can qualitatively be explained by the difference in injection barriers from each of the metals to p6P. To a first approximation the barrier can be estimated as the difference between the work function of the metal and either the LUMO or HOMO level – depending on whether the current is carried by electrons or holes. In p6P the LUMO and HOMO levels are approximately at 2.9 eV and 6.0 eV, respectively, [39]. The work functions of Al and Au are 4.3 and 5.1 eV [40]. This causes a barrier of approximately 0.9 eV for hole injection from Au while aluminium presents a barrier of 1.4 eV for electron injection. This agrees with the larger onset voltage observed for the Al-contacted device; see Fig. 12.13a.

An OLED is a two-carrier device in which the anode and the cathode should provide efficient injection of holes and electrons, respectively. Investigating the injection properties of a contact material can be done in hole-only or electron-only devices [26]. Similar current–voltage characteristics of a hole-only and an electron-only device indicate a good combination of hole- and electron-injecting materials for OLED operation. p6P thin film OLEDs with Al cathode and Indium–Tin–Oxide (ITO) anode have been reported [41, 42]. ITO has a work function of approximately 4.7 eV which suggests injection of both electrons and holes. However, since both barriers are high, the efficiency is relatively low [41]. Therefore, electrode combinations with lower barriers are preferred, for example Au and Mg. This can introduce other problems. Low-work function metals generally oxidize easily and therefore have a short lifetime. Instead of using low work function metals, additional layers can be used as discussed in the previous section [32, 41].

One way of realizing nanofiber prototype devices with asymmetric contacts is via the double shadow masking technique described above [38] as shown in Fig. 12.12. The first prototype device fabricated this way had Au and Ti electrodes. As expected, the onset voltages are different in the forward and reverse directions, however, no electroluminescence could be observed. This suggests that the electron and hole currents are significantly different and a more efficient electron injection electrode should be sought for the realization of OLEDs based on p6P nanofibers.

12.4 Nanofiber Mechanics

The technological use of nanofibers for electrooptical devices requires well-controlled detachment from the growth substrate and transfer to a desired position. If manipulation tools are used, such as described in [43, 44], this causes mechanical stress that can be potentially harmful to the nanofibers due to their soft and fragile nature. Thus, an understanding of their mechanical properties and their behavior during manipulation can aid in the development of methods for gentle nanofiber handling. Here we show how the mechanical properties of individual nanofibers can be examined through atomic force microscopy-based manipulation and how a route for individual nanofiber transfer to a target substrate can be devised.

In p6P nanofibers the molecules are arranged in a herringbone structure within stacked layers, where the primary intralayer interaction is Coulombic whereas the interaction between the layers is due to van der Waals forces [45]. Given the similarity of the molecular packing in many oligomeric crystals such as the oligophenylenes [46], oligothiophenes [47], and oligoacenes [47], which are all arranged in a herringbone structure in stacked layers, the results shown here for p6P are expected to be representative for the mechanical properties of this type of materials.

12.4.1 2-D Manipulation

The intrinsic mechanical properties of an individual organic nanofiber and its interaction strength with a supporting substrate can be probed with AFM-based techniques [48]. In addition, AFM probes or other microfabricated manipulation tools can be used to controllably slide nanofibers along a surface and thereby enable the assembly of complex structures [43].

The transfer of the nanofibers from the growth substrate is again accomplished by dispersing the nanofibers in liquid and drop-casting them onto the desired substrate [14]. On silicon dioxide the adhesion is strong, which facilitates mechanical experiments with effectively fixed nanofibers. In contrast, using a low-adhesive surface makes it possible to move the nanofibers.

Rupture Properties

A versatile method of investigating the rupture properties of a nanofiber is to use the tip of an AFM as a tool to slice through a nanofiber while measuring the applied force [49].

This is most conveniently done using an AFM equipped with nanomanipulation features such as the Nanomanipulator software and a haptic interface for advanced manipulation and visualization [50]. By imaging the surface in tapping mode, a suitable nanofiber is located. The AFM is then switched to manipulation mode, where the tip performs a horizontal motion in contact mode along a particular path, while the lateral force is recorded. Subsequent image acquisition then shows the result of the manipulation. Figure 12.14 shows a supported nanofiber before and after two manipulation sequences together with the lateral force experienced by the AFM tip during manipulation, shown in Fig. 12.14c.

Both lateral force curves exhibit a significant peak with maximum values of $\simeq$440 nN(I) and $\simeq$510 nN(II), respectively. The onset of the peaks correspond to the position where the tip initially touched the nanofiber. By estimating and subtracting the contribution from the static friction between the nanofiber and the substrate, it is possible to estimate the shear force, which, together with the cross-sectional area of the nanofiber, gives the rupture shear stress. The nanofiber cross-sectional area is found by deconvolution of the topographic data with the estimated tip profile. This results in rupture shear stress values of 1.7×10^7 Pa(I) and 2.0×10^7 Pa(II), respectively. A similar experiment performed on a doubly clamped suspended nanofiber, where the tip probed the free-hanging part and where friction was therefore eliminated, showed a comparable rupture shear stress of 1.5×10^7 Pa (not shown) [48].

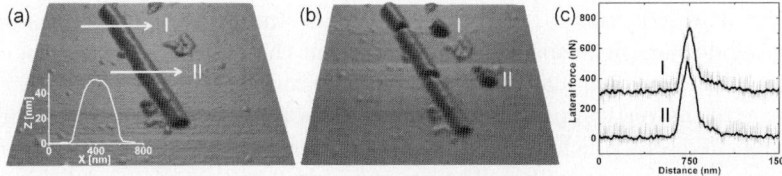

Fig. 12.14. (a) A 4.9 μm × 4.9 μm AFM image of a nanofiber supported by a silicon dioxide surface. The *arrows* indicate the trajectories followed by the tip during the subsequent manipulations. *Inset* shows cross-sectional profile as measured, i.e., it overestimates the nanofiber width due to tip convolution effects. (b) The same nanofiber after two manipulations during which the AFM tip has sliced through the nanofiber. Notice the cut-out pieces I and II. (c) Lateral force measured during manipulation I and II, respectively, with curve I shifted vertically for clarity. The normal force was kept constant at 150 nN during manipulation. Reprinted with permission from [48]

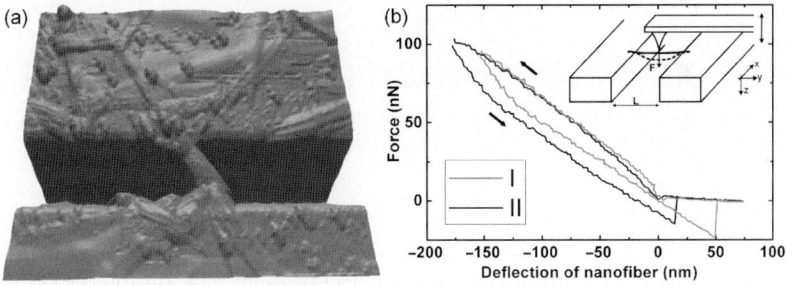

Fig. 12.15. (a) A 14.4 μm × 14.4 μm AFM image of a nanofiber suspended across a 3 μm wide trench. (b) Force–deflection curves for two subsequent pushing and retraction cycles. The *arrows* indicate the direction of the tip motion. *Inset* shows schematic drawing of the experimental procedure where the AFM tip is positioned on the midpoint of the suspended part of the nanofiber and the lowered thereby deflecting the nanofiber. Reprinted with permission from [48]

Elastic Properties

By a similar procedure, the elastic properties were probed by vertically deflecting a suspended nanofiber with the AFM tip. Here the AFM tip is positioned on top of the midpoint of the suspended part of a nanofiber and is then lowered and raised several times. This causes a deflection of the nanofiber, and based on the force vs. deflection data, the Young's modulus can be estimated.

The force-deflection curves in Fig. 12.15 were measured on a suspended nanofiber and show an approximately linear behavior in the region where the tip is in contact with the nanofiber. The onset value, where the force begins to rise, is very similar in both cases, which indicates an almost reversible deformation and therefore only little plastic deformation.

Since the deflection is small compared to the nanofiber height, the appropriate model for the suspended nanofiber is in this case that of a suspended rod that bends rather than stretches in response to a point load. However, in addition to the vertical point load F there could be a built-in tension T_0 in the nanofiber as well, which will alter the force-deflection response compared to that of a stress free nanofiber. The nanofiber response can be modeled by [51]

$$EI\frac{d^3u}{dy^3} - T\frac{du}{dy} = -\frac{F}{2}. \qquad (12.2)$$

E is Young's modulus, I is the plane moment of inertia, u is the deflection, and T is the overall tension, which is the sum of any built-in tension T_0 and the additional tension caused by the deflection. Solving (12.2) and fitting with the measured force–distance data provides an empirical expression for the Young's modulus in terms of the unknown built-in tension T_0

$$E = -1.6 \times 10^{15}\,\mathrm{m}^{-2} \times T_0 + 6.5 \times 10^8\,\mathrm{Pa}. \qquad (12.3)$$

If a negligible built-in tension is assumed, this predicts a Young's modulus of 0.65 GPa. This value represents an upper bound, since any built-in tension would mean that the measured data correspond to a softer material.

Sliding Properties

The significant adhesion and friction experienced by the nanofiber on a silicon dioxide surface renders motion along the surface almost impossible. This necessitates a surface where the friction is much lower, since otherwise the nanofiber will break. AFM-based manipulation experiments, where the tip is used as a tool to push and slide the nanofiber, have shown that on a (tridecafluoro-1,1,2,2-tetrahydrooctyl)trichlorosilane-coated silicon substrate it is indeed possible to translate and rotate micron-sized pieces of a nanofiber without breaking these [48]. Figure 12.16 shows AFM images before and after the tip has been used to move a nanofiber along with the necessary force.

As seen in Fig. 12.16 a nanofiber can be moved along a surface, given that the friction force with the substrate is low. In microscopic systems, friction scales with contact area [52]. In the case of nanofibers, measurements on different fibers indicate that the friction within measurement error is not dependent on interface area, which is more similar to the properties of macroscopic systems.

Also to be noted is the pollution of the surface, which is assumed to be caused by organic remnants from the drop-casting application of the nanofibers. The nanofiber wipes away these remains and leaves behind a smoother surface with corrugations of just a few nanometers – almost an order of magnitude smaller than before. This indicates an intimate contact between the nanofiber and the surface.

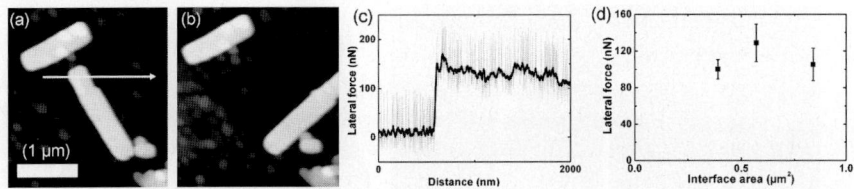

Fig. 12.16. (a) A 3.2 μm × 3.5 μm AFM image of a nanofiber on a (tridecafluoro-1,1,2,2-tetrahydrooctyl)trichlorosilane-coated silicon substrate. The *arrow* indicates the trajectory which the AFM tip followed during the subsequent manipulation. (b) Same region after manipulation. Note how the nanofiber rotated in response. (c) Lateral force experienced by the AFM tip during manipulation. (d) Lateral forces values obtained from several manipulation sequences on different nanofibers. Reprinted with permission from [48]

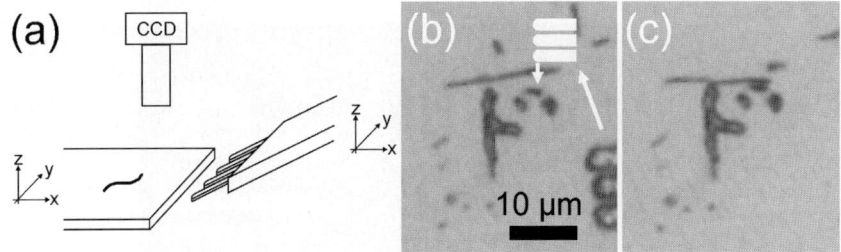

Fig. 12.17. Construction of an artificial "T" from p6P nanofibers. (a) Sketch of the experimental setup. To the *left* is the sample holder mounted on a moveable stage. To the *right* the manipulation tool is also mounted on a XYZ-stage driven by piezoelectric transducers. The manipulation is observed through an objective lens with a long working distance. (b) Microscope image from the manipulation setup showing the construction of the "T." Note the array of SiO_2 "fingers", which constitute the manipulation tool. The tool is moved along the directions indicated by the *arrows* and is used to position the upper horizontal bar of the "T." (c) The finished "T." Reprinted with permission from [43]

Assembly of Complex Structures

While the AFM has the advantage of providing quantitative data of the exerted force, it is less practical in the assembly of more complicated structures. This is due to the fact that it is a rather time-consuming method which does not allow observation during manipulation but only before and after. Also the very local, point-like nature of the applied force makes it difficult to control whether the nanofiber rotates or translates. These problems can be overcome by using a micromanipulation system where a manipulation tool is operated through a piezoelectric XYZ-stage, while at the same time the manipulation area can be observed in an optical microscope.

Figure 12.17 shows two microscope images acquired during the construction of the letter "T" from two nanofibers. On the right-hand side of (b)

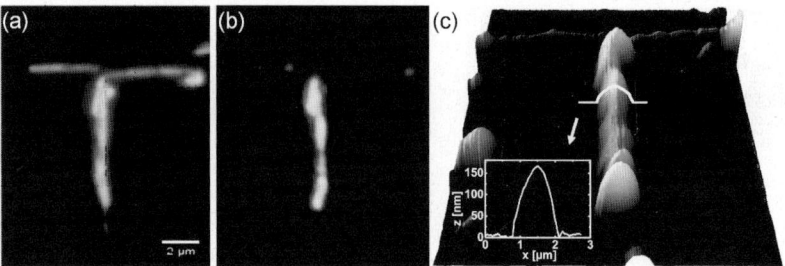

Fig. 12.18. An artificial "T" constructed from p6P nanofibers. (**a**) Fluorescence microscope image excited with light of 365 nm wave length and observed unpolarized. (**b**) Similar to (**a**) but observed with horizontal polarization (**c**) AFM image of the "T." *Inset* shows cross-sectional profile along the *indicated line*. Reprinted with permission from [43]

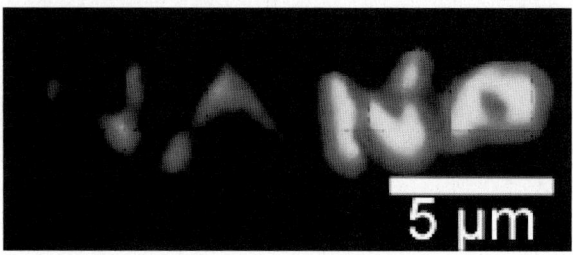

Fig. 12.19. The word "NANO" constructed from p6P nanofibers. Reprinted with permission from [43]

the manipulation tool is seen, which is a microfabricated array of SiO_2 cantilevers.

The fluorescence image in Fig. 12.18 clearly shows that the "T" consists of several micrometer long pieces of nanofiber. By observing only the light with horizontal polarization as shown in Fig. 12.18b it can be inferred that the manipulation is not causing severe damage to the internal structure of the nanofiber, i.e., the polarization-dependence of the fluorescence is retained. The width of the legs of the "T" in Fig. 12.18 is found by AFM to be of the order of some hundred nanometers.

More complex patterns can also be constructed. As an example, Fig. 12.19 shows the word "NANO" built from several nanofibers. This method provides a way to construct prototype devices based on soft organic nanofibers. For large scale fabrication a more efficient method along the lines of fluidic assembly must, however, be sought.

12.4.2 3-D Manipulation

More complicated structures based on individual nanofibers could become feasible through 3-dimensional manipulation, where an individual nanofiber

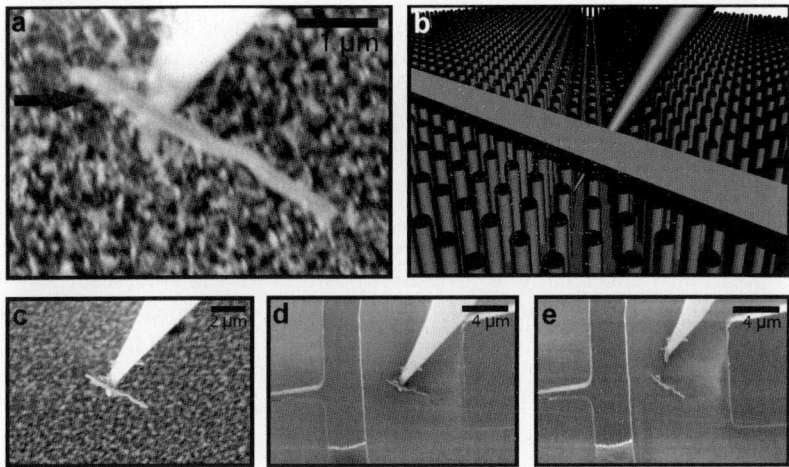

Fig. 12.20. 3-D manipulation of a p6P nanofiber. (**a**) SEM image showing a p6P nanofiber (*indicated with arrow*) resting on top of a nanotube forest. A tungsten tip is pushed into the nanotube forest beneath the p6P nanofiber. (**b**) Conceptual illustration of the method whereby a tip is lowered beneath the nanofiber, which is supported by small pillars, to facilitate easy pickup. (**c**) By pulling the tungsten tip up the nanofiber is released from the substrate and adheres to the tip. (**d**) The nanofiber is then transferred to its target location, where it is released (**e**). Reprinted with permission from [44]

could be picked up and placed elsewhere. This could enable the integration of nanofibers in prototype microsystems. A main obstacle against 3-dimensional manipulation is the difficulty in lifting up the nanofiber, i.e., overcoming the adhesion with the surface without breaking the nanofiber. One way of overcoming this problem is to use a special, low-adhesive substrate as a "manipulation workbench." An example of such a substrate is a "nanotube forest" as shown schematically in Fig. 12.20b. It consists of a large array of vertically aligned carbon nanotubes that act as pillars supporting the nanofiber, which is lying on top. Since the nanofiber is only supported in relatively few, discrete points and since the nanotube forest is mechanically compliant, it is possible for a manipulation tool to penetrate into the forest, beneath the nanofiber, and gently lift it up vertically without breaking it [44].

The carbon nanotube forest is fabricated through plasma-enhanced chemical vapor deposition from a carbon feed-stock gas on an oxidized silicon substrate coated with a suitable catalyst material [44]. The electric field generating the plasma during fabrication causes the nanotubes to align vertically as needed. Following the fabrication of the nanotube forest, the p6P nanofibers are dispersed here through drop-casting of a liquid dispersion. From this substrate it is now possible to pick up an individual nanofiber.

The manipulation tool is an electrochemically etched tungsten tip with a tip radius-of-curvature of a few tens of nanometers. This tool is mounted on a precision XYZ stage driven by piezoelectric actuators capable of motion in very fine steps. One option is to carry out the manipulation inside the vacuum chamber of an SEM. This makes the observation during manipulation rather easy due to the high resolution of the SEM compared to an optical microscope. Figure 12.20 shows a series of SEM images where the tungsten tip is used to lift up a nanofiber, transfer it to a desired position, and releasing it.

Using this method together with the shadow mask technique described earlier, it is possible to electrically contact individual nanofibers that have been positioned specifically at a desired position in a microsystem [44]. One option is to perform the manipulation in ambient conditions using an optical microscope for observation which despite the limited resolution still allows sufficient view to manipulate individual nanofibers. The in situ growth methods used for carbon nanotubes [53] and inorganic nanowires [54] are not applicable here due to the different assembly mechanism of organic crystals.

12.5 Conclusions

In this chapter measurements on optical, electrical, and mechanical properties of individual organic nanofibers have been discussed. Quantitative measurements on nanofibers embedded in liquids have revealed highly anisotropic angular intensity distributions of emitted light both along the angles ϕ perpendicular to the long nanofiber axes and Θ along it. When rotated around their long axes it was found that the intensity of emitted light changed with a $\cos^2(\phi)$ function, which is consistent with the quasicrystalline nature of the fibers, i.e., the fact that they are made of rows of well-aligned dipole emitters. In addition to that waveguiding along the longitudinal axis seems to dictate the emission properties.

The anisotropy of light emission together with the well-polarized nature of the light makes these nanofibers especially useful for further applications as light sources in submicron-scaled photonic devices.

The electrical investigations on individual nanofiber devices with well-defined submicron gaps first led to a possible new route for fabricating nanofiber devices with different cathode and anode materials. Quantitatively the investigations have shown current–voltage characteristics that indicate that Au has significantly better charge injection than Al. Therefore a cathode material with lower work function is needed to make balanced electron and hole currents in a future nanofiber-based OLED.

The mechanical properties have been investigated by atomic force microscopy-based manipulation. The nanofibers exhibit a rupture shear stress of about 2×10^7 Pa and a Young's modulus of 0.65 GPa. Despite their soft and fragile nature, pieces of individual nanofibers can be translated along a surface and assembled into complex structures using microfabricated tools. Using a special manipulation substrate, such as a carbon nanotube forest,

3-dimensional manipulation becomes possible which can provide accurate positioning on any future target substrate.

Acknowledgment

H.-G. Rubahn is indebted to his collaborator F. Balzer. He thanks the Danish research foundations FNU and FTP. J.K.-H. and P.B. thank their collaborators K. Gjerde, D. Engstrøm and C.H. Clausen.

References

1. M. Schiek, A. Lützen, R. Koch, K. Al-Shamery, F. Balzer, R. Frese, H.G. Rubahn, Appl. Phys. Lett. **86**, 153107 (2005)
2. M. Schiek, A. Lützen, K. Al-Shamery, F. Balzer, H.G. Rubahn, Cryst. Growth Design **7**(2), 229 (2007)
3. F. Balzer, H.G. Rubahn, Appl. Phys. Lett. **79**, 3860 (2001)
4. F. Balzer, H.G. Rubahn, Adv. Funct. Mater. **15**, 17 (2005)
5. H. Yanagi, T. Ohara, T. Morikawa, Adv. Mat. **13**, 1452 (2001)
6. F. Balzer, V. Bordo, A. Simonsen, H.G. Rubahn, Appl. Phys. Lett. **82**, 10 (2003)
7. F. Balzer, V. Bordo, A. Simonsen, H.G. Rubahn, Phys. Rev. B **67**, 115408 (2003)
8. F. Quochi, F. Cordella, A. Mura, G. Bongiovanni, F. Balzer, H.G. Rubahn, Appl. Phys. Lett. **88**, 041106 (2006)
9. F. Balzer, K. Al Shamery, R. Neuendorf, H.G. Rubahn, Chem. Phys. Lett. **368**, 307 (2003)
10. J. Brewer, M. Schiek, A. Lützen, K. Al-Shamery, H.G. Rubahn, Nano Lett. **6**, 2656 (2006)
11. L. Marty, V. Bouchiat, A. Bonnot, M. Chaumont, T. Fournier, S. Decossas, S. Roche, Microelectron. Eng. **61–62**, 485 (2002)
12. A. Briseno, J. Aizenberg, Y.J. Han, R. Penkala, H. Moon, A. Lovinger, C. Kloc, Z. Bao, J. Am. Chem. Soc. **127**, 12164 (2005)
13. F. Balzer, H.G. Rubahn. Patent PA 2004 01676
14. J. Brewer, C. Maibohm, L. Jozefowski, L. Bagatolli, H.G. Rubahn, Nanotechnology **16**, 2396 (2005)
15. A. Tsumura, H. Koezuka, T. Ando, Appl. Phys. Lett. **49**, 1210 (1986)
16. S. Nelson, Y.Y. Lin, D. Gundlach, T. Jackson, Appl. Phys. Lett. **72**, 1854 (1998)
17. V. Podzorov, V. Pudalov, M. Gershenson, Appl. Phys. Lett. **82**, 1739 (2003)
18. V. Sundar, J. Zaumseil, V. Podzorov, E. Menard, R. Willett, T. Someya, M. Gershenson, J. Rogers, Science **303**, 1644 (2004)
19. R. de Boer, M. Gershenson, A. Morpurgo, V. Podzorov, Phys. Stat. Sol. (A) **201**, 1302 (2004)
20. C. Tang, S. VanSlyke, Appl. Phys. Lett. **51**, 913 (1987)
21. M. Era, T. Tsutsui, S. Saito, Appl. Phys. Lett. **67**, 2436 (1995)
22. F. Boroumand, P. Fry, D. Lidzey, Nano Lett. **5**, 67 (2005)
23. H. Yamamoto, J. Wilkinson, J. Long, K. Bussman, J. Christodoulides, Z. Kafafi, Nano Lett. **5**, 2485 (2005)

24. H. Ishii, K. Sugiyama, E. Ito, K. Seki, Adv. Mater. **11**, 605 (1999)
25. M. Lambert, P. Mark, *Current Injection in Solids* (Academic, New York, 1970)
26. I. Parker, J. Appl. Phys. **75**, 1656 (1994)
27. P. Davids, I. Campbell, D. Smith, J. Appl. Phys. **82**, 6319 (1997)
28. V. Arkhipov, E. Emelianova, Y. Tak, H. Bässler, J. Appl. Phys. **84**, 848 (1998)
29. U. Wolf, V. Arkhipov, H. Bässler, Phys. Rev. B **59**, 7507 (1999)
30. Y. Shen, A. Hosseini, M. Wong, G. Malliaras, Chem. Phys. Chem. **5**, 16 (2004)
31. A. Hepp, H. Heil, W. Weise, M. Ahles, R. Schmechel, H. von Seggern, Phys. Rev. Lett. **91**, 157406 (2003)
32. M. Pfeiffer, S. Forrest, in *Nanoelectronics and Information*, ed. by R. Wasser (Wiley-VCH, Berlin, 2005)
33. J. Kjelstrup-Hansen, H. Henrichsen, P. Bøgild, H.G. Rubahn, Thin Solid Films **515**, 827 (2006)
34. J. Kjelstrup-Hansen, S. Dohn, D. Nørgaard Madsen, K. Mølhave, P. Bøggild, J. Nanosci. Nanotechnol. **6**, 1995 (2006)
35. R. de Boer, M. Jochemsen, T. Klapwijk, A. Morpurgo, J. Niemax, A. Tripathi, J. Pflaum, J. Appl. Phys. **95**, 1196 (2004)
36. J. Geurst, phys. stat. sol. **15**, 107 (1966)
37. D. Gundlach, Y.Y. Lin, T. Jackson, D. Schlom, Appl. Phys. Lett. **71**, 3853 (1997)
38. H. Henrichsen, J. Kjelstrup-Hansen, D. Engstrøm, C. Clausen, P. Bøggild, H.G. Rubahn, Org. Electron. **8**, 540 (2007)
39. N. Koch, E. Zojer, A. Rajagopal, J. Ghjisen, R. Johnson, G. Leising, J.J. Pireaux, Adv. Funct. Mat. **11**, 51 (2001)
40. H. Michaelson, J. Appl. Phys. **48**, 4729 (1977)
41. F. Meghdadi, S. Tasch, B. Winkler, W. Fischer, F. Stelzer, G. Leising, Synth. Met. **85**, 1441 (1997)
42. N. Koch, A. Pogantsch, E. List, G. Leising, R. Blyth, M. Ramsey, F. Netzer, Appl. Phys. Lett. **74**, 2909 (1999)
43. J. Kjelstrup-Hansen, P. Bøggild, J. Hvam, A. Majcher, H.G. Rubahn, Phys. Stat. Sol. (a) **203**, 1459 (2006)
44. K. Gjerde, J. Kjelstrup-Hansen, C. Clausen, K. Teo, W. Milne, H.G. Rubahn, P. Bøggild, Nanotechnology **17**, 4917 (2006)
45. R. Resel, Thin Solid Films **433**, 1 (2003)
46. P. Puschnig, K. Hummer, C. Ambrosch-Draxl, G. Heimel, M. Oehzelt, R. Resel, Phys. Rev. B **67**, 235321 (2003)
47. J. Brédas, D. Beljonne, J. Cornil, J. Calbert, Z. Shuai, R. Silbey, Synth. Met. **125**, 107 (2001)
48. J. Kjelstrup-Hansen, O. Hansen, H.G. Rubahn, P. Bøggild, Small **2**, 660 (2006)
49. M. Guthold, W. Liu, B. Stephens, S. Lord, R. Hantgan, D. Erie, R. Taylor, Jr., R. Superfine, Biophys. J. **87**, 4226 (2004)
50. R. Taylor II, J. Chen, S. Okimoto, N. Llopis-Artime, V. Chi, F. Brooks, M. Falvo, M. Paulson, P. Thiansathaporn, D. Glick, S. Washburn, R. Superfine, Proc. IEEE Vis. Conf. (1997)
51. L. Landau, E. Lifshitz, *Theory of Elasticity, Course of Theoretical Physics*, vol. 7, 3rd edn. (Butterworth-Heinemann, Oxford, 1986)
52. P. Sheehan, C. Lieber, Science **272**, 1158 (1996)
53. N. Franklin, Y. Li, R. Chen, A. Javey, H. Dai, Appl. Phys. Lett. **79**, 4571 (2001)
54. R. He, D. Gao, R. Fan, A. Hochbaum, C. Carraro, R. Maboudian, P. Yang, Adv. Mater. **17**, 2098 (2005)

13

Device Treatment of Organic Nanofibers: Embedding, Detaching, and Cutting

H. Sturm and H.-G. Rubahn

13.1 Introduction

It has been demonstrated thoroughly that organic molecules such as phenylenes can form long, quasisingle crystalline aggregates ("nanofibers" or "nanoneedles") of parallel to the surface-oriented molecules on cleaved muscovite mica upon vapor deposition (organic molecular beam epitaxy, OMBE). Phenylenes of that kind are rodlike molecules of usually between four and six benzene rings, which emit polarized blue light after UV excitation below 400 nm. Via functionalization with, e.g., methyl oxide or chlorine end groups the emission spectra can be significantly modified. The surface grown nanofibers are all mutually parallel oriented because of strong electric dipole fields on the mica surface and a quasi-heteroepitaxial relationship between adsorbate and substrate [1,2]. That way domains of parallel aggregates up to square centimeter size are grown.

Whereas individual aggregates can reach lengths of several hundred micrometers, their widths and heights are a few ten to a few hundred nanometers as measured by atomic force microscopy (AFM). Widths, heights, lengths as well as the number density of the aggregates on the surface can be controlled nearly independently of each other by choosing appropriate process parameters during deposition in vacuum.

Organic nanofibers are of special interest for the generation of single- or multifiber optoelectronic devices such as evanescent wave biosensors or organic nanolasers since waveguiding along *para*-hexaphenylene-based fibers [3] as well as lasing [4] have been demonstrated recently. However, the application potential of nanofibers for micro- and nanooptical devices relies heavily on ways to manipulate them after the original growth process. This chapter discusses some of the progress obtained in this respect with special emphasis on the following issues:

1. Coating of the mica-grown nanofibers with inorganic or polymeric films.
2. Transfer of detached nanofibers to polymer surfaces.

3. Procedures to cut or manipulate otherwise the length of the nanofibers. Especially the use of the fibers as interconnecting optical elements asks for light transfer through the end faces of the fibers. Assembling an organic nanolaser from a single fiber requires well-defined end faces, too, which may eventually be covered with a thin metallic or dielectric layer to build a resonator structure.

There are also good reasons for coating experiments. First of all, bleaching of the fibers chromophores is a serious drawback in their optical performance. Actually most of the fibers properties depend on their chemical integrity. Chromophores, i.e., the group of atoms of a molecular entity in which the electronic transition for a given spectral absorption is localized, undergo bleaching due to a photon-induced chemical reaction. Absorbing a photon, the chromophore or fluorophore transfers the energy from the ground state to an excited singlet state. Depending on the absorbing functional group, a rather long living, metastable excited triplet state can be formed. Therefore, this high-energetic molecular arrangement has time to undergo chemical reactions. Since the triplet states are of radical nature possessing unpaired electrons, all other free radicals in the surroundings, e.g., oxygen, react with some probability and lead to a new covalent bond which modifies the molecule. For a single crystal in a solid-state reaction, single damaged chromophores may just change the absorption spectra or the luminescence intensity. However even the mechanical cohesion of the fibers is at risk if the bleaching process continues and more and more atoms are rearranged and covalently bound to the incoming reaction partners. Hence, coating the nanofibers with films either to reduce the diffusion of reactants or just to change the local chemical environment of the chromophores should increase the durability of the nanofibers and improve their application potential.

The transfer of the organic nanofibers to polymer surfaces or the entire embedding into polymers is an important step to achieve more flexibility in using the organic nanofibers. Polymers are substrates or hosts for molecules which can be chosen so as to serve as a chemically inert or optically indifferent environment as well as a partner offering its own chromophores for optical interaction. Using electrically conductive polymers, even polymer electrodes become conceivable. Furthermore, some polymers can be highly elastic and flexible, others are rigid and stiff or are able to keep a bending deformation due to plasticity, stressing embedded nanofibers mechanically. Due to this high flexibility in performance, polymers represent a smart material for embedding of organic nanofibers.

Serving as embedding material, polymers can be used as a support for handling and positioning. Polymer embeddings can protect the nanofibers from contamination and protect the user or the environment against the molecules used to form the nanofibers. Nanofiber safety might become an important issue even if the number of nanofibers in an application is limited and even if the dose rate is small due to the low solubility of the molecules in aqueous media.

With an increasing number of types of tailor-made molecular building blocks the probability of an enhanced solubility increases, because the tuning of the optical properties is performed via adding end groups. The maximum allowable concentrations of nanofibers itself, e.g., at a workplace, and regardless of their chemical and physical nature, are still a matter of discussion and a subject of research, more than ever if the underlying molecules belong to the class of condensed aromatic rings. Photobleaching of device-integrated nanofibers, e.g., might result in new chemical species containing oxygen and with a low molecular weight, i.e., high diffusivity. Polymers containing the appropriate reaction partners open the chance to trap fled fragments of the nanofibers via covalent binding or dimerization of π-systems. The impact that nanohazard might have is to date totally unclear but certainly asks for increasing attention.

13.2 Coating of Organic Nanofibers on Mica

13.2.1 Parameters Related to the Embedding of Organic Nanofibers: Thermal Conductivity and Thermal Expansion

In general the growth substrate muscovite mica is assumed to limit the range of applicability of nanofibers in devices. However, mica as a substrate of the nanofibers installed in a device can also have advantages. In the case of any photonic excitation of a nanofiber deposited on mica, a part of the energy absorbed by the substrate or as a result of dissipation mechanisms in the fiber itself is converted to heat. To increase the lifetime of the optically active nanofiber and to minimize the influence of the temperature dependence of optical and electrical fiber properties, a good heat sink for the fibers is essential. The bulk value of *thermal conductivity* of mica is reported to be 0.24–$2.23\,\mathrm{Wm^{-1}K^{-1}}$ [5]. For comparison, this value ranges from 0.04 to $0.35\,\mathrm{Wm^{-1}K^{-1}}$ for polymers [5]. However, mica single crystals, namely muscovite and biotite, show a strong anisotropy due to the fact that they are sheet silicates: the in-plane thermal conductivity is found to be 3.14–$5.10\,\mathrm{Wm^{-1}K^{-1}}$ [6], whereas perpendicular to the sheets a value of 0.52–$0.84\,\mathrm{Wm^{-1}K^{-1}}$ [6] is reported.

We can conclude, that the thermal conductivity of mica in the basal plane is about four to ten times higher than that for polymers and that the basal plane of an external heat sink would be five times more efficient compared to a heat sink perpendicular to this plane assuming the same geometry. A similar observation can be made for the *thermal expansion* coefficients. The thermal expansion coefficients of muscovite mica for the three crystal axes varies from 9.9 to $13.8 \times 10^{-6}\,\mathrm{K^{-1}}$ [7]. From literature experimental data on thermal expansion coefficients of any substance used so far to form the organic nanofibers, for example p-hexaphenylene (P6P), were not found [8].

Unfortunately, a mismatch in thermal expansion could lead to a debonding between nanofiber and substrate or, depending on the rupture stress of the fibers, to the deformation fibers. The latter is more reasonable due to the fact, that low moduli of organic crystals were reported [9], p-hexaphenylene is declared to show a compression modulus of 100 kbar, i.e., 10.1 MPa. Their softness is due to the weakness of the intermolecular interactions in these organic crystals compared with ionic, covalent, or metallic crystals.

The organic nanofibers we are dealing with may have lengths up to the micrometer range. Let us simplify the geometry of a fiber on a flat surface to the situation of two beams like in a bimorph, both clamped at one side and free to move over their whole length. In the case of linear thermal expansion the change in length of a crystal piece heated up from a temperature T_1 to T_2 can be calculated as

$$\Delta l = l(T_2) - l(T_1) = \alpha\, l_0 \times (T_2 - T_1) \ . \tag{13.1}$$

Assuming a temperature change of 100 K and a length l_0 of 100 nm, we find for muscovite ($\alpha \approx 10 \times 10^{-6}\,1\,\mathrm{K}^{-1}$) a Δl of 0.100 nm. In contrast, silicon dioxide SiO$_2$ will give a Δl of 0.243 nm. Even if this is more than twice as much as for muscovite, the absolute value is small enough to be neglected on the considered length scale. Unfortunately no values of α are reported for p-hexaphenylene and the known values for organic crystals vary considerably. For example, Heimel et al. [9] report for 2,5-diphenyl-1,3,4-oxadiazine (DPO) a value of $\alpha \approx 190 \times 10^{-6}\,1\,\mathrm{K}^{-1}$ and for benzophenone a value of $\alpha \approx 0.2 \times 10^{-6}\,1\,\mathrm{K}^{-1}$. Whereas benzophenones expansion can be ignored ($\Delta l \approx 2\,\mathrm{pm}$), DPO gives a Δl of 1.9 nm, which is nearly 2% of the initial length l_0. The organic nanofibers of p-hexaphenylene exhibit a structure of stacked planes similar to graphite but under some tilt toward the mica, so it can be expected, that the fibers are able to undergo a plastic deformation before they break; see also Fig. 13.9c. The cooling process after the growth on mica using substrate temperatures more than 100 K over room temperature could lead to either a compression or a stretching of the fibers, depending on their expansion coefficients. Since cracks in the fibers are observed after rapid cooling in vacuum and cannot be seen after heating of the fibers we conclude that the thermal expansion coefficient must be more similar to benzophenone as compared to DPO, i.e., small.

As a consequence, some fine tuning has to be performed before organic nanofibers go to application, at least if they are used for light emission purposes. Fine tuning includes the management of the thermal mismatch as well as the development of a harvesting procedure at elevated temperatures to receive thermally relaxed, intact nanofibers. Supposedly the use of polymers, at least noncrystalline polymers, as embedding material should reduce the mechanical stress at the polymer/fiber interface.

The next chapter covers experiments using inorganic materials for coating organic nanofibers on mica. Results concerning polymer coatings will be discussed in Sect. 13.3.2.

13.2.2 Evaporation of Silicon Oxide

Silicon monoxide, SiO (Balzers, Germany), has been evaporated using electron beam evaporation (Auto 306 Turbo, Edwards High Vacuum Intern). The film thickness has been measured using a quartz microbalance (FTM 7, Edwards) placed close to the sample without any correction by a tooling factor or any further calibration attempts.

Silicon monoxide has relative low-vapor pressure [10,11] which allows gentle evaporation due to a low thermal stress for the sample. Silicon monoxide as a solid is metastable and decomposes partially at about 400°C to Silicon (Si) and Silicon dioxide (SiO_2). This is a reversible reaction. At 1,250–1,400°C and between 1×10^{-3} and 1×10^{-4} mbar SiO is industrially synthesized via the back reaction. Gaseous SiO is stable at temperatures above 1,000°C (Table 13.1) [12]. The composition and structure of silicon monoxide depends on various growth conditions and is still a matter of research; three different models are discussed. The random bonding model [13,14] assumes a statistic distribution of silicon–silicon and silicon–oxygen bonds in an unique and amorphous phase. The random mixture model [15,16] assumes the existence of two phases consisting of a-Si (*amorphous* silicon) and a-SiO_2. Newer investigations led to the interface clusters model [17] with clusters of a-Si and a-SiO_2 smaller than 2 nm, separated by a thin layer of silicon-rich suboxides. In our case, the structure and composition is not that clear. The disproportionation of SiO starts at 400°C. It is obvious that the temperature at the sample surface did not exceed the temperature for thermal degradation of the organic nanofibers, because the fibers are still intact. However, SiO is still a metastable solid with a large ability for morphological instability as mentioned above, and the influence of a substrate partially covered with organic nanofibers is yet not known. So for the sake of clarity, the coatings used in this work are referred to as "SiO_x" and not as "SiO".

Two different SiO_x coatings were produced on nanofibers on mica with nominal thicknesses of 160 and 254 nm, respectively. The 160 nm coating was produced with the best vacuum achievable in the chamber ($\approx 1 \times 10^{-5}$ mbar), the 254 nm coating was made at $\approx 1 \times 10^{-3}$ mbar to increase the probability of a higher oxygen content in the film. The maximum evaporation rate was ≈ 3 nm s^{-1} in each case. Additionally, a 300 nm LiF coating was made using

Table 13.1. Vapor pressure of silicon monoxide (SiO) with temperature (after [12] and references cited therein)

Temperature (°C)	Vapor pressure (mbar)
1,000	1×10^{-6}
1,060	1×10^{-5}
1,130	1×10^{-4}
1,220	1×10^{-3}

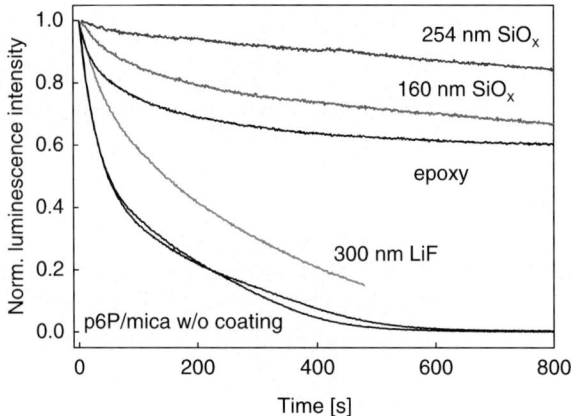

Fig. 13.1. Measured luminescence as a function of illumination time for differently coated P6P nanofiber films on mica, showing different efficiencies for the protection against oxygen driven photo bleaching. Reprinted with permission from [21]

$\approx 1 \times 10^{-5}$ mbar and $\approx 3\,\mathrm{nm\,s^{-1}}$, respectively. The epoxy coating was made of SU-8 photoresist using a spin coater with undefined thickness.

13.2.3 Antibleaching Effect with SiO$_x$ Coatings

Bleaching experiments were performed with P6P nanofibers on mica, coated with SiO$_x$, epoxy, and LiF with different thicknesses (see Fig. 13.1).

In the following only the coating with SiO$_x$ will be discussed, because the LiF coating turned out not to be efficient. The epoxy coating was not further examined, because the coating mechanism is expected to be more complicated due to fact that the negative-resist SU-8 changes, e.g., optical absorption, chemical composition, and density during exposure.

The most efficient coating was the coating with 254 nm SiO$_x$, made at $\approx 1 \times 10^{-3}$ mbar. Up to now there are no systematic investigations available whether or not the chemical composition of such a coating varies from a coating produced at $\approx 1 \times 10^{-5}$ mbar, so the thickness of the SiO$_x$ films should be regarded as the solely varied property.

13.2.4 Microscopical Analysis of Nanofibers on Mica, Covered by SiO$_x$

SiO$_x$-coated p-hexaphenylene nanofibers were exposed to ultraviolet (UV) light. Figure 13.2 presents the inspection of a spot illuminated with UV light (experimental conditions see Chap. 3). Figure 13.2a shows that p-hexaphenylene nanofibers are still in good order within the spot concerning their geometrical integrity, whereas their fluorescence emission is modified.

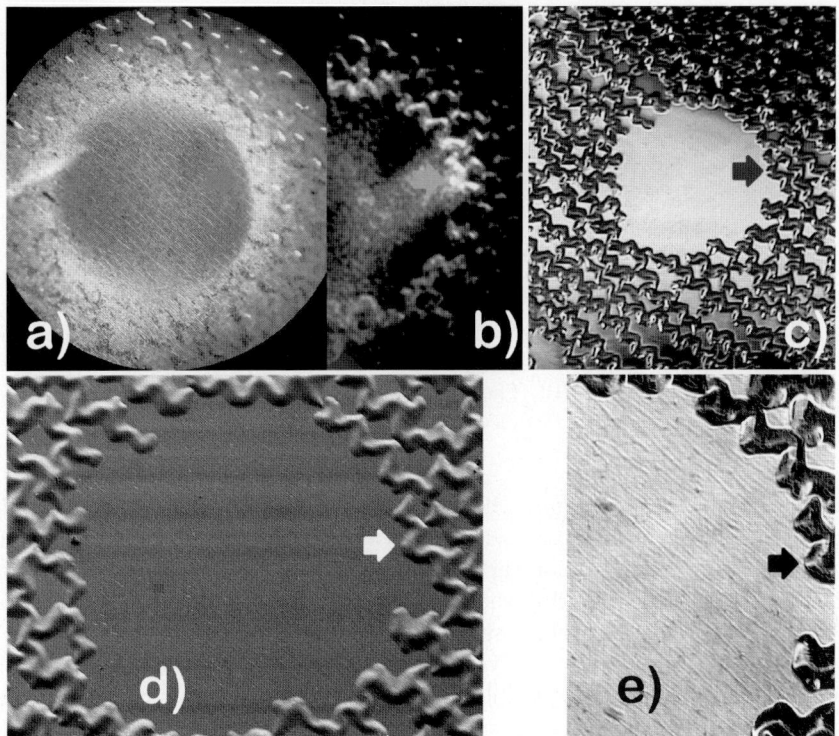

Fig. 13.2. P-hexaphenylene nanofibers, coated with 254 nm of SiO_x. In the center of image (a), the fibers were exposed after coating to UV light. Outside the exposed zone, a partial delamination of the thin SiO_x film did take place after the exposure. The morphology of the "blisters" is known as "telephone cord buckling." The general direction of the "telephone cord" lines is parallel to the direction of the nanofibers. (a) Fluorescence emission image showing the change in color and intensity of emission inside the previously illuminated spot. (b) *Red* part of (a) facilitating the identification for the alignment. (c) Light microscopy image. (d) ESEM image, GSE contrast, 0.5 mbar water pressure, 1.7 kV. (e) Light microscopy image, higher magnification than (c)

The rippled surface outside the illumination spot can be understood by examination of the light microscopy (Fig. 13.2c) and the electron microscopy images (Fig. 13.2d). Outside the illumination spot, a so-called "telephone cord buckling" occurred, recently described in a review about the physics of adhesion by Gerberich and Cordill [18]. Telephone cord buckles are frequently observed on thin, compressively stressed films presumably in order to balance the driving force for delamination and the interfacial fracture energy.

Unfortunately the mica substrate, partially covered with nanofibers and presumably covered with an ultrathin organic film in-between the fibers, is a rather complex substrate for the SiO_x film. In general, the origin of the

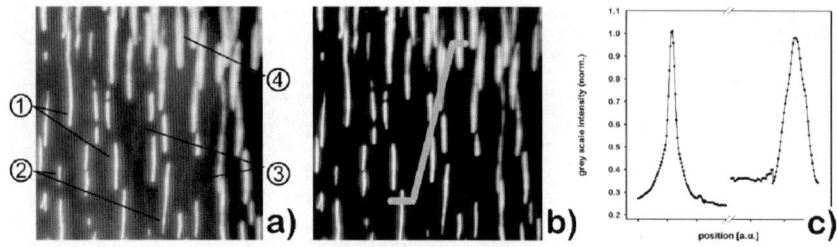

Fig. 13.3. *P*-hexaphenylene nanofibers, coated with ≈254 nm of SiO$_x$, at high magnification. (**a**) Fluorescence emission image with markers indicating moderately bleached fibers (1, 2), totally erased fibers (3) and unbleached fibers (4). (**b**) Same image converted to *gray* scale. The *green line* indicates the positions for the cross-sectional analysis given in (**c**). The analysis is restricted to the line parts between the white marked points in (**b**) (see broken axis in (**c**))

stress between substrate and the SiO$_x$ overlayer can be assumed to be the mismatch of the thermal expansion coefficients. However, that telephone cord bucklings only occur outside the illuminated zone, i.e., can be suppressed by the processes involved during the photobleaching, remains an interesting fact worth the effort of further investigations.

Figure 13.3 shows an analysis of *p*-hexaphenylene nanofibers, one unbleached and one moderately bleached. The aim is to find two fibers which show the same intensity maximum to compare the lateral distribution of the intensity. In doing so it is considered that the width of the two lines is comparable. Figure 13.3c shows that at least at the base the two chosen profiles have nearly the same width. The most interesting next step would be a spectral analysis of each position of the image. Absence of calibration hinders a detailed analysis, however, a simple color separation of the digital image may also turn out to be sufficient.

A spectral analysis via color separation shows some interesting differences between the unbleached and the bleached fibers (Fig. 13.4). For the unbleached fiber, as expected, the integrated intensities for magenta and blue are higher due to the larger widths. The bleached fibers show for all colors except for red and yellow a loss in intensity outside of the center position, nicely observable in the profile for cyan, where the intensity outside the center is even higher for the bleached fiber than for the unbleached one. Also the blowups in (d) depict this, compare, e.g., label 3 with label 4 and label 5 with label 6. The analysis also shows the unexpected result that the heavily bleached fibers (see Fig. 13.4d, blow-up labels 1, 3, 5, and 7) emit much magenta, some blue, no cyan but then again some green light. Let us keep in mind that photobleaching starts a chemical process leading to smaller molecules and that the fibers are confined between two films which can hardly be penetrated by them. This means, that the products of the bleaching process are still present and can undergo an independent absorption and emission process.

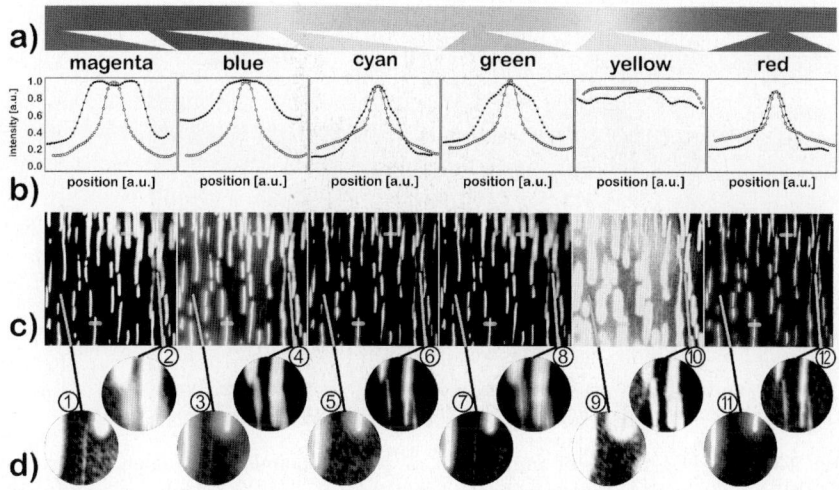

Fig. 13.4. Primitive analysis of the fluorescence emission image for two different *p*-hexaphenylene nanofibers, one unbleached, one moderately bleached. (**a**) colors, accessible using the color separation options to RGB (*red, green, blue*) and CMYK (*cyan, magenta, yellow, black*) of an image analysis program. The order of the colors from *left* to *right* follows the decrease in energy. (**b**) Intensity cross-sections at each separated color for the unbleached (*closed circles*) and bleached (*open circles*) nanofibers. The chosen fiber position is indicated in (**c**). (**c**) *Gray* scale images after color separation, the brightness indicating the intensity of the particular color. The *orange bars* correspond in length and width to the image pixels used for the calculated profiles under (**b**), the unbleached fiber is the one in the upper right corner. (**d**) Blow-ups of the images given in (**c**), magnification about twofold, contrast enhanced. In the text, so-called "strongly bleached fibers" are also mentioned. The best visible one can be found in (**d**) label 1

A reasonable conclusion is, that the *moderately bleached fibers* degrade easier from outside to inside in the sense that the emission of the light decreases stronger at the borders. Since we observe only a two-dimensional emission pattern similar to a projection instead of the three-dimensional distribution of emission, we should also see an intensity decrease in the center of the fibers. This is not the case. Outside of the fiber center, light is emitted with lower energy compared to the nonbleached fiber. The existence of new chemical species with a red-shifted fluorescence emission could explain this phenomenon, presumably produced by the chemical degradation in the course of the photobleaching process. The *strongly bleached fibers* (not further marked in the images, but visible as fine lines between the moderately bleached fibers), assumed to have approximately the same average width and height, show a different result. They are all very narrow and emit very little light only at magenta and green, but not in-between at cyan (see Fig. 13.4, label 5), very dim at yellow (label 9), and not at all at red (label 11). This might be explained

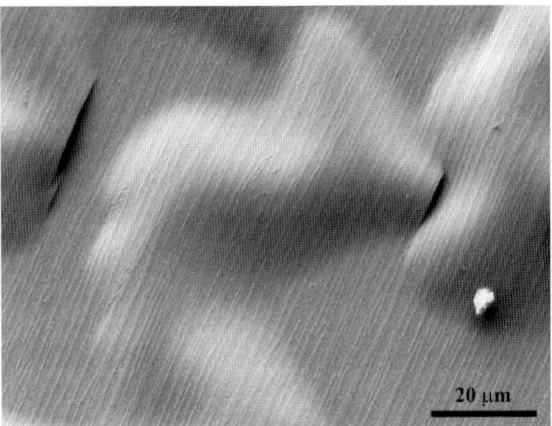

Fig. 13.5. ESEM image of nanofibers of *p*-hexaphenylene on mica, covered by SiO_x. The image was taken outside the bleached zone, so the telephone cord buckles appear. The different trajectories of the detected electrons and ions provide pseudo-topographical information (*shading effect*). Acceleration voltage 1.7 kV, partial water pressure 1.1 mbar, tilt 30.8°

by assuming that either some of the molecules did escape through the coating or that the new chemical species formed exhibit nearly no chromophores any longer. A diffusion of the molecules might be possible for the case that the SiO_x film is no longer intact or that the molecules are intercalated between the crystal planes of the mica at imperfect sites of the layered mica crystal. Facing this question, inspections of the SiO_x-covered nanofibers using an environmental scanning electron microscope (ESEM) seem to be advisable, which allows inspection of insulating substrates.

Figure 13.5 shows some delaminations of the SiO_x layer outside the zone exposed to the bleaching. The nanofibers are also clearly visible. Additionally, three cracks in the surface can be found which might allow molecular diffusion. Interestingly, these cracks are aligned parallel to the direction of the nanofibers and thus parallel to the major direction of the buckle lines. From that it is concluded that the organic nanofibers influence the mechanical stability of the SiO_x layer.

Next the question arises why the nanofibers are visible in this image although the electron penetration depth at 1.7 kV is presumably smaller that the coating thickness. In order to answer this question, a Monte Carlo Simulation of the electron trajectories at different acceleration voltages has been performed.

Figure 13.6 shows the results of three Monte Carlo simulations for electron trajectories in the layered system water vapor/SiO_x/muscovite. Note that the value of 254 nm thickness for the SiO_x that has been read out from the deposition machinery has been increased to 376 nm since direct measurements using electron microscopy indicate this latter value.

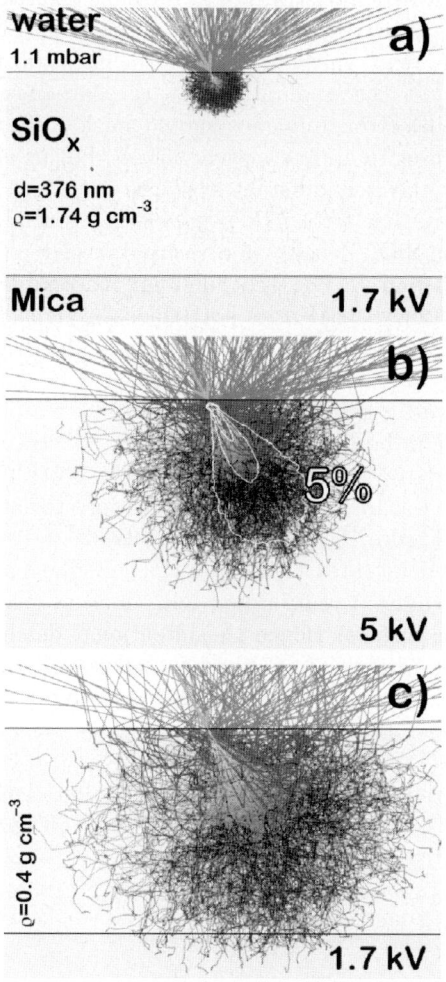

Fig. 13.6. Monte Carlo Simulations of the electron trajectories using the CASINO software version 2.42 NT [19]. (**a**) The simulated sample consists of (from *top* to *bottom*) water (plotted thickness 124 nm) at 1.1 mbar (8.1348×10^{-4} g cm^{-3}), SiO$_x$ (assumed to be SiO, thickness 376 nm) and a mica substrate. The acceleration voltage is 1.7 kV, angle of incidence 30.8° (see Fig. 13.5) The direction of the incoming electrons is along the *green* arrow. The *red* trajectories in water vapor are backscattered electrons. The trajectories within the solid are color-coded according to their energy. Minimum energy 0.03 keV, 1,000 trajectories plotted. (**b**) Same conditions as before, but at 5 kV acceleration voltage. Additionally, equi-energy lines (*white*) are plotted, showing the energy decrease with penetration depth (50%, 25%, 10% and 5% of 5 keV, based on 9,000 calculated trajectories). (**c**) Same conditions as (**a**), but with a virtual density of 0.4 g cm^{-3} for the SiO$_x$. Further calculation parameters see [19]

Figure 13.6a clearly demonstrates that it can be excluded that the electrons penetrating at 1.7 kV acceleration voltage through an SiO_x layer of 376 nm thickness are able to reach the nanofibers on the mica substrates. Even at 5 kV (Fig. 13.6b) and under the assumption that the nanofibers possess heights of about 100 nm, the electron influence should be low, because the energy of the electrons decreases to approximately 250 eV. Let us now assume that the density of the SiO_x layer is different from the literature values. Simulations carried out in a way that for 1.7 kV some electrons would at least reach the nanofibers with the SiO_x density as a free parameter result in a density for the SiO_x film of $0.4 \,\mathrm{g\,cm^{-3}}$. This is beyond any reasonable value and it is thus concluded that (a) the ESEM does not look "through" the coating and (b) the topography of the coated nanofibers mirrors the original topography of the uncoated substrate.

Let us now investigate whether is possible to remove the SiO_x film to check if the fibers can be freed from the growth substrate without rupturing them and to inspect the inner morphology of the protective coating. For this purpose the sample was imposed at several positions to external pressure using a small metal tip. Figure 13.7 shows the results.

Figure 13.7a demonstrates, that either the nanofibers are removed from the mica substrate by chance totally (lower left panel) or they stay geometrically intact at the mica surface. Hence the intentional delamination of such an SiO_x film is not an appropriate method for harvesting these nanofibers from the mica substrate. However, there are interesting morphological details: The

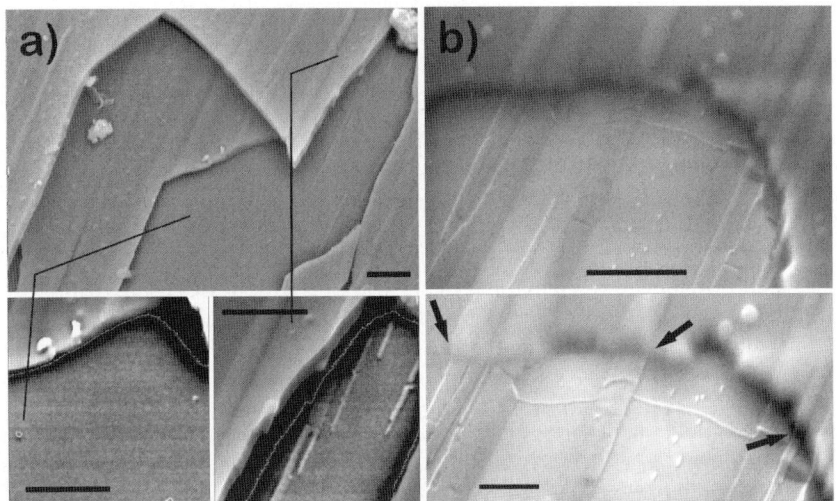

Fig. 13.7. (a) SiO_x film cracked on top of a blister. Two details are magnified, filtered, and contrast enhanced (the white lines are artefacts). All bars 2 μm. ESEM-mode (9.1 kV, 1.1 mbar water, tilt 30.8°, gaseous detection device GSE). (b) SiO_x film cracked at a position besides a blister. The lower image has been background filtered using the WSxM software [20]. All bars 2 μm. SEM-mode (14 kV, high vacuum, sample gold sputtered, tilt 30.8° backscattered electron detector BSE)

width of the fibers on top of the SiO_x film is about a factor of two to five times larger on top of the SiO_x film than the width of the free nanofibers itself. As Fig. 13.7b clearly shows, this larger width is not linearly increasing with increasing thickness of the film, but it appears immediately after a few nanometers of SiO_x film have been deposited. The side parts of the wide fibers in SiO_x are visible throughout the whole film thickness and their angle to the mica plane is close to 90°. The small prolate dots often observed in between the fibers have the same orientation as the fibers and can be regarded as their seed crystals. These dots show essentially the same behavior as the fibers and their widths on the surface is strongly increased compared with the size of the dots on mica, too. This effect is not correlated to the crystal orientation of the mica, because the change of the aspect ratio of the seed crystals from a small dash to a nearly round shape means that the effect is isotropic.

The organic molecules are brought to the mica surface by evaporation onto a heated mica substrate with a carefully tuned temperature, allowing diffusional surface motion. After a seed crystal has been formed and is large enough for further growing, a molecule coming close enough to the crystal will be assembled into it due to an energetic benefit. The organic crystals grow in height and width, consuming all molecules in the local proximity. At the end of the evaporation and after the temperature has slowly decreased, three different areas can be distinguished (1) areas coated with organic molecules in the crystal state (nanofibers of different length and thickness down to small seed crystals), (2) areas coated with organic molecules not yet assembled to the crystalline state, and (3) a depletion zone close to the nanofibers and seed crystals. This depletion zone just reflects the fact that close to a crystal the probability for assembly is highest. Hence, the nanofibers and seed crystals are surrounded by a free and not contaminated mica surface. For the SiO molecules impinging the surface during the next evaporation step, these three different areas lead to different film growth modes. It is reasonable to assume that the SiO molecules adsorb best at the uncoated mica surface, whereas the sticking probability is low at the organic surfaces. Due to the kinetic energy of the incoming SiO molecules, organic molecules outside the crystal lattice are mobilized and displaced from the surface. It cannot be excluded that even molecules being a part of the nanofibers are involved, but in general at least perfect crystals are much more stable than amorphous films. Figure 13.8 depicts this model of the SiO deposition process.

The model presented in Fig. 13.8 is slightly refined from a version published recently [21]. Figure 13.9 presents other examples where the interfaces failed as predicted. Figure 13.9a shows the SiO_x film from the backside, where replicas of the nanofibers can be identified. Figure 13.9b shows that the nanofibers coated with SiO_x which was grown in the depletion zone are more stable than the SiO_x grown over the contaminated mica surface. This result encourages further experiments with SiO_x coatings, because it proves that the easily deformable organic nanofibers (see Fig. 13.9c) can be mechanically stabilized by such a coating and that it should be possible to harvest them for further use in applications.

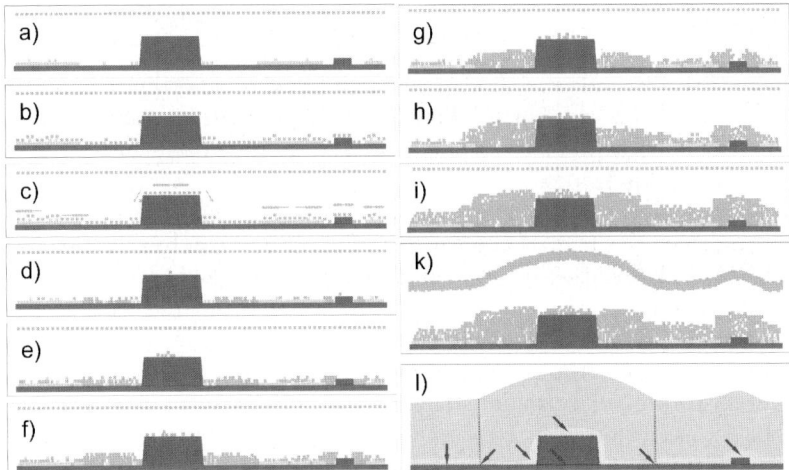

Fig. 13.8. Model for the SiO deposition on nanofiber-coated mica. (**a**) Three areas are visible: The crystalline fiber and a seed crystal (*dark blue*), the depletion zone surrounding them and areas with organic molecules, i.e., a wetting layer (*cyan*) which failed to become part of a crystal and thus contaminate the mica substrate (*orange*). SiO molecules are color-coded grey. (**b**) Uniform distribution of the SiO. (**c**) Diffusion and adsorption starts, increasing the SiO thickness in the depletion zone. (**d**) Layer by layer is deposited. (**e**) Some organic molecules are detached from the mica. (**f**) Within the depletion zone, the SiO film thickness increases more rapidly as on top of the organic crystals. (**g**) The seed crystal is nearly covered. From now on, the growth rate there is assumed to be the same as for the depletion zone. The film might have crystal failures. (**h**) and (**i**) Remaining organic molecules disturb the build-up of the film outside the depletion zone. (**k**) End of deposition, intermediate layers are indicated by a continuous *gray line*. (**l**) The depletion zones modified the growth process of the SiO_x film in a way that even over the long range of 376 nm thickness the morphology remains changed (drawings roughly to scale). The width of this modified zone remains constant from the mica substrate to the surface, because the nanofiber sides are steep. The *red arrows* point to the positions of internal stress or badly adhering interfaces. *Dotted red lines* point to the observed crack failures (see Figs. 13.7 and 13.9)

13.3 Parameters Related to the Embedding of Organic Nanofibers: Preparation of Polymer Films

13.3.1 Motivation for Encapsulation of Nanofibers in Polymers

The goal is to develop recipes for the fabrication of working prototypes and demonstrators, based on organic nanofibers embedded in or deposited on polymers, which act as supports and hosts. Compatibility, interface properties, internal mechanical stress, and the optical properties of the polymer have to be carefully tuned to ensure the successful nondestructive transfer

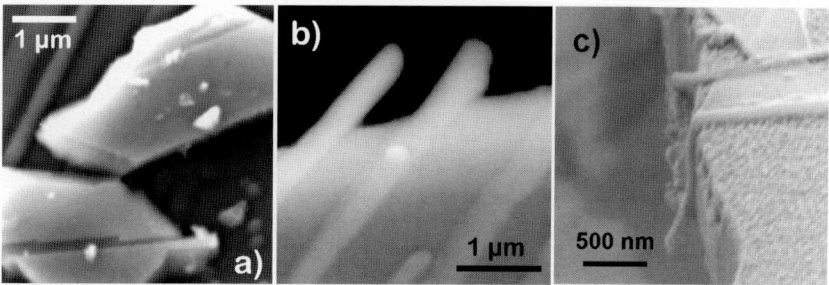

Fig. 13.9. (a) ESEM image of the backside of a delaminated SiO$_x$ film showing replicas of the nanofiber (9.1 kV, 1.1 mbar water pressure, tilt 30.8°, GSE detector). (b) ESEM image of delaminated nanofibers stabilized by the SiO$_x$ coating (14.0 kV, Au coated, 0.5 mbar water pressure, GSE detector) (c) SEM image of two uncoated, transferred nanofibers at the border of a silicon substrate (H. Henrichsen, J. Kjelstrup-Hansen, private communication, 2005)

to the polymer and to retain or to engineer the functionality of the organic nanofiber. A deeper understanding of the interaction between the organic nanofiber and polymer solutions or melts is of essential importance and might be obtained by future in situ observations of wetting in ESEM/AFM hybrid systems. The control of the length of a nanofiber can be achieved by a careful (dielectrophoresis) selection process combined with cryo-cutting of polymer embedded fibers with a diamond knife. Flat or acute angles can be obtained by cutting following oriented embedding. For single organic nanofibers length control, we favor AFM cutting techniques like dynamic plowing lithography in the case where the integrity of the chromophores at the ends of the fibers must be guaranteed.

An important target are recipes for the transfer of nanofibers from their growth substrate onto or into a solid, transparent, and flexible corpus, adaptable to form building-blocks. These building-blocks should fulfil several requirements: The handling of the building-blocks should be easy and the organic nanofibers should be protected against pollution, oxygen, and mechanical stress. Electroactive or optically active functional elements formed from organic materials suffer usually from contamination by aggressive molecules like oxygen, ozone, nitrogen oxides, and sulphur oxide. As already proven, a protection against oxygen is especially important, because the generation of excited states during optical stimulation in the presence of oxygen increases the photobleaching effect. Furthermore, the improved handling of the building-blocks and the protection against mechanical stress enable experiments and testing of device prototypes in a laboratory environment, where nanometer accuracy is unusual or impossible.

As another important benefit, an encapsulation of the organic nanofibers hinders monomer residues, clusters, and parts of the nanofibers to get into the environment. The potential hazard of a single soft organic nanofiber might be assumed to be negligible compared to asbestos or carbon nanotubes. However, appearing in larger amounts handling of organic compounds always asks for special safety precautions.

13.3.2 Essential and Desirable Polymer Properties, Preparation Strategies

Though low adhesion to the substrate and good adhesion to the organic nanofibers are necessary, the polymer chosen and hence the preparation procedure have to fulfil additional requirements. For example, noncrystalline polymers are preferred due to their transparency. Depending on the preparation technique, semicrystalline polymers can exhibit large crystallites which scatter light. Beside the classical high-transparent polymers like polystyrene (PS) and polymethylmethacrylate (PMMA) or relatives, an interesting new amorphous technical polymer is available, namely a statistical copolymer of hexafluoroethylene (PTFE) and 2,2,-bistrifluoromethyl-4,5-difluoro-1,3-dioxole. As an example for desirable properties, some key features of this polymer (Teflon® AF, DuPont, USA) are, e.g., the transparency at 200 nm wavelength ($\approx$98% for a 1.6 µm film), the chemical stability (similar to PTFE), hydrophobicity, a low dielectric constant ($\approx$1.9) and refractive index ($\approx$1.3), a low surface free energy, high temperature stability (glass transition temperature 240–360°C, depending on the molecular weight), a low dissipation factor even at high frequencies ($\approx$0.002 at 2 GHz) and its solubility (hydrofluorocarbons). In this case, a solvent casting must be used to produce the polymer film. Several polymers are already tested and some of them are promising candidates, following the two different strategies:

- Preparation of a polymer melt on the nanofiber-coated mica substrate followed by a lift-off or
- Preparation of a solvent casted film on the nanofiber-coated mica substrate followed by a lift-off

Several experiments demonstrate the usefulness of melt preparation. Two major difficulties must be overcome, which are the temperature stability of the nanofibers and the adhesion of the polymer to the mica surface and the nanofiber, respectively. Although the stability of hexaphenylene nanofibers exceeds 170°C and the transfer to a matrix of large molecular weight polystyrene (PS) was successful at this temperature, further optimization is needed. Instead of using a melt, a mixture of an epoxy and an amine hardener was used at first, curing close to room temperature.

Figure 13.10a shows a simple control after the application of a thermoset polymer droplet onto the mica sheet covered with nanofibers. Though the nanofibers are still intact, the sample is useless, because the lift-off process failed (Fig. 13.10b). The wide range of different epoxies, the possibility to vary the composition, and the chemical nature of epoxy and hardener leave this group of polymers still a promising candidate due to the low-temperature

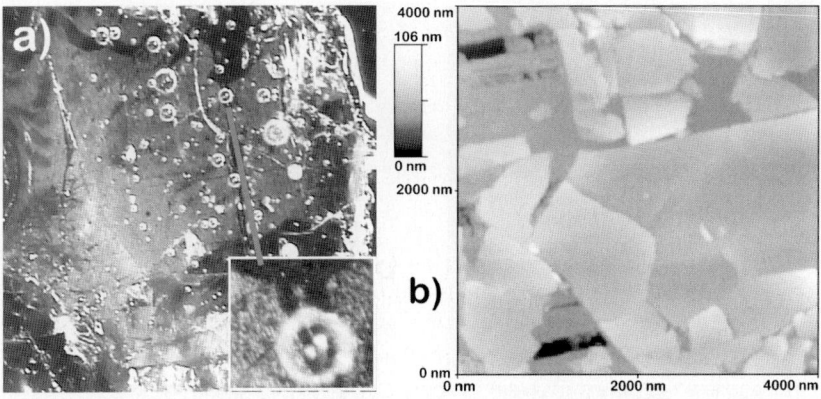

Fig. 13.10. (a) Epoxy cured on mica covered with P6P nanofibers after lift-off. The *blue* fluorescence light shows that the nanofibers remained mainly intact (see also the parallel lines in the *inset*). The image width is about 1.5 mm. (b) The AFM topography image of this surface (TopoMetrix Discoverer, contact mode) shows that the lift-off process failed. The whole surface is covered with broken mica sheets still adhering to the epoxy and covering the nanofibers

stress during preparation. The use of thermosets is additionally advantageous, because a low preparation temperature decreases the mechanical stress provoked by the mismatch of the thermal expansion coefficients.

The assumption is, supported by measurements of topography and materials contrast (friction) as depicted in Fig. 13.11c, d, that the contact to hot polystyrene leads to partial singling and the formation of nanodots. These nanodots still show some blue fluorescence. The nanodots show the same (high) friction as the other material except of the grooves. It is reasonable to assume that PS is stiffer than the P6P fibers or films. Therefore we can conclude that the grooves are polystyrene (PS) and the nanofibers were destroyed while in contact with the molten PS. To decrease the temperature for deposition, PS with a lower molecular weight should be used in the future. The stability of the P6P nanocrystals is reduced, because polystyrene possesses phenyl rings with a π-system similar to P6P, so some solubility of the P6P in a PS melt can be assumed.

The next and last example shows the result after the lift-off of a Teflon® AF, film, made from perfluorocarbon solution (FC-77, 3M, USA) by drying at room temperature. Figure 13.12 shows the results of a transfer to a polymer film casted from solution. Due to the influence of the solvent or due to the stresses caused by the high density changes during drying, many P6P nanofibers are detached from their original position. The low stiffness of the polymer film led additionally to a higher bending during the lift-off process, which could be avoided in the future by using an additional support enforcing the polymer film. In principle, the preparation of solvent-casted films should imply less problems due to the fact that the chemically unmodified organic nanofibers are not soluble.

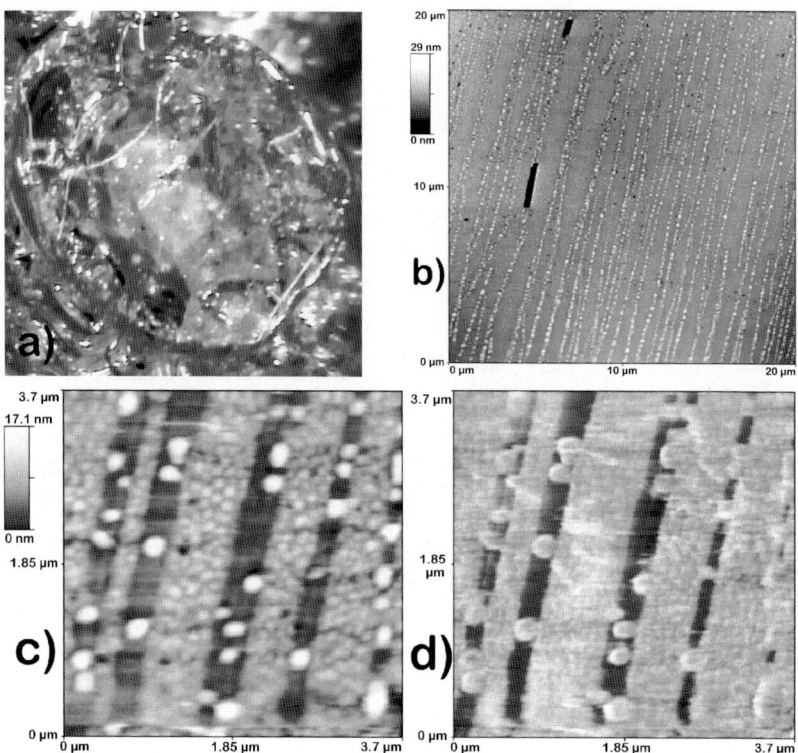

Fig. 13.11. (a) Droplet of polystyrene on mica covered with P6P nanofibers after lift-off, deposition temperature 170°C±10°C. The *blue* fluorescence light shows that at least the chromophores of the nanofibers remained intact. The image width is about 1.5 mm. (b) AFM topography (TopoMetrix Discoverer, contact mode). The nanofibers are destroyed and chains of pearl-like features lay in a groove. (c) AFM topography, showing the grooves of approximately 10 nm depth and some globuli inside. (d) The friction contrast (lateral force) simultaneously measured with (c) shows, that the grooves exhibit a lower friction coefficient. This is a strong hint that the materials composition of the grooves is different from the rest. During scanning, the pearls can be pushed by the tip if the contact force is increased

13.4 Cutting of Nanofibers

The main problem in cutting or shaping the end faces of the fibers is their extremely small dimensions with heights around 50 nm and widths of the order of several hundred nanometers. These exclude any direct mechanical cutting methods and point to either cutting in embedding environments or optical cutting methods. Test experiments have shown that electron beam cutting is useless since the irradiated nanofiber parts are no longer luminescing but still cannot be removed from the surface.

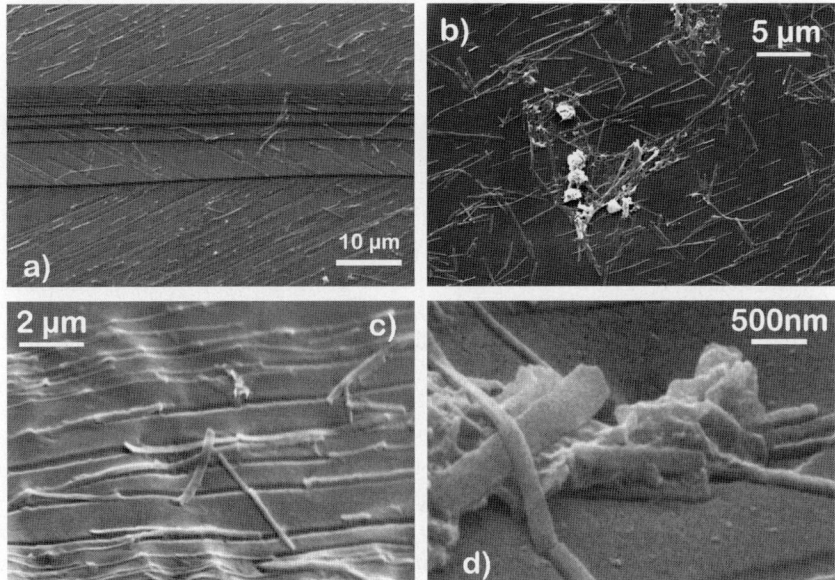

Fig. 13.12. P6P nanofibers embedded in solvent-casted Teflon® AF. All images at 20 kV, sample Au coated, secondary electron detector device. (**a**) P6P nanofibers on polymer film. Some of the fibers are still laying in their impressed grooves, some are displaced during the lift-off. Additionally, replica of steps at the mica substrate are visible in the polymer film. (**b**) Same as before, but with smaller number of imprinted grooves. (**c**) Blow-up of (**b**) to demonstrate that the nanofibers are able to follow a bending of the polymer film during lift-off without detaching. (**d**) Nanofiber exhibiting plastic deformation

Microfabrication by laser ablation on the other hand has become a strong and flexible tool for the generation of components or devices for a variety of applications in microelectronics, micromechanics, microfluidics, and microoptics. In many cases the ablative removal of material is applied to fabricate holes, channels, or three dimensionally shaped surfaces. Microlenses, waveguides, and gratings are examples for optical elements, which can be fabricated this way.

Generally, the quality of the generated structures increases when reducing the ablating laser wavelength. Besides providing better optical resolution, this is mainly due to stronger light absorption leading to a more confined interaction volume. Thus transparent polymers like PMMA or polycarbonate can be processed with much higher precision at a laser wavelength of 193 nm compared to, e.g., 308 nm. This should also be true for the phenylenes that the present nanofibers are made of. However, at a wavelength of 193 nm even substrate materials that are transparent in the visible spectral range like glass, ceramics, or mica start to absorb strongly, so that their ablation or damage thresholds are easily reached. In the case of removing a film or nanoaggregate from a substrate, this leads to the existence of a sometimes narrow "processing window," where the fluence is sufficient to ablate the layer, but not so high that the substrate is damaged.

Ablation was performed with an ArF excimer laser emitting at 193 nm in a mask projection set-up. The masks were imaged with 25× demagnification (Schwarzschild reflective objective, numerical aperture 0.4) on the sample resulting in irradiation spots of $100 \times 100\,\mu m^2$ and $10 \times 6\,\mu m^2$, respectively. The fluence was controlled using a variable dielectric attenuator, measuring the pulse energy by a pyroelectric Joulemeter. As masks we have used crossed razor blades of various distances.

Fluorescence images of samples irradiated with different fluences by single UV laser shots are presented in Fig. 13.13. As seen, at $100\,mJ\,cm^{-2}$ a clear ablation pattern is obtained. Again, there are no significant bleaching effects on the remaining fiber pieces.

A direct comparison of ablated and nonablated areas with the maximum possible optical resolution is shown in Fig. 13.14. Whereas the border region of the nanofibers is undefined preceding the ablation process, it becomes straightened after the ablation. There is no hint for a melting of the nanofibers and only very weak bleaching at the very ends of the fibers.

A closer look at the border region separating ablated and nonablated fiber parts is given in Fig. 13.14c with the help of an AFM image. In between the fibers lots of small islands are located, which are remaining from the growth process of the needles. Apparently even these small P6P aggregates are completely removed by the ablation process and the fiber ends are all lined up and possess a predefined angle of 67° with respect to the fibers long axes. This angle could be adjusted easily, e.g., to obtain Brewster angle cut fibers.

As for the horizontal cutting angle definition up to now a minimum width of 300–400 nm for the cutting region has been obtained. Whether there is a deteriorating effect of this weak definition on the effective acceptance angles of the nanofibers still has to be investigated. Possible routes to improvement are tilt-angle projection, preadsorption of an ultrathin light-absorbing layer, immersion-medium enhanced projection, or encapsulation of the nanofibers.

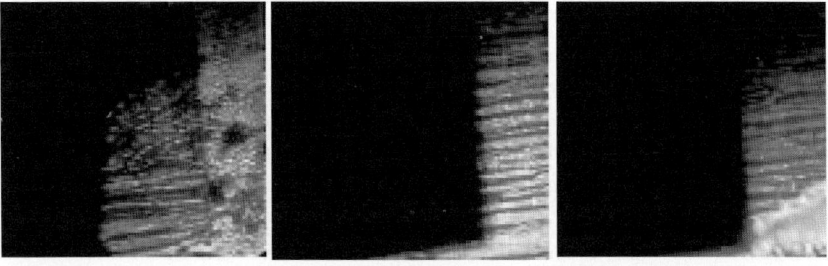

Fig. 13.13. Epifluorescence micrographs ($100 \times 100\,\mu m^2$) of hexaphenylene nanofibers on mica, irradiated by single UV pulses (193 nm) with $60\,mJ\,cm^{-2}$ (left-hand side), $80\,mJ\,cm^{-2}$ (*middle*), and $100\,mJ\,cm^{-2}$ (right-hand side)

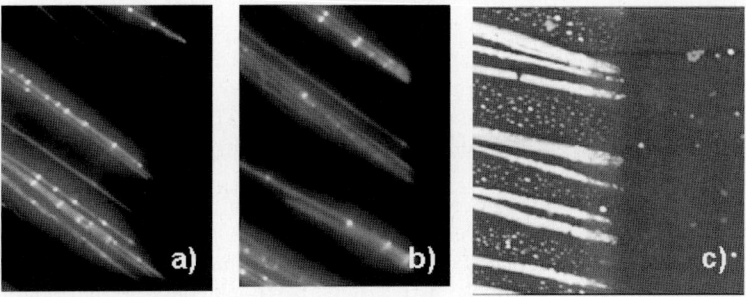

Fig. 13.14. (a), (b) Epifluorescence micrographs ($100 \times 60\,\mu m^2$) of hexaphenylene nanofibers on mica, not irradiated (left-hand side) and irradiated by two $100\,mJ\,cm^{-2}$ UV pulses (193 nm). The fibers are clearly cut on the right-hand side. (c) AFM image ($7 \times 7\,\mu m^2$) of an array of organic nanofibers irradiated with 193 nm UV laser light. Note that the ablated area has been cleaned not only from the nanofibers, but also from the small cluster in between them and from the wetting layer

13.5 Conclusions

In this chapter the state of the art of glass coating, polymer embedding, and laser cutting of organic nanofibers is reviewed. It is demonstrated that coating of nanofibers with an SiO_x layer of appropriate thickness reduces significantly the photooxidation process and thus the bleaching of nanofibers. Extended studies have revealed details of the coating process which led to a model for the glass embedding. Polymer embedding is still in its infantile stage, but first results are rather promising, especially showing that it is possible to embed nanofibers in a polymer layer and simultaneously remove them from the growth substrate.

Cutting of bare nanofibers to arbitrary lengths and angles is difficult, even using ultraviolet laser ablation. The cutting steepness seems to be limited to a few hundred nanometers which is much worse as compared to the "as-grown" fibers, where the steepness is of the order of 100 nm. There is also an obvious problem with debris deposition on top of the cut fibers. It is rather safe to predict that a combination of the embedding technology discussed in the first part of this chapter and laser cutting will lead to significantly improved results.

Acknowledgment

H.-G. Ruban is indebted to the Danish research foundations SNF (21-03-0469) and STVF (26-04-0253) as well as the European Communitys Human Potential Programme under contract HPRN-CT-2002-00304, "FASTNet", for financial support.

References

1. F. Balzer, H.G. Rubahn, Appl. Phys. Lett. **79**, 3860 (2001)
2. F. Balzer, H.G. Rubahn, Adv. Funct. Mater. **15**, 17 (2005)
3. F. Balzer, V. Bordo, A. Simonsen, H.G. Rubahn, Appl. Phys. Lett. **82**, 10 (2003)
4. F. Quochi, F. Cordella, A. Mura, G. Bongiovanni, F. Balzer, H.G. Rubahn, Appl. Phys. Lett. **88**, 041106 (2006)
5. NIST Standard Reference Database 81 (http://srdata.nist.gov/insulation) and The Physics Hypertextbook, http://hypertextbook.com/thermal/conduction, and Landolt-Börnstein 1983
6. C. Clauser, E. Huenges, in *Rock Physics and Phase Relations – A Handbook of Physical Constants*, ed. by T. Ahrens (AGU reference shelf 3, American Geophysical Union, Washington DC, 1995), p. 119, and Landolt-Börnstein 1983
7. Y. Fei, in *Mineral Physics and Crystallography – A Handbook of Physical Constants*, ed. by T. Ahrens (AGU reference shelf 1, American Geophysical Union, Washington DC, 1995), p. 29
8. R. Hammond, K. Pencheva, K. Roberts, P. Mougin, D. Wilkinson, J. Appl. Cryst. **38**, 1038 (2005)
9. G. Heimel, P. Puschnig, M. Oehzelt, K. Hummer, B. Koppelhuber-Bitschnau, F. Porsch, C. Ambrosch-Draxl, R. Resel, J. Phys.: Cond. Matt. **15**, 3375 (2003)
10. X. Huang, K. Terashima, K. Hoshikawa, Jpn. J. Appl. Phys. **38**, L1153 (1999)
11. T. Carlberg, J. Electrochem. Soc. **133**, 1940 (1986)
12. F. Kolb, Wachstum und Charakterisierung von Siliziumnanodrähten. Ph.D. thesis, Martin-Luther Universität Halle-Wittenberg (2005). Electronic document identification urn:nbn:de:gbv:3-000008655
13. H. Philipp, J. Phys. Chem. Solids **32**, 1935 (1971)
14. H. Philipp, J. Non-Cryst. Solids **8–10**, 627 (1972)
15. G. Brady, J. Phys. Chem. **63**, 1119 (1959)
16. R. Temkin, J. Non-Cryst. Solids **17**, 215 (1975)
17. A. Hohl, T. Wieder, P. Van Aken, T. Weirich, G. Denninger, M. Vidal, S. Oswald, C. Deneke, J. Mayer, H. Fuess, J. Non-Cryst. Solids **320**, 255 (2003)
18. W. Gerberich, M. Cordill, Rep. Prog. Phys. **69**, 2157 (2006)
19. D. Drouin, A. Couture, R. Gauvin, P. Hovington, P. Horny, H. Demers, *CASINO, monte CArlo SImulation of electroN trajectory in sOlids* (Univ. of Sherbrooke, Sherbrooke, Quebec, Canada). http://www.gel.usherbrooke.ca/casino/. Additional calculation parameters chosen: Total and partial cross section: Mott by interpolation, effective section ionisation: Casnati, ionisation potential: Joy and Luo 1989, random number generator: Press et al. 1986, direct cosin: Soum et al. 1979, dE/ds calculation: Joy and Luo 1989
20. Used procedure: "flatten filter" using paths crossing the steps followed by several line fits. The WSxM free software: http://www.nanotec.es
21. C. Maibohm, J. Brewer, H. Sturm, F. Balzer, H.G. Rubahn, J. Appl. Phys. **100**, 054304 (2006)

Index

a-Si, 329
a-SiO$_2$, 329

ablation
   ArF laser, 344
   threshold, 343
absorption, 34
absorption spectra
   polarized, 270
activation energy, 51
adhesion, 60, 316, 321, 331, 340
adsorption sites, 168
AES, 8
AFM, 16–18, 302
   cutting, 339
   enforced agglomeration, 51
   manipulation, 316
Ag(111), 51, 168, 170, 201
agglomeration, 4, 52
   AFM enforced, 51
aggregate band, 287
Al, 170
   (111), 10, 168, 170, 179, 180, 265, 283
   (111) oxidized, 175, 186
Al$_2$O$_3$, 289
Al(111), 157
alkali halides, 31, 49, 174
alkanethiol, 37
alkylthiole, 167
amorphous silicon, 329
amplified spontaneous emission, *see* ASE
anchoring, quaterphenyl, 214

Anderson localization, 89
angular resolved UPS, *see* ARUPS
anisotropic
   properties, organic layers, 264
   charge transport, 119
   current–voltage characteristics, 292
   electroluminescence, 278
   epitaxial growth, 171–178, 271, 277, 282
   sticking, 166, 172
annealing, 50, 114, 288
anthracene, 124
antibleaching, 330
ArF laser ablation, 344
ARUPS, 9, 170
ASE, 239, 280
   narrowing, 241
atomic force microscopy, *see* AFM
Au, 58, 283
   (110), 170
   (111), 31, 37, 157, 170, 197–202, 265, 283
   (111)–C, 284
   foil, 31, 197, 203
   polycrystalline, 197, 203
Auger electron spectroscopy, *see* AES
azimuthal alignment, 137, 170

back-scattering, 249
band alignment, 165, 179, 182
   local, 184
band structure, 9, 10, 178
BCB, 266, 289

beam damage, 15, 37, 207
benzene, 9, 10, 168, 180, 199
biosensor, 325
biphenyl, 11, 69, 198
birefringence, 236
bithiophene, 11
bleaching, 326, 330, 344
blue light
　polarised, 77
blue shift, 78, 224
bonding
　π, 170
　covalent, 283
　ionic, 283
　van der Waals, 283
bottom-up, 68
Bragg's law, 126
breaks, 21, 225, 249
bright field method, TEM, 135
broadening of a peak, 133
building block principle, 12
bulk limited current, 310
bulk organic single crystal, 263
bulk SHG, 81

C_{60}, 92, 265, 288
　Ba-doped, 92, 98
　on mica, 95
C–C stretching vibration, 219
carbon nanotubes, 321
carrier transport, 278
centrosymmetry, 22, 80, 124, 125
CH–π–interaction, 68
charge
　carrier mobility, 239, 263, 291, 292,
　　294, 311
　　minimum, 314
　carrier transport, 285
　injection, 179
　transfer, 98
　transport, 310
　transport anisotropy, 119
　trapping, 182
charging, 11, 37
chemisorption, 4
chemosensing, 256
cleaving steps, 7
close packing principle, 121
CLP4, 51, 73, 301

clusters, 36, 43, 45, 54, 248, 337, 344
　AFM induced agglomeration, 51
　agglomeration, 60
　critical number density, 50, 51, 109
　critical size, 51
　height and width, 47
　number density, 48, 50
　strain, 134
CNP4, 73
coating, 325, 327
　SiO_x, 330
coherent propagation model, 251
cohesion, 58, 326
colour separation, 332
columnar growth mode, 6
commensurable type of epitaxy, 137
commensurate, 169, 173
commensurism, 39
compression modulus, 328
conjugated organic molecules, 89
conjugated polymers, 89
contact mode, 16, 106
contact plane, 137, 141
contact resistance, 291
contacts, 310, 311, 313
contamination, 339
coronene, 121
Coulombic interaction, 315
covalent bonds, 167, 283, 326
cracks, 328
critical
　angle, 131
　cluster size, 51
　nucleus, 5
　number density, 50, 51, 109
　terrace width, 206
cryo-cutting, 339
crystallisation, 4
crystallite size, 132, 133, 148
　lateral and vertical, 134
crystallite strain, 134
crystallographic order, 137
Cu
　(110), 151, 168–170, 182
　(110)-(2×1)O, 169, 170, 175, 182
　-phthalocyanine, 122
current-voltage characteristics, 292
cutting, 326, 339, 342
Cyclotene^{TM}, see BCB

damage threshold, 343
dangling bonds, 6
dark field microscopy, 21, 78
   TEM, 136, 149
defect emission, 221
defects, 43, 168, 313
   dimer-like, 288
   grain boundary, 292
degradation, 11
   laser-induced, 244
delamination, 334
dendritic islands, 53, 54, 122, 204
density functional theory, *see* DFT
denuded zone, *see* depletion zone
depletion zone, 36, 74, 110, 174
deposition rate, 49, 51, 54
device
   multi fiber, 325
   single fiber, 311, 325
DFT, 12, 69
divinyltetramethyldisiloxane-bis(benzocyclobutane), *see* BCB
dichroic ratio, 36, 273, 280
dielectric tensor, 226
diffraction
   angle, 126
   angle, in-plane, 132
   conditions, 126
   method, TEM, 135
   pattern, 135
diffusion, 334, 337
   anisotropy, 171, 174
   preferential, 166
dimers, 285, 288
dioctahedral mica, 35
dipole
   -assisted self-assembly, 302
   assisted alignment, 41
   induced interaction, 272
   induced-dipole interaction, 240
disc-shaped molecules, 121
domains, 36, 43
doping, 289
drop-casting, 321
dynamic plowing lithography, 339

EBSD, 204
electrodes, 293, 311

electroluminescence, 311
   anisotropic, 278
   polarized, 274
electron
   beam cutting, 342
   diffraction, 120
   injection, 291
   mobility, 291, 314
   push/pull groups, 78, 80
   ray path, 135
   trajectories, 334
electron backscatter diffraction, *see*
   EBSD
electronic properties
   anisotropic, 264
electronic structure, 178
elongated islands, 122
embedding, 302, 326, 327, 338
encapsulation, 338, 339
energy barrier, 310
enhanced pole density, *see* EPD
environmental scanning electron
   microscope, *see* ESEM
EPD, 131
epitaxial
   growth, 104
   matrix, 39, 141, 145
   relationship, 131, 137, 140
epitaxy, 39, 41, 43, 90, 137
   alignment of distinct directions, 137
   alignment of initially adsorbed
      molecules, 137
   anisotropic, 171–178, 271
   commensurable type, 137
   grammar, 140
   line-on-line, 142
   point-on-line, 142
   point-on-point, 141
epoxy, 330, 340
ESEM, 334
evanescent field, 225
evanescent waves, 21
Ewald sphere, 135
exciton, 244, 271, 285, 311
   –exciton scattering, 221
   emission, 219
   spectrum, 286
   transitions, 77

350   Index

Förster energy transfer, 288
Fermi level, 9
  alignment, 179
fiber bundles, 236
film growth, 4
fluorene, 122
fluorescence, 34, 330
  microscopy, 21
  polarized, 31, 36, 41, 54, 278
  spectrum, 77, 219
    time resolved, 224
  tuneable, 77
Frank–van der Merwe mode, 6
friction, 318, 341
fullerenes, 265
functionalization, 325

GaAs, 33, 51, 101
  (001), 271
gain, 253
  measurement, 252
  saturation, 252
GID, 173, 189
glass, 81, 101, 167
glass coating, 330
grain boundaries, 119
grammar of epitaxy, 140
grazing incidence diffraction, see GID
green emission, 224, 286
grooves, 35
growth conditions, 137, 177
growth model, 51

Hall effect, 94
HAS, 37
helium atom scattering, see HAS
herringbone
  layer, 148
  structure, 32, 120, 264, 315
heteroepitaxy
  organic–organic, 185–190
heterostructures
  Au–mica, 58
  organic–organic, 185–190
hexacene, 124
hexafluoroethylene, see PTFE
hexaphenylene, see sexiphenyl, 31, 32, 36, 67, 196, 201, 219, 301, 303
high resolution TEM, 150

highest occupied molecular orbital, see HOMO
hole mobility, 309, 314
HOMO, 40, 179–182, 310
HOMO–LUMO gap, 69, 82
HOPG, 197
Hot-Wall Epitaxy, see HWE
Hot-Wall-Beam Epitaxy, see HWBE
HWBE, 90
HWE, 90, 241, 265
hydrophilic, 35, 56
hydrophobic, 56
hyperpolarisability, 69, 82

in-plane
  diffraction angle, 132
  order, 140
  rocking curve, 132
incoherent hopping, 309
infrared reflection, 271
infrared spectroscopy, 274
injection
  barrier, 311
  limited current ILC, 310
  limited regime, 310
instability, 73
integral breadth, 133
interaction
  π–π, 49, 172
  CH–π, 68
  Coulombic, 32, 315
  dipole induced, 272
  dipole induced-dipole, 240
  molecule–molecule, 167
  molecule–substrate, 167
  van der Waals, 33, 58, 121, 137, 201, 289, 315
interface dipole, 180, 311
interface layer, see wetting layer, 215
intermittent contact mode, 16
intramolecular forces, 201
ionic bonds, 278, 283
IR spectroscopy, 274
islands, 36, 74, 75, 149
ITO, 167, 265, 278

KAP, 31, 34
KCl, 33, 94, 101, 240, 265
  (001),(100), 138, 277
  steps, 281

kinetics, 45
kinks, 74
Knudsen cell, 31, 73, 197, 241, 283

L scan, 104, 127
laser, 285, 301
   -induced surface heating, 276
   ablation, 343
laser scanning microscope, *see* LSM
lasing, 239, 308, 325
   coherent random, 241
   electrical pumping, 256
   ensemble averaged, 242
   incoherent random, 241
   optical pumping, 239, 242
   random, 240
   single fiber, 247
lateral force, 316
lateral size, 134
lattice match, 137, 189
Laue-equations, 126
LEED, 35, 37, 158, 196
   multi-channel plate, 37
level alignment, 179
LiF, 34, 330
lift-off process, 303, 340
light emission
   angular, 304
   spatial anisotropy, 307
light scattering, 21, 239, 240, 249
line profile analysis, 132, 151
line-on-line coincidence, 142
local field enhancement, 82
low energy electron diffraction, *see* LEED
Löwenstein's rule, 35
lowest unoccupied molecular orbital, *see* LUMO
LSM, two-photon, 303, 304
luminescence, 21
   clamping, 250
   lifetime, 221
   quantum yield, 239
LUMO, 40, 310
lying molecules, 9, 16, 33, 34, 44, 49, 51, 54, 57, 68, 105, 112, 139, 149, 155, 156, 170, 172, 175, 181, 185–187, 189, 278, 279, 282

Malus law, 236
mask-shadowing, 101, 278
mass transfer, 303
MBE, 8, 31, 90, 93, 207, 271
mechanical stress, 339
memory circuits, 263
metal films, 167
metallic substrates, 283
mica, *see* muscovite, 31, 58, 72, 93, 265, 271, 293, 303, 325
   $2M_1$ polytype, 73
   cleavage step, 36
   electric fields, 73
   grooves, 35, 73
microfabrication, 311
micromanipulation, 312, 319
micropipette, 303, 311
microrings, 56, 240
microstrain, 132, 133
MOCLP4, 78
MOCNP4, 78
model function, 132
molecular beam epitaxy, *see* MBE
molecular conformation, 181
molecular strings, 168
molecule
   –molecule interaction, 167
   –substrate interaction, 167, 201
   azimuthal alignment, *see* azimuthal alignment, 170
molecules
   edge-on aligned, 139
   end-on orientation, 139
   flat-on orientation, 140
   lying, *see* lying molecules, 9, 16, 33, 34, 44, 49, 51, 54, 57, 68, 105, 112, 139, 149, 155, 156, 170, 172, 175, 181, 185–187, 189, 278, 279, 282
   orientation, 228, 234
   upright, 9, 16, 43, 49, 53, 57, 156, 173, 181, 276, 278, 282
MONHP4, 78
monolayer, 131, 165, 167
Monte Carlo simulation, 52, 334
MOP4, 73, 301, 303, 304
morphology, 13, 145
mosaicity, 128
Mott–Gurney theory, 294, 310, 313
multi-fiber devices, 325

multiphoton effects, 22
muscovite, *see* mica, 34, 36, 53, 221, 240
   steps, 43

n-type doping, 99
NaCl, 33, 34, 94, 227
nano-needle, *see* nanofiber
nanochains, 122
nanofiber, 21, 36, 171
   2d manipulation, 316
   3d manipulation, 320
   arrays, 303
   assembly, 319
   coating, 325, 327
   cutting, 326, 342
   elastic properties, 317
   embedding, 326
   encapsulation, 338
   manipulation, 325
   mechanics, 315
   protection, 326
   rupture, 316
   sliding, 318
   transfer, 311, 315, 325, 326
   transferred, 81
Nanofiber A/S, 303
nanomanipulation, 316
nanostructured interfaces, 182–184
nanowire
   silicon, 311
near edge X-ray absorption fine structure spectroscopy, *see* NEXAFS
near field microscopy, 24, 231
   apertureless , 27
needle
   breaks, 21, 225, 249
   bundles, 236
   growth, 51
      temperature dependence, 45
      thickness dependence, 46
   height, 47
   length distribution, 47
   manipulation, 206
   strain, 134
   twinned, 41
   width, 47
needles, 31, 36, 74, 75, 149, 150, 272, 276, 301, 325
   number density, 50

neutron diffraction, 15, 120
NEXAFS, 12, 158, 170, 172, 186, 187, 196
Ni, 51
   (110), 9, 14, 168
   (110)-(2 × 1)O, 167
NMeP4, 73
non-contact mode, 16
nonlinear optics, 22, 80, 301
   push/pull groups, 80
nucleation, 4, 175
   and growth, 4
   theory, 49, 51

octahedral site, 35, 43
octathiophene, 125
OFET, 68, 166, 195, 263, 266, 288, 292, 309
   sexiphenyl, 293
off-needles, 76
Ohmic conduction, 310
OLED, 68, 166, 263, 309, 315
   polarized, 264, 285, 309
oligo-acenes, 124, 315
oligo-phenylenes, 67, 124, 264, 315
   methoxy-substituted, 69
oligo-phenylenevinylene, 285
oligo-thiophene dioxide, 255
oligo-thiophenes, 12, 34, 53, 67, 125, 264, 315
   lattice constants, 32
OMBE, 265, 283, 325
OMDB, 241
optical
   anisotropy, 119
   cutting, 342
   properties
      anisotropic, 264
   resolution, 22
   response
      dynamic, 220
      static, 219
   transition, 40
organic electronics, 89, 263
organic field-effect transistor, *see* OFET
organic light-emitting diode, *see* OLED
organic molecular beam deposition, *see* OMBD

organic molecular beam epitaxy, *see* OMBE
organic solar cells, 263
out-of-plane order, 128, 138

packing forces, 125
para-phenylenes, 31, 36
  functionalized
    CLP4, 51, 73, 301
    CNP4, 73
    MOCLP4, 78
    MOCNP4, 78
    MONHP4, 78
    MOP4, 73, 301, 303, 304
    NMeP4, 73
  lattice constants, 32
  molecular conformation, 124
  non-symmetrically functionalized, 69, 78
  symmetrically functionalized, 69, 71, 73
  synthesis, 70
para-sexiphenyl, 93
  islands and nano-fibers, 106
  needles, 101
  needles and islands, 102
  on KCl, 102
  on mica, 106
  rearrangement of islands, 109
PAX, 184
peak broadening, 133
Pearson VII function, 132
pentacene, 8, 67, 121, 124, 169, 179, 200, 264
persistence length, 58
perylene, 67, 121, 122
perylene-3,4,9,10-tetracarboxylic-3,4,9,10-dianhydride, *see* PTCDA
PES, 9
phase contrast, 17
phlogopite, *see* mica, 34, 35, 43
photobleaching, 231, 327, 339
photoelectron spectroscopy, *see* PES
photoemission of adsorbed Xenon, *see* PAX
photoexcitation
  excitonic, 220
photoluminescence
  polarized, 270, 274, 285
  steady state, 285
  time resolved, 285, 287
  time-resolved, 271
photonic sensing, 255
phthalocyanines, 121
physisorption, 4, 90
π bonding, 170
π–π interaction, 49, 172
π–π^* absorption, 274
π–conjugated polymers, 263
π–conjugation, 67, 68
π^*-resonance, 199
planarity, 265
plasmon excitation, 27
plasmonics, 27
plastic deformation, 317, 328
plate-like crystals, 123
PMMA, 340, 343
point-on-line coincidence, 39, 142
point-on-point commensurism, 141
polarizability, 41
polarization, 77, 79, 320
polarized
  absorption, 40
  blue light, 36
  emission, 40, 224
  fluorescence, 236
  fluorescence microscopy, 41, 55, 57
  two-photon microscopy, 57
pole figure technique, 101, 130, 142
poles, 131
pollution, 339
poly(p-phenylenevinylene), 255
poly(2,5-thienylene vinylene), *see* PTV
polycarbonate, 343
polymer
  conjugated, 89
  films, 338
  melt, 340
  substrate, 326
polymerization, 207
polymethyl-methacrylate, *see* PMMA
polymorphism, 34, 122, 152, 278
polystyrene, 340, 341
positioning, 302
powder
  pattern, 128
  three-dimensional, 128
  two-dimensional, 137

processing window, 344
Pseudo Voigt function, 132
Pt(111), 199
PTCDA, 51
PTFE, 340
PTV, 268
purification, 8
push-back, 180
push/pull groups, 78, 80
pyrene, 121, 122

quadrupolar electrostatic forces, 121
quartz microbalance, 8
quaterphenylene, 69, 70, 124, 196, 283
   α-phase, 198
   β-phase, 198
   epitaxy, 284
   needle-like chains, 284
   needles, 203
   plate-like crystallites, 284
   tetragonal islands, 284
   twisted conformation, 201
quaterthiophene, 53, 125
quinquethiophene, 125

radiationless transition, 221
radio frequency identification device, see RFID
Raman gain amplification, 240
random lasing, 240, 288
reabsorption, 225
reciprocal
   breadth, 133, 151
   lattice rods, 131
   space, 126
   space map, 129, 139
   space point, 135
red shift, 34, 78, 83, 225, 288
reflectance difference spectroscopy, RDS, 90, 170
reflection high energy electron diffraction, see RHEED
resolution limit, 21
RFID, 195
Rh(111), 199
RHEED, 90
rocking curve, 94, 128, 276
   in-plane, 132
rodlike molecules, 121

rubbed layers, 101
rubrene, 201, 309
rupture shear stress, 316
rutile, see TiO_2, 142

S-S annihilation, 256
SAED, 37
safety, 326, 340
sandwich herringbone crystal structure, 120
scan
   γ/ω-, 132
   ω-, 132
   $\theta/2\theta$-, 127
   $q_\parallel$-, 132
   $q_\perp$-, 132
   q_x-, 129
   q_z-, 127
   L-, 127
   specular, 127
scanning electron microscope, see SEM
scanning near field optical microscope, see SNOM
scanning tunneling microscopy, see STM
scattering vector, 126
Scherrer equation, 133
Schlagfigur, 35, 39
Schottky–Mott limit, 179
SCLC, 293, 310
second harmonic generation, see SHG
seed crystals, 337
seeding, 177
selected area electron diffraction, see SAED
self-assembly, 165, 167
self-waveguiding, 102
SEM, 196, 204
semiconductor
   n-type, 268, 288
   p-type, 267
septiphenyl, 124
septithiophene, 125
sexiphenyl, see hexaphenylene, 8, 14, 119, 122, 124, 167, 169, 170, 175, 177–179, 186, 187, 240, 271
   crystallites, 275
   islands, 283
   molecule orientation, 276

needles, 174, 278
  on isotropic substrates, 152
  on KCl(100), 138, 146, 149
  on mica, 150
  on sexithiophene, 189, 190
  on TiO$_2$(110), 142
  standing molecules, 276, 278
sexithienyl, 285
sexithiophene, 8, 51, 121, 125, 169, 175, 179, 186
  on Cu(110), 151
  on sexiphenyl, 186–190
  on TiO$_2$, 147, 174
shadow masking, 311, 322
shear force, 26
sheet silicate, 34
SHG, 22, 81
Si(111), 175
silica, 34
silicate, 34
silicon, 34
  amorphous, 329
  nanowire, 311
single molecule detection, 27
single-fiber devices, 325
SiO, 329
SiO$_2$, 167, 289
SiO$_x$, 329, 330
size parameter, 133
smart material, 326
SNOM, 24, 231
space charge, 291, 310
space-charge limited current, see SCLC
spatial distribution, 130
spectral lines
  width, 219
specular scan, 127
spirophenyl, 240
spontaneous emission, 243
stacked crystal structure, 120
stamp, elastomeric, 309
step
  bunching, 43
  density, 205
step flow mode, 6
steps, 43
sticking anisotropy, 166, 172
STM, 14, 171, 196
  imaging artifacts, 208
  tip created defects, 214

strain, 133
strain energy, 58
Stranski–Krastanov mode, 6, 51
stress, 45, 332
structural defects
  dimers, 285
structure factor, 127
substrate roughness, 68
substrate screening, 9
superstructure matrix, 39, 141
surface
  waves, 21
  diffraction, 131
  electric field, 35
  energy, 56, 68, 175
  mobility, 74
  reconstruction, 182
  science, 7
  second harmonic generation, 81
  steps, 272
  unit cell, 139, 141
surface reconstruction, 167, 170
susceptibility, 82
Suzuki cross-coupling reaction, 71
swallow wings, 76

tapping mode, 16, 94
TDS, 106, 112, 196
TED, 37, 275
Teflon® AF, 340, 341
telephone cord buckling, 331
TEM, 15, 120, 134, 275
  bright field method, 135
  dark field method, 136, 149
  diffraction method, 135
  high resolution, 150
templates
  inorganic, 166, 168
  nanostructures, 182
  organic, 190
terphenyl, 124
terraced islands, 122, 149
terraces, 7
terthiophene, 125
tetracene, 124, 314
tetrahedra, 35
thermal
  conductivity, 327
  displacement, 198

expansion, 327
  mismatch, 332
  libration, 125
thermal desorption spectroscopy, see TDS
thermodynamic growth, 171
thin film growth, 263
thin film phase, 124
thiophene/p-phenylene co-oligomers, 255
three-dimensional powder, 128
threshold length, 110
TiO_2, 31, 142, 147, 170, 179, 187
  (110), 175, 177, 178, 282
    sputtered, 175
  (110)-(1×1), 169, 170
TM–mode, 226
tomography, 304
torsional modes, 198
total reflection, 131
TPI SNOM, 234
transfer, 81, 302, 311, 315, 325, 326, 341
transfer matrix, 251
transistor
  ambipolar, 268
transition dipole, 40
transition dipole moment, 200, 202, 279
transmission electron diffraction, see TED
transmission electron microscopy, see TEM
trap, 313
trioctahedral mica, 35
triplet state, 326
tunneling, see STM
  photons, 24
twist angle, 182, 198
two-dimensional powders, 137
two-photon
  intensity near field microscopy, see TPI SNOM
  luminescence, 81
  microscopy, 22, 228
    polarized, 57

UHV, 7, 165, 187, 197

ultra high vacuum, see UHV
ultramicroscopy, 21
ultrashort laserpulses, 23, 81, 221, 242, 303
ultraviolet photoemission spectroscopy, see UPS
upright molecules, 9, 16, 43, 49, 53, 57, 156, 173, 181, 276, 278, 282
UPS, 9, 178, 179, 182
UV-vis absorption spectrum, 273

van der Pauw geometry, 94
van der Waals
  bonds, 167, 170, 272, 278, 283
  epitaxy, 90
  forces, 4, 16
  interactions, 33, 58, 121, 137, 201, 289, 315
vertical size, 134
vibronic
  bands, 244
  progression, 219, 243, 286
  structure, 274
vicinal surface, 169
Voigt function, 132
Volmer–Weber mode, 6, 58

walking sticks, 79
wave vector, 135
waveguide, 225, 232, 280, 301, 308, 325
wetting, 339
wetting layer, 51, 54, 68, 81, 112, 170, 178, 186, 215
  defect, 113
Williamson–Hall analysis, 134
work function, 9, 178–184, 311, 314
worms, 77

X-ray diffraction, see XRD
X-ray photoemission spectroscopy, see XPS
XPS, 8, 9
XRD, 15, 37, 94, 104, 120, 271, 275

Young's modulus, 317

Springer Series in
MATERIALS SCIENCE

Editors: R. Hull   R. M. Osgood, Jr.   J. Parisi   H. Warlimont

40 **Reference Materials
in Analytical Chemistry**
A Guide for Selection and Use
Editor: A. Zschunke

41 **Organic Electronic Materials**
Conjugated Polymers and Low
Molecular Weight Organic Solids
Editors: R. Farchioni and G. Grosso

42 **Raman Scattering
in Materials Science**
Editors: W. H. Weber and R. Merlin

43 **The Atomistic Nature
of Crystal Growth**
By B. Mutaftschiev

44 **Thermodynamic Basis
of Crystal Growth**
$P-T-X$ Phase Equilibrium
and Non-Stoichiometry
By J. Greenberg

45 **Thermoelectrics**
Basic Principles
and New Materials Developments
By G. S. Nolas, J. Sharp,
and H. J. Goldsmid

46 **Fundamental Aspects
of Silicon Oxidation**
Editor: Y. J. Chabal

47 **Disorder and Order
in Strongly
Nonstoichiometric Compounds**
Transition Metal Carbides,
Nitrides and Oxides
By A. I. Gusev, A. A. Rempel,
and A. J. Magerl

48 **The Glass Transition**
Relaxation Dynamics
in Liquids and Disordered Materials
By E. Donth

49 **Alkali Halides**
A Handbook of Physical Properties
By D. B. Sirdeshmukh, L. Sirdeshmukh,
and K. G. Subhadra

50 **High-Resolution Imaging
and Spectrometry of Materials**
Editors: F. Ernst and M. Rühle

51 **Point Defects in Semiconductors
and Insulators**
Determination of Atomic
and Electronic Structure
from Paramagnetic Hyperfine
Interactions
By J.-M. Spaeth and H. Overhof

52 **Polymer Films
with Embedded Metal Nanoparticles**
By A. Heilmann

53 **Nanocrystalline Ceramics**
Synthesis and Structure
By M. Winterer

54 **Electronic Structure and Magnetism
of Complex Materials**
Editors: D.J. Singh and
D. A. Papaconstantopoulos

55 **Quasicrystals**
An Introduction to Structure,
Physical Properties and Applications
Editors: J.-B. Suck, M. Schreiber,
and P. Häussler

56 **SiO_2 in Si Microdevices**
By M. Itsumi

57 **Radiation Effects
in Advanced Semiconductor Materials
and Devices**
By C. Claeys and E. Simoen

58 **Functional Thin Films
and Functional Materials**
New Concepts and Technologies
Editor: D. Shi

59 **Dielectric Properties of Porous Media**
By S.O. Gladkov

60 **Organic Photovoltaics**
Concepts and Realization
Editors: C. Brabec, V. Dyakonov, J. Parisi
and N. Sariciftci

61 **Fatigue in Ferroelectric Ceramics
and Related Issues**
By D.C. Lupascu

62 **Epitaxy**
Physical Principles
and Technical Implementation
By M.A. Herman, W. Richter, and H. Sitter

Springer Series in
MATERIALS SCIENCE

Editors: R. Hull   R. M. Osgood, Jr.   J. Parisi   H. Warlimont

63 **Fundamentals
of Ion-Irradiated Polymers**
By D. Fink

64 **Morphology Control of Materials
and Nanoparticles**
Advanced Materials Processing
and Characterization
Editors: Y. Waseda and A. Muramatsu

65 **Transport Processes
in Ion-Irradiated Polymers**
By D. Fink

66 **Multiphased Ceramic Materials**
Processing and Potential
Editors: W.-H. Tuan and J.-K. Guo

67 **Nondestructive
Materials Characterization**
With Applications to Aerospace Materials
Editors: N.G.H. Meyendorf, P.B. Nagy,
and S.I. Rokhlin

68 **Diffraction Analysis
of the Microstructure of Materials**
Editors: E.J. Mittemeijer and P. Scardi

69 **Chemical–Mechanical Planarization
of Semiconductor Materials**
Editor: M.R. Oliver

70 **Applications of the Isotopic Effect
in Solids**
By V.G. Plekhanov

71 **Dissipative Phenomena
in Condensed Matter**
Some Applications
By S. Dattagupta and S. Puri

72 **Predictive Simulation
of Semiconductor Processing**
Status and Challenges
Editors: J. Dabrowski and E.R. Weber

73 **SiC Power Materials**
Devices and Applications
Editor: Z.C. Feng

74 **Plastic Deformation
in Nanocrystalline Materials**
By M.Yu. Gutkin and I.A. Ovid'ko

75 **Wafer Bonding**
Applications and Technology
Editors: M. Alexe and U. Gösele

76 **Spirally Anisotropic Composites**
By G.E. Freger, V.N. Kestelman,
and D.G. Freger

77 **Impurities Confined
in Quantum Structures**
By P.O. Holtz and Q.X. Zhao

78 **Macromolecular Nanostructured
Materials**
Editors: N. Ueyama and A. Harada

79 **Magnetism and Structure
in Functional Materials**
Editors: A. Planes, L. Mañosa,
and A. Saxena

80 **Micro- and Macro-Properties of Solids**
Thermal, Mechanical
and Dielectric Properties
By D.B. Sirdeshmukh, L. Sirdeshmukh,
and K.G. Subhadra

81 **Metallopolymer Nanocomposites**
By A.D. Pomogailo and V.N. Kestelman

82 **Plastics for Corrosion Inhibition**
By V.A. Goldade, L.S. Pinchuk,
A.V. Makarevich and V.N. Kestelman

83 **Spectroscopic Properties of Rare Earths
in Optical Materials**
Editors: G. Liu and B. Jacquier

84 **Hartree–Fock–Slater Method
for Materials Science**
The DV–X Alpha Method for Design
and Characterization of Materials
Editors: H. Adachi, T. Mukoyama,
and J. Kawai

85 **Lifetime Spectroscopy**
A Method of Defect Characterization
in Silicon for Photovoltaic Applications
By S. Rein

86 **Wide-Gap Chalcopyrites**
Editors: S. Siebentritt and U. Rau

87 **Micro- and Nanostructured Glasses**
By D. Hülsenberg and A. Harnisch